Godfrey C. Onwubolu (Ed.)

Hybrid Self-Organizing Modeling Systems

Studies in Computational Intelligence, Volume 211

Editor-in-Chief
Prof. Janusz Kacprzyk
Systems Research Institute
Polish Academy of Sciences
ul. Newelska 6
01-447 Warsaw
Poland
E-mail: kacprzyk@ibspan.waw.pl

Further volumes of this series can be found on our homepage: springer.com

Vol. 190. K.R. Venugopal, K.G. Srinivasa and L.M. Patnaik
Soft Computing for Data Mining Applications, 2009
ISBN 978-3-642-00192-5

Vol. 191. Zong Woo Geem (Ed.)
Music-Inspired Harmony Search Algorithm, 2009
ISBN 978-3-642-00184-0

Vol. 192. Agus Budiyono, Bambang Riyanto and Endra Joelianto (Eds.)
Intelligent Unmanned Systems: Theory and Applications, 2009
ISBN 978-3-642-00263-2

Vol. 193. Raymond Chiong (Ed.)
Nature-Inspired Algorithms for Optimisation, 2009
ISBN 978-3-642-00266-3

Vol. 194. Ian Dempsey, Michael O'Neill and Anthony Brabazon (Eds.)
Foundations in Grammatical Evolution for Dynamic Environments, 2009
ISBN 978-3-642-00313-4

Vol. 195. Vivek Bannore and Leszek Swierkowski
Iterative-Interpolation Super-Resolution Image Reconstruction:
A Computationally Efficient Technique, 2009
ISBN 978-3-642-00384-4

Vol. 196. Valentina Emilia Balas, János Fodor and Annamária R. Várkonyi-Kóczy (Eds.)
Soft Computing Based Modeling in Intelligent Systems, 2009
ISBN 978-3-642-00447-6

Vol. 197. Mauro Birattari
Tuning Metaheuristics, 2009
ISBN 978-3-642-00482-7

Vol. 198. Efrén Mezura-Montes (Ed.)
Constraint-Handling in Evolutionary Optimization, 2009
ISBN 978-3-642-00618-0

Vol. 199. Kazumi Nakamatsu, Gloria Phillips-Wren, Lakhmi C. Jain, and Robert J. Howlett (Eds.)
New Advances in Intelligent Decision Technologies, 2009
ISBN 978-3-642-00908-2

Vol. 200. Dimitri Plemenos and Georgios Miaoulis *Visual Complexity and Intelligent Computer Graphics Techniques Enhancements,* 2009
ISBN 978-3-642-01258-7

Vol. 201. Aboul-Ella Hassanien, Ajith Abraham, Athanasios V. Vasilakos, and Witold Pedrycz (Eds.)
Foundations of Computational Intelligence Volume 1, 2009
ISBN 978-3-642-01081-1

Vol. 202. Aboul-Ella Hassanien, Ajith Abraham, and Francisco Herrera (Eds.)
Foundations of Computational Intelligence Volume 2, 2009
ISBN 978-3-642-01532-8

Vol. 203. Ajith Abraham, Aboul-Ella Hassanien, Patrick Siarry, and Andries Engelbrecht (Eds.)
Foundations of Computational Intelligence Volume 3, 2009
ISBN 978-3-642-01084-2

Vol. 204. Ajith Abraham, Aboul-Ella Hassanien, and André Ponce de Leon F. de Carvalho (Eds.)
Foundations of Computational Intelligence Volume 4, 2009
ISBN 978-3-642-01087-3

Vol. 205. Ajith Abraham, Aboul-Ella Hassanien, and Václav Snášel (Eds.)
Foundations of Computational Intelligence Volume 5, 2009
ISBN 978-3-642-01535-9

Vol. 206. Ajith Abraham, Aboul-Ella Hassanien, André Ponce de Leon F. de Carvalho, and Václav Snášel (Eds.)
Foundations of Computational Intelligence Volume 6, 2009
ISBN 978-3-642-01090-3

Vol. 207. Santo Fortunato, Giuseppe Mangioni, Ronaldo Menezes, and Vincenzo Nicosia (Eds.)
Complex Networks, 2009
ISBN 978-3-642-01205-1

Vol. 208. Roger Lee, Gongzu Hu, and Huaikou Miao (Eds.)
Computer and Information Science 2009, 2009
ISBN 978-3-642-01208-2

Vol. 209. Roger Lee and Naohiro Ishii (Eds.)
Software Engineering, Artificial Intelligence, Networking and Parallel/Distributed Computing, 2009
ISBN 978-3-642-01202-0

Vol. 210. Andrew Lewis, Sanaz Mostaghim, and Marcus Randall (Eds.)
Biologically-Inspired Optimisation Methods, 2009
ISBN 978-3-642-01261-7

Vol. 211. Godfrey C. Onwubolu (Ed.)
Hybrid Self-Organizing Modeling Systems, 2009
ISBN 978-3-642-01529-8

Godfrey C. Onwubolu (Ed.)

Hybrid Self-Organizing Modeling Systems

Springer

Godfrey Onwubolu
Knowledge Management & Mining Inc.
Richmond Hill, Ontario
Canada
E-mail: onwubolu_g@dsgm.ca

ISBN 978-3-642-01529-8 e-ISBN 978-3-642-01530-4

DOI 10.1007/978-3-642-01530-4

Studies in Computational Intelligence ISSN 1860949X

Library of Congress Control Number: Applied for

Typeset & Cover Design: Scientific Publishing Services Pvt. Ltd., Chennai, India.

Printed in acid-free paper

9 8 7 6 5 4 3 2 1

springer.com

This book is dedicated entirely to God Almighty for His sovereignty in creation. Every bit of knowledge that I bring together into a book manuscript enables me to see and appreciate more the greatness of Almighty God!

Foreword

Models form the basis of any decision. They are used in different context and for different purposes: for identification, prediction, classification, or control of complex systems. Modeling is done theory-driven by logical-mathematical methods or data-driven based on observational data of the system and some algorithm or software for analyzing this data. Today, this approach is summarized as Data Mining.

There are many Data Mining algorithms known like Artificial Neural Networks, Bayesian Networks, Decision Trees, Support Vector Machines. This book focuses on another method: the Group Method of Data Handling. Although this methodology has not yet been well recognized in the international science community as a very powerful mathematical modeling and knowledge extraction technology, it has a long history.

Developed in 1968 by the Ukrainian scientist A.G. Ivakhnenko it combines the black-box approach and the connectionism of Artificial Neural Networks with well-proven Statistical Learning methods and with more behavioral-justified elements of inductive self-organization. Over the past 40 years it has been improving and evolving, first by works in the field of what was known in the U.S.A. as Adaptive Learning Networks in the 1970s and 1980s and later by significant contributions from scientists from Japan, China, Ukraine, Germany. Many papers and books have been published on this modeling technology, the vast majority of them in Ukrainian and Russian language.

The unique feature of the self-organizing modeling approach of GMDH is that it allows, systematically and autonomously, developing optimal complex models by performing both parameter *and* structure identification. It inductively builds the model structure or composition of terms or network topology, automatically. This is possible, because self-organizing modeling closely links model accuracy to model complexity. It introduces the concept of an optimal complex model as a model that optimally balances model quality on a learning data set and its generalization power on new, not previously seen data with respect to the data's noise level and the purpose of modeling (prediction, modeling, control, etc.). This has been the key idea for solving the basic problem of experimental systems analysis of avoiding overfitted models based on the data's information, only, and this makes advanced implementations of this algorithm a so powerful, efficient and easy-to-use knowledge extraction tool.

Today, there is a spectrum of self-organizing modeling algorithms, which are all summarized as GMDH algorithms: Different flavors of the initial parametric GMDH algorithm for developing linear and nonlinear regression models, but also a number of self-organizing non-parametric algorithms for solving pattern recognition, clustering, or fuzzy modeling problems. This book adds to this spectrum a new element by combining GMDH with other state-of-the-art Soft Computing and Computational Intelligence methods and algorithms.

I wish this book, and hopefully many other new publications on GMDH, helps bringing this widely underestimated knowledge extraction technology back into peoples' minds as a first-class tool for modeling and for solving complex real-world problems.

KnowledgeMiner Software, Berlin, Germany Frank Lemke
January 2009

Preface

Background
The Group Method of Data Handling (**GMDH**) which Ivakhnenko introduced is a typical inductive modeling method that is built on principles of self-organization. Since then, inductive modeling has been developing and applied to complex systems in several key areas such as prediction, modeling, clusterization, system identification, as well as data mining and knowledge extraction technologies, to several fields such as social science, science, engineering, medicine, etc. There have been more recent developments such as utilization of Genetic Programming, Genetic Algorithms, Differential Evolution, Particle Swarm Optimization and other Computational Intelligence approaches or the idea of Active Neurons and multileveled self-organization to build models from data.

Since its introduction, attempts have been made to publicize the theory, algorithms, applications, solutions, and new developments of GMDH. A dedicated website on GMDH is perhaps the most useful resources centre available to researchers and practitioners to finding published papers. Historically, there was the first International Conference on Inductive Modeling (ICIM'2002) in Lviv, Ukraine, in May 2002. Following its success, there was an initial Workshop that took place in Kyiv, Ukraine, in July 2005. Then more recently, the 2nd International Workshop on Inductive Modeling (IWIM07) was held in Prague on September 23-26, 2007. The series of conferences and workshops have been the only international forum that focuses on theory, algorithms, applications, solutions, and new developments of GMDH. The motivation of these conference and workshops was to analyze the state-of-the-art of modeling methods that inductively generate models from data, to discuss concepts of an automated knowledge discovery workflow, to share new ideas on model validation and visualization, to present novel applications in different areas, and to give inspiration and background on how inductive modeling can evolve and contribute given the current global challenges.

To date, there are very few books written in English language that describe the traditional GMDH. Hybridizing the classical GMDH with computational intelligence methods is a new idea. The main purpose of this book therefore is to present the work done by the originators of a number of hybrid GMDH variants, in which the classical GMDH has been hybridized with various computational intelligence methods. These hybrids are presented in such a manner that readers are able understand how they are realized and the benefits of hybridization are highlighted by showing much superior results that are obtained using hybrid GMDH variants when compared to the classical GMDH. Virtually all the hybrid GMDH architectures discussed in this book build on the multi-layer GMDH (well known as MIA-GMDH), with integration with computational intelligence methods. The book also presents a framework, in which 'self-organization modeling' is emphasized. Consequently, it is anticipated that the two separate domains of neural network (NN) and GMDH which have been used for modeling could be brought together under one bigger umbrella, and a framework or standard could be realized for hybridizing these self-organizing modeling systems.

Chapter authors' background: Chapter authors are to the best of my knowledge the originators or closely related to the originators of the above mentioned hybrid inductive modeling approaches. Hence, this book will be one of the leading books in hybrid inductive modeling approaches.

Organization of the Chapters

The Editor of the book, Godfrey Onwubolu, presents "Hybrid Computational Intelligence and GMDH Systems" in Chapter 1 in order to give an overview of the book and the context of computational intelligence in hybridization with GMDH. Hitoshi Iba, the originator of Hybrid Genetic Programming and GMDH System presents STROGANOFF in chapter 2. Nader Nariman-zadeh and Ali Jamali, the originators of Hybrid Genetic Algorithm and GMDH System present chapter 3. Godfrey Onwubolu, the originator of Hybrid Differential Evolution and GMDH System presents chapter 4, which is the kernel of *Knowledge Management & Mining* (KMM) software that he has developed. Anuraganand Sharma and Godfrey Onwubolu, the originators of Hybrid Particle Swarm Optimization and GMDH System present chapter 5. Pavel Kordik the originator of GAME - Hybrid Self-Organizing Modeling System based on GMDH presents chapter 6.

Audience: The book will be an instructional material for senior undergraduate and entry-point graduate students in computer science, cybernetics, applied mathematics, statistics, engineering, bioinformatics, who are working in the areas of machine learning, artificial intelligence, complex system modeling and analysis, neural networks, and optimization. Researchers who want to know how to realize effect hybrids of classical modeling approaches together with computational methods will find this book a very useful handbook and the starting point. The book will be a resource handbook and material for practitioners who want to apply methods that work on real-life problems to their challenging applications.

Canada
January 2009

Godfrey C. Onwubolu

Acknowledgements

This is my first book in the area of group method of data handling (GMDH) and there was a journey to get to the point of writing this book. In this regard, I want to first, thank all those who helped me to understand the fundamentals of GMDH presented in this book. My association with key leaders in the area was instrumental to my grasping the concepts of GMDH. Then, my association with the Inductive Modeling community significantly helped me to have more insight into this exciting technology for modeling complex real-life systems. For this, I thank Professor Volodymyr Stepashko, Frank Lemke, and Assistant Professor Pavel Kordik whom I have closely worked with, in organizing Inductive Modeling workshops and conference. I also thank all members of the Inductive Modeling community whom I have been in touch with regarding various subjects of GMDH.

I am also particularly thankful to the authors of the various chapters of this book for their contributions to the hybrid variants of GMDH that they have worked on, and their co-operation throughout the preparation of the book.

Donald Davendra and I have worked on Differential Evolution (DE) for several years since when I supervised him for his undergraduate and graduates theses on the subject. The polished DE code from his work that I supervised formed the basis for the hybrid GMDH-DE that I proposed.

There was a logistic problem in bringing this book to a completion. Two chapters were written in LateX format while all others were written in Word format. It was impossible to get authors to switch from one form into the other so as to a have a unified format of LateX or Word. The compromise was for the book editor to convert all the chapters in Word into LateX. It was not an interesting experience for me and I had to approach Donald Davendra to bail me out from the challenging situation. He agreed despite his extremely busy schedule on completion of his PhD thesis; I really appreciate his help in this regard.

My association with Springer-Verlag over the years has been extremely pleasant. I have enjoyed working on this book project with Dr. Thomas

Ditzinger and Heather King and their other colleagues at Springer-Verlag, Heidelberg, Germany; they are pleasant people to work with.

My beloved wife, Ngozi, gave me all the support that I needed to bring this book project to completion. I commenced this project while we were still living in Fiji and it was concluded while we moved to settle permanently in Canada; this period of relocation was extremely challenging for me conclude this book project, but my wife supported me to sail through. I am really thankful to our children, Chukujindu, Chinwe and Chinedu who were settling down with us at the time of our relocation for their usual tolerance, as well as Chioma and Chineye, although they were living in another Province.

Contents

Hybrid Computational Intelligence and GMDH Systems 1
Godfrey Onwubolu
1 Introduction 1
2 Group Method of Data Handling (GMDH) Networks 2
2.1 GMDH Layers 3
2.2 GMDH Nodes 3
2.3 GMDH Connections 4
2.4 GMDH Network 5
2.5 Regularized Model Selection 5
2.6 GMDH Algorithm 6
2.7 Advantages of GMDH Technique 7
2.8 Limitations of GMDH Technique 7
3 Rationale for Hybrid Systems 11
4 Computational Intelligence 12
4.1 What Is Intelligence? 13
4.2 Can Computers Be Intelligent? 13
4.3 Computational Intelligence Paradigms 14
5 Hybrid GMDH Systems 21
6 Conclusion 23
References 23

Hybrid Genetic Programming and GMDH System: STROGANOFF 27
Iba Hitoshi
1 Introduction 27
2 Background 28
2.1 System Identification Problems 28
2.2 Difficulties with Traditional GP 29
2.3 Numerical Approach to GP 31
3 Principles of STROGANOFF 31

3.1 STROGANOFF Algorithm 31
3.2 GMDH Process in STROGANOFF.............. 32
3.3 Crossover in STROGANOFF 34
3.4 Mutation in STROGANOFF 35
3.5 Fitness Evaluation in STROGANOFF 36
3.6 Overall Flow of STROGANOFF 36
3.7 Recombination Guidance in STROGANOFF 38
4 Numerical Problems with STROGANOFF............... 40
4.1 Time Series Prediction with STROGANOFF 40
4.2 Comparison with a Traditional GP 43
4.3 Statistical Comparison of STROGANOFF and a Traditional GP 46
5 Symbolic Problems with STROGANOFF................ 52
5.1 Extension of STROGANOFF 52
5.2 Symbolic Regression........................... 53
6 Applying STROGANOFF to Computational Finances 59
6.1 Predicting Stock Market Data 60
6.2 Developping Day-Trading Rules................. 68
7 Inductive Genetic Programming........................ 72
7.1 Polynomial Neural Networks 73
7.2 PNN Approaches 74
7.3 Basic IGP Framework 77
7.4 PNN vs. Linear ARMA Models 79
7.5 PNN vs. Neural Network Models............... 80
8 Discussion.. 83
8.1 Comparison of STROGANOFF and Traditional GP 83
8.2 Genetic Programming with Local Hill Climbing ... 84
8.3 Limitations and Further Extensions of STROGANOFF 89
8.4 Applicability to computational finances 90
9 Conclusion 91

Hybrid Genetic Algorithm and GMDH System 99
Nader Nariman-zadeh and Jamali Ali
1 Introduction.. 99
2 Modelling Using GMDH-Type Neural Networks 101
3 Hybrid Genetic/SVD Design of GMDH-Type Neural Networks... 103
3.1 Application of SVD in the Design of GMDH-Type Networks 103
3.2 Application of SVD in the Design of GMDH-Type Networks 103
4 Single-Objective Hybrid Genetic Design of GMDH-Type Neural Networks Modelling and Prediction of Complex Processes.. 108

4.1 Application to the Modelling and Prediction of Level Variations of the Caspian Sea 108
4.2 Application to the Modelling and Prediction of the Explosive Cutting Process 111
5 Multi-objective Hybrid Genetic Design of GMDH-Type Neural Networks Modelling and Prediction of Complex Processes 115
5.1 Multi-objective Optimization 116
5.2 Multi-objective Uniform-Diversity Genetic Algorithm (MUGA) 117
5.3 Multi-objective Genetic Design of GMDH-Type Neural Networks for a Variable Valve-Timing Spark-Ignition Engine 120
5.4 Multi-objective Genetic Design of GMDH-Type Neural Networks for a Nonlinear System 125
5.5 Multi-objective Genetic Design of GMDH-type Neural Networks for Modelling and Prediction of Explosive Cutting Process 130
6 Conclusion 136
References 137

Hybrid Differential Evolution and GMDH Systems 139
Godfrey Onwubolu
1 Introduction 139
2 Inductive Modeling: Group Method of Data Handling (GMDH) 140
2.1 GMDH Layers 140
2.2 GMDH Nodes 140
2.3 GMDH Connections 142
2.4 GMDH Network 142
2.5 Advantages of GMDH Technique 143
2.6 Limitations of GMDH Technique 143
3 Classical Differential Evolution Algorithm 144
3.1 The Steps Involved in Classical Differential Evolution 145
3.2 Ten different Working Strategies in Differential Evolution 147
3.3 Discrete Differential Evolution 148
3.4 Permutative Population 149
3.5 Forward Transformation 149
3.6 Backward Transformation 150
3.7 Recursive Mutation 150
3.8 Discrete Differential Evolution (DDE) 151
3.9 Enhanced Differential Evolution (EDE) 152
4 The Hybrid Differential Evolution And GMDH System.... 152

4.1 Structural Optimization: Representation of Encoding Strategy of Each Partial Descriptor (PD) 154
4.2 Parametric Optimization: Coefficient Estimation of the Polynomial Corresponding to the Selected Node (PN) ... 158
4.3 Framework of the Design Procedure of the DE-GMDH Hybrid System ... 163
4.4 The Hybrid DE-GMDH Algorithm ... 165
5 DE-GMDH Mechanics Illustrated ... 168
6 Applications of the DE-GMDH Hybrid System ... 177
6.1 DE-GMDH for Modeling the Tool-Wear Problem ... 177
6.2 Exchange Rates Forecasting Using the DE-GMDH Paradigms ... 181
6.3 Gas Furnace Experimentation Using the DE-GMDH Learning Network ... 186
6.4 CPU Time Cost of the DE-GMDH Algorithm ... 187
7 Conclusions ... 188
References ... 189

Hybrid Particle Swarm Optimization and GMDH System ... 193
Anurag Sharma and Godfrey Onwubolu
1 Introduction ... 193
2 The Group Method of Data Handling (GMDH) ... 194
2.1 Overview of Traditional GMDH ... 194
2.2 Drawbacks of Traditional GMDH ... 197
3 Particle Swarm Optimization Algorithm ... 198
3.1 Explosion Control ... 200
3.2 Particle Swarm Optimization Operators ... 200
3.3 Particle Swarm Optimization Neighborhood ... 202
4 The Proposed Hybrid PSO-GMDH Algorithm ... 206
4.1 Overview ... 206
4.2 Technical View ... 209
5 Experimentation ... 221
5.1 Tool Wear Problem ... 221
5.2 Gas Furnace Problem ... 226
6 Conclusion ... 229
References ... 230

GAME – Hybrid Self-Organizing Modeling System Based on GMDH ... 233
Pavel Kordík
1 Introduction ... 233
1.1 Self-Organizing Modelling ... 234
2 Group of Adaptive Models Evolution (GAME) ... 237
2.1 The Concept of the Algorithm ... 237

2.2 Contributions of the GAME Algorithm 241
2.3 Optimization of GAME Neurons 245
2.4 Optimization Methods (Setting Up Coefficients) 249
2.5 Combining Optimization Methods 255
2.6 Structural Innovations 258
2.7 Genetic Algorithm 259
2.8 Ensemble Techniques in GAME 266
3 Benchmarking the GAME Method 269
3.1 Summary of Results 271
4 Case Studies – Data Mining Using GAME 271
4.1 Fetal Weight Prediction Formulae Extracted from GAME 271
5 The FAKE GAME Project 273
5.1 The Goal of the FAKE GAME Environment 275
References 278

Author Index 281

Contributors

Iba Hitoshi
Department of Information and Communication Engineering, Faculty of Engineering, University of Tokyo, 7-3-1 Hongo, Bunkyo-ku, Tokyo, 113-8656, Japan
e-mail: iba@iba.k.u-tokyo.ac.jp

Nariman-zadeh Nader
Department of Mechanical Engineering, University of Guilan, PO Box 3756, Rasht, Iran
e-mail: nnzadeh@guilan.ac.ir

Jamali Ali
Department of Mechanical Engineering, University of Guilan, PO Box 3756, Rasht, Iran.

Godfrey Onwubolu
Knowledge Management & Mining, Richmond Hill, Canada
e-mail: onwubolu_g@dsgm.ca

Pavel Kordk
Department of Computer Science and Engineering, FEE, Czech Technical University, Prague, Czech
e-mail: kordikp@fel.cvut.cz

Anurag Sharma
School of School of Computing, Information System, Mathematical Sciences and Statistics, Faculty of Science & Technology, The University of the South Pacific, Private Bag, Suva, Fiji
e-mail: sharma_au@usp.ac.fj

Hybrid Computational Intelligence and GMDH Systems

Godfrey Onwubolu

Abstract. The multilayer GMDH is known to often under-performs on non-parametric regression tasks, while time series modeling GMDH exhibits a tendency to find very complex polynomials that cannot model well future, unseen oscillations of the series. In order to alleviate the problems associated with standard GMDH approach, a number of researchers have attempted to hybridize GMDH with some evolutionary optimization techniques. This is the central theme of this book. This chapter prepares the groundwork for hybridizing computational intelligence methods with standard GMDH in order to realize more robust and flexible hybrids for solving complex, real-world problems which currently cannot be solved using standard GMDH approach.

1 Introduction

Hybridization of intelligent systems is a promising research field of modern artificial intelligence concerned with the development of the next generation of intelligent systems. A fundamental stimulus to the investigations of Hybrid Intelligent Systems (HIS) is the awareness amongst practitioners and in the academic communities that combined and integrated approaches will be necessary if the remaining tough problems in artificial intelligence are to be solved. Recently, hybrid intelligent systems are becoming popular due to their capabilities in handling many real world complex problems, involving imprecision, uncertainty and vagueness, high-dimensionality. However, the popularity of HIS is well known in the neural network (NN) domain but not in the group method of data handling (GMDH) domain. Therefore, this book aims to unify the Hybrid Self-Organizing Modeling Systems (HSOMS) by including NN and GMDH hybrid systems. Self-organization is synonymous with inductiveness; hence, we refer to this knowledge base as self-organizing modeling or

Godfrey Onwubolu
Knowledge Management & Mining, Richmond Hill, Canada
e-mail: onwubolu_g@dsgm.ca

G.C. Onwubolu (Ed.): Hybrid Self-Organizing Modeling Systems, SCI 211, pp. 1–26.
springerlink.com

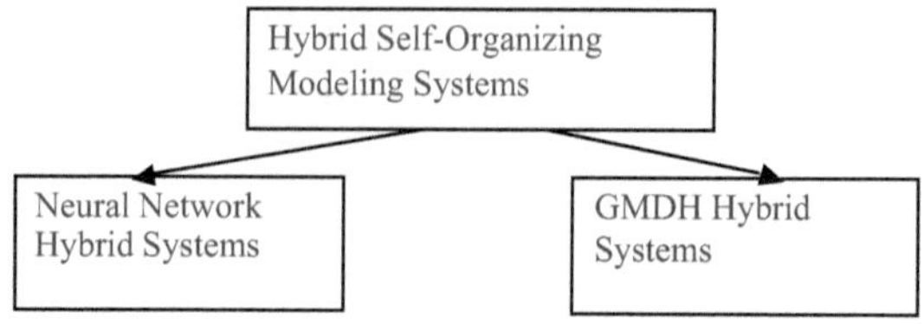

Fig. 1 Framework of Hybrid Self-Organizing Modeling Systems (HSOMS)

inductive modeling. This new framework is shown in Figure 1 and it could be seen that it is a bigger umbrella than the current context in which Hybrid Intelligent Systems (HIS) is used.

This proposed framework has a number of advantages:

- Neural Network Hybrid Systems (NNHS) may not be able to solve most real world complex problems on their own;
- GMDH Hybrid Systems (GMDHHS) may not be able to solve most real world complex problems on their own;
- HSOMS encompasses the spectra of NN and GMDH and consequently may be robust in solving most real world complex problems
- More possibilities exist in integrating NN and GMDH horizontally rather than viewing them as parallel approaches; consequently, both horizontal and vertical connections are possible.

2 Group Method of Data Handling (GMDH) Networks

The causality relationship between the inputs and the output of a multiple inputs single output self-organizing network can be represented by an infinite Volterra-Kolmogorov-Gabor (VKG) polynomial of the form 1:

$$y_n = a_0 + \sum_{i=1}^{M} a_i x_i + \sum_{i=1}^{M}\sum_{j=1}^{M} a_{ij} x_i x_j + \sum_{i=1}^{M}\sum_{j=1}^{M}\sum_{k=1}^{M} a_{ijk} x_i x_j x_k \ldots \quad (1)$$

where $X = (x_1, x_2, ..., x_M)$ is the vector of input variables and $X = (x_1, x_2, ..., x_M)$ is the vector of coefficients or weights.

This is the discrete-time analogue of a continuous time Volterra series and can be used to approximate any stationary random sequence of physical measurements. Ivakhnenko showed that the VKG series can be expressed as a cascade of second order polynomials using only pairs of variables[1] [2]. The corresponding network can be constructed from simple polynomial and delay elements. As the learning procedure evolves, branches that do not contribute significantly to the specific output can be pruned, thereby allowing only the dominant causal relationship to evolve. The multilayer GMDH network algorithm constructs hierarchical cascades of bivariate activation polynomials in the nodes, and variables in the leaves. The activation polynomial outcomes are fed forward to their parent nodes, where partial polynomial models are made. Thus, the algorithm produces high-order multivariate polynomials by composing simple and tractable activation polynomial allocated in the hidden nodes of the network.

In neural network idiom, the higher-order polynomial networks grown by the GMDH algorithm are essentially feed-forward, multi-layered neural networks. The nodes are hidden units, the leaves are inputs, and the activation polynomial coefficients are weights. The weights arriving at a particular hidden node are estimated by ordinary least squares (OLS) fitting.

2.1 GMDH Layers

When constructing a GMDH network, all combinations of the inputs are generated and sent into the first layer of the network. The outputs from this layer are then classified and selected for input into the next layer with all combinations of the selected outputs being sent into layer 2. This process is continued as long as each subsequent $layer_{(n+1)}$ produces a better result than $layer_{(n)}$. When $layer_{(n+1)}$ is found to not be as good as the $layer_{(n)}$ the process is stopped.

2.2 GMDH Nodes

Self-organizing networks are constructed from elemental polynomial neurons each of which possesses only a *pair* of dissimilar inputs (x_i, x_j). Each layer consists of nodes generated to take a specific pair of the combination of inputs as its source. Each node produces a set of coefficients a_i where $i\varepsilon\{0,1,2,3,...,m\}$ such that equation 2 is estimated using the set of *training* data. This equation is tested for fit by determining the mean square error of the predicted $\hat{y}$ and actual y values as shown in equation 3 using the set of *testing* data.

$$\hat{y}_n = a_0 + a_1 x_{i_n} + +a_2 x_{j_n} + a_3 x_{i_n} x_{j_n} + a_4 x_{i_n}^2 + a_5 x_{j_n}^2 \tag{2}$$

$$e = \sum_{n=1}^{N} (\hat{y}_n - y_n)^2 \tag{3}$$

In determining the values of a that would produce the “best fit”, the partial derivatives of equation 3 are taken with respect to each constant value a_i and set equal to zero.

$$\frac{\partial e}{\partial a_i} = 0 \tag{4}$$

Expanding equation 4 results in the following system of equations that are solved using the training data set.

$$\sum_{n=1}^{N} y = \sum_{n=1}^{N} a_0 + a_1 x_i + +a_2 x_j + a_3 x_i x_j + a_4 x_i^2 + a_5 x_j^2 \tag{5}$$

$$\sum_{n=1}^{N} yx_i = \sum_{n=1}^{N} a_0x_i + a_1x_i^2 + +a_2x_ix_j + a_3x_i^2x_j + a_4x_i^3 + a_5x_ix_j^2 \quad (6)$$

$$\sum_{n=1}^{N} yx_j = \sum_{n=1}^{N} a_0x_j + a_1x_ix_j + +a_2x_j^2 + a_3x_ix_j^2 + a_4x_i^2x_j + a_5x_j^3 \quad (7)$$

$$\sum_{n=1}^{N} yx_ix_j = \sum_{n=1}^{N} a_0x_ix_j + a_1x_i^2x_j + +a_2x_ix_j^2 + a_3x_i^2x_j^2 + a_4x_i^3x_j + a_5x_ix_j^3 \quad (8)$$

$$\sum_{n=1}^{N} yx_i^2 = \sum_{n=1}^{N} a_0x_i^2 + a_1x_i^3 + +a_2x_i^2x_j + a_3x_i^3x_j + a_4x_i^4 + a_5x_i^2x_j^2 \quad (9)$$

$$\sum_{n=1}^{N} yx_j^2 = \sum_{n=1}^{N} a_0x_j^2 + a_1x_ix_j^2 + +a_2x_j^3 + a_3x_ix_j^3 + a_4x_i^2x_j^2 + a_5x_j^4 \quad (10)$$

The equations can be simplified using matrix mathematics as follows.

$$Y = \begin{pmatrix} 1 & x_i & x_j & x_ix_j & x_i^2 & x_j^2 \end{pmatrix} \quad (11)$$

$$X = Y^TY \quad (12)$$

$$X = \begin{pmatrix} 1 & x_i & x_j & x_ix_j & x_i^2 & x_j^2 \\ x_i & x_i^2 & x_ix_j & x_i^2x_j & x_i^3 & x_ix_j^2 \\ x_j & x_ix_j & x_j^2 & x_ix_j^2 & x_i^2x_j & x_j^3 \\ x_ix_j & x_i^2x_j & x_ix_j^2 & x_i^2x_j^2 & x_i^3x_j & x_ix_j^3 \\ x_i^2 & x_i^3 & x_i^2x_j & x_i^3x_j & x_i^4 & x_i^2x_j^2 \\ x_j^2 & x_ix_j^2 & x_j^3 & x_ix_j^3 & x_i^2x_j^2 & x_j^4 \end{pmatrix} \quad (13)$$

$$\mathrm{a} = \begin{pmatrix} a_0 & a_1 & a_2 & a_3 & a_4 & a_5 \end{pmatrix} \quad (14)$$

$$b = (yY)^T \quad (15)$$

This system of equations then can be written as:

$$\sum_{n=1}^{N} aX = \sum_{n=1}^{N} b \quad (16)$$

The node is now responsible for evaluating all inputs of x_{i_n}, x_{j_n}, y_n data values in **a** and **b** for the *training* set of data. Solving the system of equations results in **x** being the node's computed set of coefficients. Using these coefficients in equation 2, the node then computes its error by processing the set of *testing* data in equations 2 and 3. The error is the measure of fit that this node achieved.

2.3 GMDH Connections

A GMDH layer sorts its nodes based on the error produced, saving the best N nodes. The generated y_n values (classifications) of each node become one set of inputs to be used by the next layer when it combines all outputs from the previous layer's nodes assigning them to the new layer's nodes (See Figure 2). The layer must remember the nodes that were saved so that other data submitted to the network will follow the same generated path to the output.

2.4 *GMDH Network*

When the GMDH network is completed, there is a set of original inputs that filtered through the layers to the optimal output node. This is the computational network that is to be used in computing predictions (in our application, classifications are implied).

The best nodes in the input layer (starred nodes in Figure 2) are retained and form the input to the next layer. The inputs for layer 1 are formed by taking all combinations of the surviving output approximations from the input layer nodes. It is seen that at each layer the order of the polynomial approximation is increased by two. The layer 2 best nodes for approximating the system output are retained and form the layer 3 inputs. This process is repeated until the current layer's best approximation is inferior to the previous layer's best approximation.

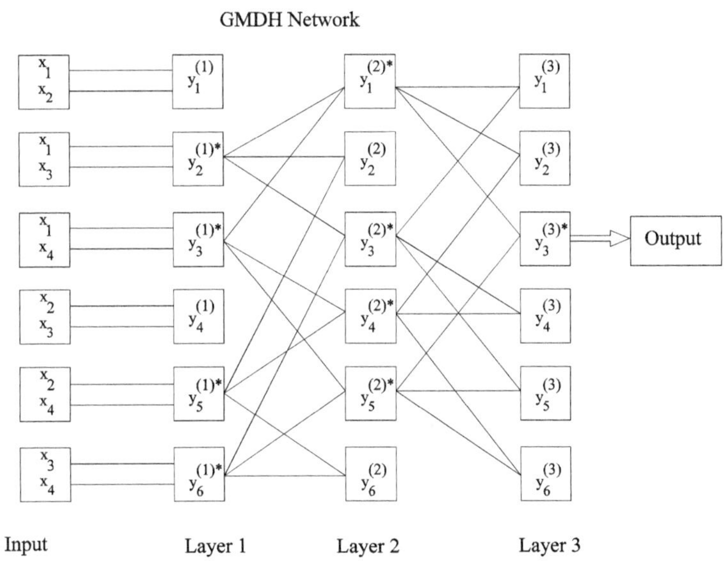

Fig. 2 GMDH forward feed functional network

2.5 *Regularized Model Selection*

A model selection criterion is necessary to achieve over-fitting avoidance; that is to pursue construction of not only accurate but also predictive networks. The model selection criterion is essential since it guides the construction of the network topology, and so influences the quality of the induced function model. Two primary issues in the design of a model selection function for over-fitting avoidance are:

1. favoring more fit networks by incorporating a mean-squared-error sub-criterion; and
2. tolerating smoother network mappings having higher generalization potential by incorporating a regularization sub-criterion.

Knowing that a large weight in a term significantly affects the polynomial surface curvature in the dimensions determined by the term variables, a *correcting smoothness sub-criterion* that accounts for the weights' magnitude is accommodated in a *regularized average error* (RAE) as

$$RAE = (1/n_t)\left(\sum_{i=1}^{n_t}(y_i - F(x_i,x_k))^2 + \alpha\sum_{j=1}^{W} a_j^2\right) \tag{17}$$

where α is regularization parameter whose proper values are found using statistical techniques [3], a_j are the weights or coefficients, such that $1 \leq j \leq W$ and $F(x_i,x_k) = h(x_i,x_k)a$. This formula 17 is known as weight decay regularization [4] [5] and it requires the usage of regularized least square (RLS) fitting method for estimating the weights

$$a = \left(X^T X + \alpha I\right)^{-1} X^T y \tag{18}$$

where **a** is the coefficients vector. The regularized least square is also called the ridge regression [6]. The $\alpha \geq 0$ is a parameter to control the amounts of shrinkage. Consequently, the advantage of regularization approach is that since the regression was used as a building block, the regularization techniques can be easily incorporated and provide more stable and meaningful solutions, especially when there exist a large amount of input variables [6].

2.6 GMDH Algorithm

This section gives the steps involved in the basic GMDH algorithm as shown in Figure 3.

Initialization
Given a data series $\partial = \{(x_{ij},y_i)\}\quad i = 1,2,...,n;\ \ j = 1,2,...,m$; where the number of training data is n_t and the number of testing data is n_c such that $n_t + n_t = n$. Let the layer label be $l = 1$, the lowest error be $\varepsilon = MaxInt$ and the activation polynomials expressed as $p(x_i,x_k) = a_0 + a_1x_i + a_2x_k + a_3x_ix_k + a_4x_i^2 + +a_5x_k^2$ or $p(x_i,x_k) = h(x_i,x_k)a \Rightarrow H[h_1,h_2,...,h_N]^T$.

Network construction and weight training

Step 1: Make all $c = \binom{m}{2}$ combinations of variables (x_i,x_k), $l \leq i,j \leq r$.

Step 2: Make a polynomial $p_c^l(x_i,x_k)$ from each combination

2.1 Estimate its coefficients a_c by OLS fitting: $a_c = \left(H^T H\right)^{-1} H^T y$

2.2 Evaluate the error or external criterion (EC) of the polynomial: $p_c^l(x_i, x_k) = ha_c$ $EC_c = (1/n_t)\sum_{i=1}^{n_t}(y_i - p_c^l(x_i, x_k))^2$

2.3 Compute the model selection criterion using the regularized average error (RAE): $RAE_c = f(EC_c)$

Step 3: Order the polynomials with respect to their RAE_c, and choose r of these with lower criterion values.

Step 4: Consider the lowest error from this layer: $\varepsilon^{l+1} = \min\{RAE_c\}$.

Step 5: If $\varepsilon^{l+1} > \varepsilon$ then terminate, else set $\varepsilon = \varepsilon^{l+1}$ and continue

Step 6: The polynomial outputs become current variables: $x_c \equiv p_c^l$

Step 7: Repeat the construction and training step with $l = l + 1$.

2.7 Advantages of GMDH Technique

The advantage of using pairs of input is that only six weights (coefficients) have to be computed for each neuron. The number of neurons in each layer increases approximately as the square of the number of inputs. During each training cycle, the synaptic weights of each neuron that minimize the error norm between predicted and measured values are computed and those branches that contribute least to the output of the neuron are discarded, the remaining branches being retained and their synaptic weights kept unchanged thereafter. A new layer is subsequently added and the procedure is repeated until the specified termination conditions are met.

There could be summarized that the GMDH-type polynomial networks influence the contemporary artificial neural network algorithms with several other advantages [7]:

1. they offer adaptive network representations that can be tailored to the given task;
2. they learn the weights rapidly in a single step by standard OLS fitting which eliminates the need to search for their values, and which guarantees finding locally good weights due to the reliability of the fitting technique;
3. these polynomial networks feature sparse connectivity which means that the best discovered networks can be trained fast.

2.8 Limitations of GMDH Technique

Although standard GMDH provides for a systematic procedure of system modeling and prediction, it has also a number of shortcomings. Anastasakis and Mort [8] have carried out a comprehensive study of the shortcomings of GMDH, mong the most problematic can be stated:

Selection of Input Arguments

One of the main features of GMDH is its ability to objectively select the most appropriate input arguments amongst a set of candidates. However, the identification of these candidate input arguments is not straightforward and may affect its performance [9].

Inaccuracies in Parameter Estimation

The method of least square estimates is the most popular method to calculate the coefficients of partial descriptions. If the data matrix is well defined its estimates will be accurate however, in the majority of real world systems the data matrix is ill-defined and the least squares biased. Duffy et al. [10] propose the utilization of stepwise multiple regression techniques as well as the re-estimation of all the terms in the final equation using both training and testing set of data. The reasons of inadequacy in least square estimates are explained by Sarychev in [11]. According to that the problem is based on the false assumption that the distribution of the error vector is normal and the author argues that the assumption of a binary exponential distribution is more suitable. This argument is justified by the different nature of the error distributions in different selected intermediate variables, the simple structure of the partial descriptions in the first layer with respect to the true model and the difference of the contribution of the individual descriptions in the previous layer to the total change in the output of the current layer.

Multicollinearity

Another problem found exclusively in multilayer algorithm, which affects the stability of coefficients, is that of multicollinearity. The selected variables in one layer may be highly correlated to those selected in previous layers, which will result to the appearance of multilayerness error. Duffy and Franklin [10] attempt to solve the problem by applying a stepwise multiple regression technique for the formulation of partial descriptions in place of least squares. The ridge regression analysis is another effective approach for stabilizing the coefficients of models and solves the multicollinearity phenomenon [12].

Reduction of Complexity

Another shortcoming found GMDH approach is a tendency to generate quite complex polynomial (since the complexity of the network increases with each training and selection cycle through addition of new layers) for relatively simple systems (data input); also, an inclination to producing overly complex network (model) when dealing with highly nonlinear systems owing to its limited generic structure (quadratic two-variable polynomial). Ivakhnenko [13] claims that if the number of selected models in every layer is as large as possible the optimum model will be never lost. On the other hand, following that procedure the complexity of the model as well as its computation time is increased. Triseyev [14] reduces the complexity of GMDH algorithms by following a different approach for the selection of intermediate variables, which is based on the diversity of variables criterion and the structural number of partial descriptions. Parker et al. [15] in order to avoid an increase of the model order use second order polynomial in the first layer but only linear forms in subsequent layers.

Multiplicative-Additive GMDH Algorithm

It was mentioned above that the form of partial descriptions might affect the model complexity. The choice of partial descriptions is closely related to the field of applications [16]. The different types of descriptions and different complexing methods have been driven in a wide range of GMDH algorithms. Generally, partial descriptions of parametric polynomial models can be divided into four main categories according to the combination of their terms. Additive where new terms are added to the partial descriptions, multiplicative of unit power of factor, generalized multiplicative-additive and descriptions where the power of factors is replaced by a number p which can be either be pre-specified or not [17].

Formulas of Partial Descriptions
Despite the wide range of partial descriptions the majority of the researchers follows the argument that Volterra series are capable of identifying any non-linear system and therefore have adopted polynomial partial descriptions similar to Ivakhnenko polynomial [18]-[20]. However, due to the complexity of the model and the requirement of including the theory behind the object, many modifications have been designed in order to adapt to system's properties. Duffy et al. [10] in order to increase the spectrum of partial descriptions in every layer, introduce the linear combination of all input variables as an additional partial description to the second order polynomial. Ikeda et al. [21] proposed the introduction of each input variable into a polynomial prior to their application in the partial generators in the expense of increasing complexity. Park et al. [22] propose a wide range of partial descriptions like, linear, quadratic, cubic, bilinear, bicubic, trilinear and tricubic.

Overfitting
A consequence of complexity is the over-fitting problem and poor generalization. The partition of the data into two subsamples and the selection of the optimum model according to its accuracy on an unknown set of data may ensure the good generalization. However, the large number of parameters in the final model could create over-fitting problems and therefore techniques, which eliminate the number of parameters, should be adopted. Mehra in [23] adopts the application of stepwise regression method for parameter estimation that is capable of eliminating the multicollinearity problem as well. Additionally, the Stern estimator is proposed for parameter identification but it is based on Akaike' s Information Criterion.

Partition of Data
The objectiveness of GMDH algorithm is based on the utilization of an external criterion to select the optimum model, which requires the partition of the data. The subsamples should cover the operating regions of the system and have similar properties in order to avoid poor generalization. The requirement of splitting data into two groups will lead to different models for different subsamples and researchers have investigated a number of techniques to overcome it [24]. A simple technique will include the most recent observations on the checking set with the rest data being in the training set. Another technique may involve the variance of the data, where a mix of low and high variance data will be included in both subsamples. Duffy et al. in [10] propose two different approaches, which ensure a proper distribution of the data in both sets. The first suggest a fixed selection of pattern such as putting

alternative points in time in the training and testing set. The second ensures a better spread of data and based on a random function which binary output (0-1) will indicate the data used in the training and checking set.

Low Accuracy in GMDH Method

Although a number of modifications have been tested and proved to improve the accuracy of GMDH. However, in many cases and particularly in applications of long range prediction the GMDH was inaccurate. Ivakhnenko in [25] recognizing this failure of GMDH, summarizes its causes in the existence of a short delta form correlation between output and predictors, the insufficient functional variety of the model candidates, the immoderate use of a sequence of external criteria for choosing the optimal complexity and the over-complication of individual models. In addition, GMDH has been primarily developed for the solution of small and modest problems, which is not the case for real world systems. The application of correlation analysis prior to GMDH algorithm as well as the development of a combined criterion in the place of external criterion could solve these problems and therefore improve accuracy. Another cause of low accuracy is the possibility of eliminating important variables during the sorting out procedure. GMDH is geared to minimize the mean square error of the resulting model so it takes into account average tendencies only. Any variable, which causes the function values to fall out that average tendency will be characterized as noise and therefore eliminated despite its importance. Also other research studies revealed that on time series modeling GMDH exhibits a tendency to find very complex polynomials that cannot model well future, unseen oscillations of the series [7]. Experimental studies revealed that the multilayer GMDH often underperforms on non-parametric regression tasks [30].

GMDH Algorithm for Discrete Process

The majority of GMDH algorithms has been developed for continuous variables and cannot be applied to binary or discrete problems. A rebinarization technique, which will be used for the transition from binary to continuous attributes with the subsequent use of well known GMDH algorithms, is a potential solution. Ivakhnenko et al. in [26] introduce such an algorithm, which reconstructs with a sufficient precision an unknown harmonic function that is represented by a binary code. The sliding control criterion is applied to improve the parameter estimation since least square estimations sometimes provide imprecise estimates. The harmonic rebinarization or rediscretization algorithm can be also applied to discrete pattern recognition problems allowing the application of parametric GMDH algorithms to find the optimum space of features, the structure of a decision rule and estimate its coefficients [27].

Model's Validation

A very important subject in every modelling procedure is that of model validation. It is significant to ensure that the selected model is adequate to reflect the causal relationships between input-output. Muller in[28] proposes the computation of the model with and without randomization as one of the solutions to that problem. On the other hand, Krotov et al. [29] present a number of criteria, which can prove the verification of the forecast. The correlation coefficient, the mean square error S of the forecast and the mean squared deviation of the predicted process from the mean value of the entire series of observations (norm) can be used. In that case the reliability of the model could be characterized by the ratio of the mean square error to the mean squared deviation.

3 Rationale for Hybrid Systems

In order to alleviate the problems associated with standard GMDH approach as discussed in the preceding section, a number of researchers have attempted to hybridize GMDH with some evolutionary optimization techniques. Amongst them, Iba *et al.* [31] presented the GP-GMDH (Genetic Programming-GMDH) algorithm and showed that it performs better than the conventional GMDH algorithm. Recent development in some of the GMDH aspects has involved the concept of Genetic Algorithms (GA). Robinson [32] points out that the disadvantages of GMDH are its fixed structure and the deterministic nature of the search for the best model. These shortcomings were fixed by using multi-objective genetic algorithm (MOGA) optimization algorithm to search the space of possible polynomials in order to optimize the performance of GMDH. The Ivakhnenko polynomial is replaced by a full four order polynomial and GA is used to identify the optimal partial description. The above modification have been characterised as Term Optimization of GMDH (TOGMDH) since it only finds the optimum terms in partial descriptions and does not alter its structure. Robinson also proposed the Structure Optimization of GMDH (SOGMDH), which optimizes both the model's terms and the structure of the final model. In that algorithm, MOGA optimization algorithm performs a wider stochastic search over a large range of possible models. SOGMDH uses the form of partial descriptions in TOGMDH but allows the evolution of the model in more than one layer allowing in that way the combination of two different partial descriptions in a later stage. Both algorithms have been tested in regression and classification tasks where SOGMDH has shown a remarkable increase in accuracy. Nariman-Zadeh *et al.* [33] proposed a hybrid of genetic algorithm (GA) and GMDH which outperforms conventional GMDH approach. Onwubolu [34] proposed a hybrid of differential evolution (DE) and GMDH and clearly showed that this framework outperforms conventional GMDH approach. Onwubolu and Sharma [35] recently proposed a hybrid of particle swarm optimization (PSO) and GMDH and showed that this framework performs appreciably well compared to the conventional GMDH approach. The Group of Adaptive Models Evolution (GAME) [36] uses neurons (units) with several possible types of transfer function (linear, polynomial, sigmoid, harmonic, preceptron net, etc.).

Further rationale for hybridization of intelligent systems is the fact that *combined* and *integrated* approaches will be necessary if the remaining tough problems (involving imprecision, uncertainty and vagueness, high-dimensionality) in artificial intelligence are to be solved.

Imprecision: This is a very difficult problem to solve and it is our opinion that neither NN nor GMDH has the capability to deal with imprecision problem. For example, the problem of missing data falls under this category. In our opinion, a preprocessor would be needed to deal with such imprecise data before the refined data is operated upon in the NN or GMDH module.

Uncertainty and vagueness: This is a feature that falls into the class of fuzzy problems because uncertain and vague information could be more easily solved when fuzzy paradigm is integrated with NN or GMDH.

High-dimensionality: This is a very difficult problem to solve and it is our opinion that neither NN nor GMDH has the capability to deal with high-dimensionality problem. NN is known to be able to solve large problems more efficiently than GMDH but that attribute does not qualify it to be classed as a method that can handle real-life high-dimensionality problems as for example common in the areas of bioinformatics and medical datasets. In our view, a preprocessor would be required to reduce high-dimensionality problems to low-dimension domain before NN or GMDH could be applied in solving many real-life problems which are usually intractable to find good solutions.

Hybridization can be extremely useful if NN or GMDH is integrated with some computational intelligence (CI) methods that could improve the learning pattern and result in finding global optimal solutions. We refer to this attribute of such CI methods as *Structural Optimization*. This book emphasizes this particular feature and each of the succeeding chapter introduces particular CI methods that have been utilized for integration with GMDH.

Hybridization is the central theme of this book. CI methods have the capabilities of enhancing structural optimization and consequently making the hybrid GMDH more effective and efficient in dealing with complex real-world systems.

It is therefore in order to first discuss the different CI methods currently in use and thereafter present an overview of GMDH hybrid systems.

4 Computational Intelligence

A major thrust in algorithmic development is the design of algorithmic models to solve increasingly complex problems. Enormous successes have been achieved through the modeling of biological and natural intelligence, resulting in so-called "intelligent systems". These intelligent algorithms include artificial neural networks, evolutionary computation, swarm intelligence, artificial immune systems, and fuzzy systems. Together with logic, deductive reasoning, expert systems, case-based reasoning and symbolic machine learning systems, these intelligent algorithms form part of the field of Artificial Intelligence (AI). Just looking at this wide variety of AI techniques, AI can be seen as a combination of several research disciplines, for example, computer science, physiology, philosophy, sociology and biology.

4.1 What Is Intelligence?

A major thrust in the algorithmic development and enhancement is the design of algorithmic models to solve increasingly complex problems and in an efficient manner. Enormous successes have been achieved through modeling of biological and natural intelligence, resulting in "intelligent systems". These intelligent algorithms include neural networks, evolutionary computing, swarm intelligence, and fuzzy systems. Together with logic, deductive reasoning, expert systems, case-based reasoning and symbolic machine learning systems, these intelligent algorithms form part of the field of Artificial Intelligence (AI) [38]. Just looking at this wide variety of AI techniques, AI can be seen as a combination of several research disciplines, for example, engineering, computer science, philosophy, sociology and biology.

There are many definitions to intelligence. Here, we prefer the definition from [38]-Intelligence can be defined as the ability to comprehend, to understand and profit from experience, to interpret intelligence, having the capacity for thought and reason (especially, to a higher degree). Other keywords that describe aspects of intelligence include creativity, skill, consciousness, emotion and intuition. Computational Intelligence (CI) is the study of adaptive mechanisms to enable or facilitate intelligent behavior in complex, uncertain and changing environments. These adaptive mechanisms include those AI paradigms that exhibit an ability to learn or adapt to new situations, to generalize, abstract, discover and associate.

4.2 Can Computers Be Intelligent?

This is a question that to this day causes more debate than the definitions of intelligence. In the mid-1900s, Alan Turing gave much thought to this question. He believed that machines could be created that would mimic the processes of the human brain [37]. Turing strongly believed that there was nothing the brain could do that a well-designed computer could not. More than fifty years later his statements are still visionary. Today, much success has been achieved in using machining learning methodologies for modeling small parts of biological neural systems; however, there are still no solutions to the complex problem of modeling intuition, consciousness and emotion-which form integral parts of human intelligence.

A more recent definition of artificial intelligence came from the IEEE Neural Networks Council of 1996: the study of how to make computers do things at which people are doing better. This is a definition that seems flawed. Most books (see [38] for example) concentrate on a sub-branch of AI, namely Computational Intelligence (CI) - the study of adaptive mechanisms to enable or facilitate intelligent behavior in complex and changing environments. These mechanisms include those AI paradigms that exhibit an ability to learn or adapt to new situations, to generalize, abstract, discover and associate. The following CI paradigms are covered: artificial neural networks, evolutionary computation, swarm intelligence, artificial immune systems, and fuzzy systems. While individual techniques from these CI paradigms have been applied successfully to solve real-world problems, the current trend is to develop hybrids of paradigms, since no one paradigm is superior to the others in all situations. In doing so, we capitalize on the respective strengths of the components of the hybrid CI system and eliminate weaknesses of individual components.

At this point it is necessary to state that there are different definitions of what constitutes CI. The classification of CI in this book follows that of [38]. For example, swarm intelligence (SI) and artificial immune systems (AIS) are classified as CI paradigms, while many researchers consider these paradigms to belong only under Artificial Life. However, both particle swarm optimization (PSO) and ant colony optimization (ACO), as treated under SI, satisfy the definition of CI given above, and are therefore included in this book as being CI techniques. The same applies to AISs.

4.3 Computational Intelligence Paradigms

This book considers five main paradigms of Computation Intelligence (CI), namely artificial neural networks (NN), evolutionary computation (EC), swarm intelligence (SI), artificial immune systems (AIS), and fuzzy systems (FS) [see Figure 3]. In addition to CI paradigms, probabilistic methods are frequently used together with CI techniques, which are also shown in the figure. Soft computing, a term coined by Lotfi Zadeh, is a different grouping of paradigms, which usually refers to the collective set of CI paradigms and probabilistic methods. The arrows indicate that techniques from different paradigms can be combined to form hybrid systems.

Each of the CI paradigms has its origins in biological systems. NNs model biological neural systems, EC models natural evolution (including genetic and behavioral evolution), SI models the social behavior of organisms living in swarms or

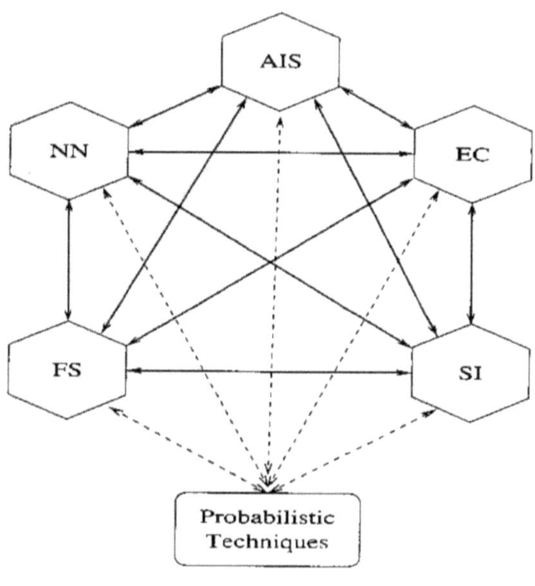

Fig. 3 Computational Intelligence Paradigms

colonies, AIS models the human immune system, and FS originated from studies of how organisms interact with their environment.

4.3.1 Artificial Neural Networks

The way that biological neurons work has intrigued neurologists for several years. It is known that neurons are connected together via synapses, and a synapse produces a chemical response to input. The biological neuron fires if the sum of all the reactions from the synapses is sufficiently large. For many years, scientists and engineers have been interested in the actions of biological neurons in order to define new models of parallel problem solving. The McCulloch-Pitts' theory [39] that treat the human brain as a computational organism is the foundation for all activities in the central nervous system and forms the basis for most neural-network models. In their work, McCulloch and Pitts model the central nervous system as neural circuits that have computational power. Each neuron sends impulses to many other neurons in a process known as divergence, receives impulses from many other neurons in a process known as convergence, and also receives impulses from *feedback paths*.

Let us briefly examine the activity that occurs at the connection between two neurons called the synaptic junction or synapse. Communication between two neurons occurs as a result of postsynaptic cell absorbing chemical substances called neurotransmitter by the presynaptic cell as shown in Figure 4. As the action potential arrives at the presynaptic membrane, the permeability of the membrane changes, resulting in influx of calcium irons. These irons cause the vesicles containing the neurotransmitters to fuse with the presynaptic membrane, resulting in their releasing neurotransmitters into the synaptic cleft. Consequently, the neurotransmitters diffuse across the synaptic cleft into the membrane of the postsynaptic membrane at certain receptor sites. The chemical-action at the receptor sites influences the permeability of the postsynaptic membrane. When positive ions enter the receptor sites, the action results in depolarisation, an effect referred to as *excitatory*. On the other hand, when negative ions enter the receptor sites, the action results in hyperpolarisation, an effect referred to as *inhibition*. Both, the excitatory and inhibition

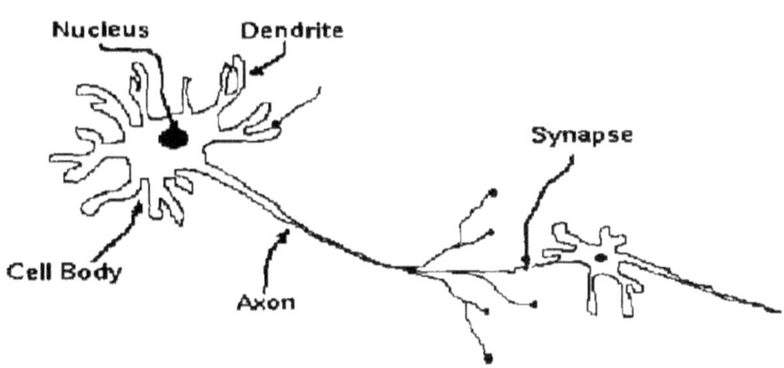

Fig. 4 A biological neuron

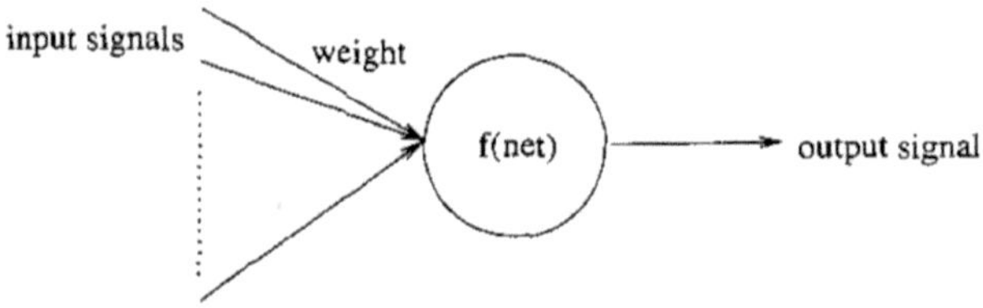

Fig. 5 An artificial neuron

actions are local actions that take place within a finite distance into the cell body and are summed up at the axon hillock. If the sum is greater than a certain threshold, an action potential is generated. If the sum is less than this threshold, an action potential in not generated.

The AN collects all incoming signals, and computes a net input signal as a function of the respective weights. The net input signal serves as input to the activation function which calculates the output signal of the AN.

The artificial neural network (ANN) is an adaptive algorithm that takes its roots in the way that biological neurons work. Neural networks are massively parallel interconnected networks of simple elements which are usually adaptive and their hierarchical organizations which are intended to interact with the objects of the real world in the same way as biological nervous systems do. The basic components of a neural network are nodes that correspond to biological synapses. The weighted-inputs to a neuron are accumulated and then passed on to an activation function that determines the nervous response. A positive weight represents an excitatory connection while a negative weight represents an inhibitory connection. In fact, the units were originally invented as an attempt to model biological neurons, hence the use of the term neural networks. A neural network is divided into layers: input layer, hidden layer(s) and the output layer as shown in Figure 6.

Neural network may be classified on the basis of the directions in which, signals flow. In the feed-forward network, signals propagate in one direction from the input neurons through intermediate neurons in the hidden layer(s) to the neurons in the output layer. In the recurrent network, signals may propagate from the output of any neuron to the input of any neuron.

Another way in which neural networks may be classified is the extent to which the user (teacher) guides the learning-process. A supervised learning neural network adjusts weights of nodes of the hidden layer(s) and output on the basis of the difference between the values of the output units and the expected values assigned by the teacher, for a given input pattern. An unsupervised learning neural network adjusts weights of the nodes and classifies the input into sets without being guided. In artificial intelligence application to most engineering problems, it is useful to implement unsupervised feed-forward or unsupervised recurrent neural networks. There are basically two major classifications of neural networks: feed forward and recurrent. A neural network is either supervised or unsupervised. Supervised, feed forward neural networks include; perceptron, Hamming network, counter-propagation network (CPN), linear associative memory (LAM), and Boltzman machine. Unsupervised, feed forward neural networks include: clustering-network and self-organising

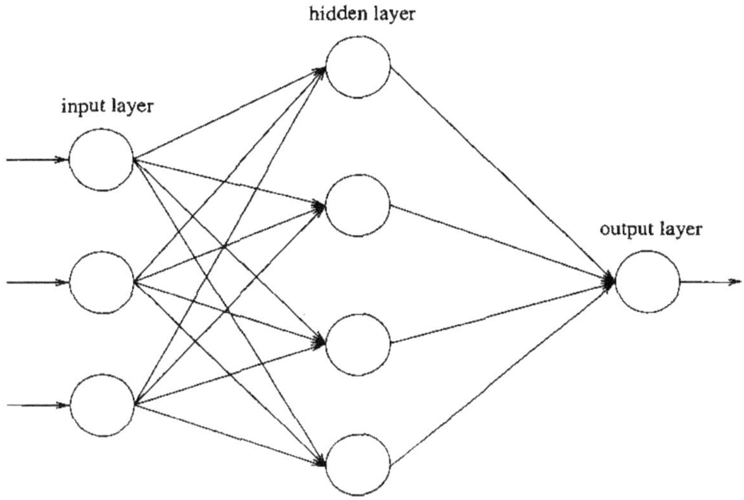

Fig. 6 An Artificial Neural Network

feature maps (SOM). Supervised, recurrent neural networks include; bi-directional associative memory (BAM), auto-associative memory, and Hopfield network. Unsupervised recurrent neural networks include; adaptive resonance theory (ART1 and ART2). An excellent summary of the taxonomy of the most important network models may be found in Huang and Zhang [40]. The details of ANN are found in Onwubolu [41].

These NN types have been used for a wide range of applications, including diagnosis of diseases, speech recognition, data mining, composing music, image processing, forecasting, robot control, credit approval, classification, pattern recognition, planning game strategies, compression, and many others.

4.3.2 Evolutionary Computation

Evolutionary computation (EC) has as its objective of survival of the fittest: the weak must give way to the strong. In natural evolution, survival is achieved through reproduction. In this concept, it is postulated that offspring, reproduced from two parents (sometimes more than two), contain genetic material of both (or all) parents - hopefully the best characteristics of each parent. Those individuals that inherit bad characteristics are weak and lose the battle to survive. This is nicely illustrated in some bird species where one hatchling manages to get more food, gets stronger, and at the end kicks out all its siblings from the nest to die.

Evolutionary algorithms use a population of individuals, where an individual is referred to as a chromosome. A chromosome defines the characteristics of individuals in the population. Each characteristic is referred to as a gene. The value of a gene is referred to as an allele. For each generation, individuals compete to reproduce offspring. Those individuals with the best survival capabilities have the best chance to reproduce. Offspring are generated by combining parts of the parents, a process referred to as crossover. Each individual in the population can also undergo mutation which alters some of the allele of the chromosome. The survival strength of an individual is measured using a fitness function which reflects the objectives and constraints of the problem to be solved. After each generation, individuals may undergo culling, or individuals may survive to the next generation (referred to as elitism). Additionally, behavioral characteristics (as encapsulated in phenotypes) can be used to influence the evolutionary process in two ways: phenotypes may influence genetic changes, and/or behavioral characteristics evolve separately.

Different classes of evolutionary algorithms (EA) have been developed:

- Genetic algorithms which model genetic evolution.
- Genetic programming which is based on genetic algorithms, but individuals are programs (represented as trees).
- Evolutionary programming which is derived from the simulation of adaptive behavior in evolution (phenotypic evolution).
- Evolution strategies which are geared toward modeling the strategy parameters that control variation in evolution, i.e. the evolution of evolution.
- Differential evolution, which is similar to genetic algorithms, differing in the reproduction mechanism used.
- Cultural evolution which models the evolution of culture of a population and how the culture influences the genetic and phenotypic evolution of individuals.
- Co-evolution where initially "dumb" individuals evolve through cooperation, or in competition with one another, acquiring the necessary characteristics to survive.

Other aspects have also been modeled. For example, mass extinction, and distributed (island) genetic algorithms, where different populations are maintained with genetic evolution taking place in each population. In addition, aspects such as migration among populations are modeled. The modeling of parasitic behavior has also contributed to improved evolutionary techniques. In this case parasites infect individuals. Those individuals that are too weak are replaced by the stronger ones. On the other hand, immunology has been used to study the evolution of viruses and how antibodies should evolve to destroy virus infections. Evolutionary computation has been used successfully in real-world applications, for example, data mining, combinatorial optimization, fault diagnosis, classification, clustering, scheduling, and time series approximation.

Generally, the main steps in EC algorithms are as follows.

- Initialize the initial generation of individuals.
- While not converged

  1. Evaluate the fitness of each individual.
  2. Select parents from the population
  3. Recombine selected parents using crossover to get offspring
  4. Mutate offspring
  5. Select new generation of populations

4.3.3 Computational Swarm Intelligence

Swarm intelligence (SI) originated from the study of colonies, or swarms of social organisms. Studies of the social behavior of organisms (individuals) in swarms prompted the design of very efficient optimization and clustering algorithms. For example, studies of the foraging behavior of ants resulted in ant colony optimization (ACO) algorithms and simulation studies of the graceful, but unpredictable, choreography of bird flocks led to the design of the particle swarm optimization algorithm.

The collective performance of social insects, such as ants, bees, wasps or termites has intrigued entomologists for several years. Their main concern is about the mechanisms that allow the individuals of the same colony to co-ordinate their activities and to favour the survival of the species. Apparently everything works out because of an underlying factor which regulates the activities of each individual. Studies have shown that this global adaptive behaviour arises from a multitude of very simple local interactions. The nature of these interactions, the treatment of the information, the differences between the solitary behaviour and the social behaviour have remained unclear for a long time. The realisation of a specific task by a colony of insects has shown that the co-ordination of the work does not depend on the insects but rather on the advancing state of the task. Co-ordination emerges from an auto-catalytic chain retroaction between stimuli and responses. An insect does not control directly its work, the whole process progresses as if each insect were guided by its work. While working an insect modifies the form of the stimulation which triggers its behaviour. This induces the emergence of a new stimulation that will trigger new reactions in the colony.

In order to illustrate the emergence of collective structures in an insect society, let us cite the example of an ant colony in search of a nearby feeding source. Initially, ants leave the nest and move randomly. When an ant discovers a feeding source, it informs its other ants belonging to the same colony by laying a temporary trail on the ground on its way back to the nest. The trail is nothing else than a chemical substance called pheromone, which guides the other ants towards the same feeding source. On their way back, the latter also lay pheromone on the ground and thus reinforce the marking of the path that leads from the nest to the discovered feeding source. The reinforcement of the marking by pheromone optimises the collection of food. All trails laid on the ground evaporate progressively as time goes by. Because of the larger elapsed time between two passages of an ant on the paths leading to remote feeding sources, trails will get undetectable faster on these paths. In the long term, ants will thus, all prefer the closest feeding source. This example shows that an ant colony converges towards an optimal solution whereas each single ant is unable to solve the problem by itself within a reasonable amount of time. In this case, as pointed out [42] 'the environment plays the role of spatio-temporal memory keeping track of the swarm past actions while selecting its own dynamic regime'.

For many years, engineers have been interested in the behaviour of social insects in order to define new models of collective problem solving. The Ant System developed recently in Dorigo [43], is an adaptive algorithm which takes its roots in the collective behaviour of an ant colony. In the co-operation phase of an ant algorithm each solution of the population is examined with the aim of updating a global memory keeping track of important structures of the set of all feasible solutions which have been successfully exploited in the past. The self-adaptation phase uses a problem-specific constructive method to create a new population of solution on the basis of the global memory. Therefore, the ants are able to optimise their paths by this process.

An ant system, as we use the term here, is a computational paradigm inspired by ants' collective contributions in solving a common problem. A similar process can be transposed to combinatorial optimisation: solutions of the problem are built using a statistics on solutions previously generated. This statistics play the role of the pheromone traces and it gives a higher weight to the best solutions. After a while, it is observed that such a procedure is able to build solutions of better quality than a procedure guided by partial objective function evaluations only. The different components of fast ant system include the memory or pheromone trail, solutions manipulation, intensification, and diversification.

Studies of ant colonies have contributed in abundance to the set of intelligent algorithms such as shortest path optimization algorithms, routing optimization in telecommunications networks, graph coloring, scheduling and solving the quadratic assignment problem, clustering and structural optimization algorithms.

Particle swarm optimization (PSO) [44] is a stochastic optimization approach, modeled on the social behavior of bird flocks, PSO is a population-based search procedure where the individuals, referred to as particles, are grouped into a swarm. Each particle in the swarm represents a candidate solution to the optimization problem. In a PSO system, each particle "flown" through the multidimensional search space, adjusting its position in search space according to its own experience and that of neighboring particles. A particle therefore makes use of the best position encountered by itself and the best position of its neighbors to position itself toward an optimum solution. The effect is that particles "fly" toward an optimum, while still searching a wide area around the current best solution. The performance of each particle (i.e. the "closeness" of a particle to the global minimum) is measured according to a predefined fitness function which is related to the problem being solved.

Applications of PSO include function approximation, clustering, optimization of mechanical structures, and solving systems of equations. Details of SI are found in Onwubolu and Babu [45].

4.3.4 Artificial Immune Systems

The natural immune system (NIS) has an amazing pattern matching ability, used to distinguish between foreign cells entering the body (referred to as non-self, or antigen) and the cells belonging to the body (referred to as self). As the NIS encounters antigen, the adaptive nature of the NIS is exhibited, with the NIS memorizing the structure of these antigen for faster future response the antigen.

In NIS research, four models of the NIS can be found:

- The classical view of the immune system is that the immune system distinguishes between self and non-self, using lymphocytes produced in the lymphoid organs. These lymphocytes "learn" to hind to antigen.
- Clonal selection theory, where an active B-Cell produces antibodies through a cloning process. The produced clones are also mutated.
- Danger theory, where the immune system has the ability to distinguish between dangerous and non-dangerous antigen.
- Network theory, where it is assumed that B-Cells form a network. When a B-Cell responds to an antigen, that B-Cell becomes activated and stimulates all other B-Cells to which it is connected in the network.

An artificial immune system (AIS) models some of the aspects of a NIS, and is mainly applied to solve pattern recognition problems, to perform classification tasks, and to cluster data. One of the main application areas of AISs is in anomaly detection, such as fraud detection, and computer virus detection.

4.3.5 Fuzzy Systems

Traditional set theory requires elements to be either part of a set or not. Similarly, binary-valued logic requires the values of parameters to be either 0 or 1, with similar constraints on the outcome of an inferencing process. Human reasoning is, however, almost always not this exact. Our observations and reasoning usually include a measure of uncertainty. For example, humans are capable of understanding the sentence: "Some Computer Science students can program in most languages". But how can a computer represent and reason with this fact?

Fuzzy sets and fuzzy logic allow what is referred to as approximate reasoning. With fuzzy sets, an element belongs to a set to a certain degree of certainty. Fuzzy logic allows reasoning with these uncertain facts to infer new facts, with a degree of certainty associated with each fact. In a sense, fuzzy sets and logic allow the modeling of common sense.

The uncertainty in fuzzy systems is referred to as non-statistical uncertainty, and should not be confused with statistical uncertainty. Statistical uncertainty is based on the laws of probability, whereas non-statistical uncertainty is based on vagueness, imprecision and/or ambiguity. Statistical uncertainty is resolved through observations. For example, when a coin is tossed we are certain what the outcome is, while before tossing the coin, we know that the probability of each outcome is 50%. Non-statistical uncertainty, or fuzziness, is an inherent property of a system and cannot be altered or resolved by observations. Fuzzy systems have been applied successfully to control systems, gear transmission and braking systems in vehicles, controlling lifts, home appliances, controlling traffic signals, and many others.

5 Hybrid GMDH Systems

Many researchers consider five main paradigms of Computation Intelligence (CI), namely:

1. Artificial Neural Networks (ANN),
2. Evolutionary Computation (EC),
3. Swarm Intelligence (SI),
4. Artificial Immune Systems (AIS), and
5. Fuzzy Systems (FS).

However, in this book, we consider that GMDH and ANN are the two well established self-organizing modeling (SOM) methods which are in use for practical problems. In this book, we present a framework for hybridizing GMDH with other components of CI as shown in Figure 7.

Therefore, the five dominant computational intelligence paradigms are GMDH/NN, evolutionary computing, swarm intelligence, artificial immune systems, and fuzzy systems as illustrated in Figure 7. For many years now, NNs have dominated the literature while GMDH has been in the "sleep mode", albeit GMDH has been found be robust for modeling and prediction of complex systems. In the model presented in this book GMDH/NN mean "GMDH or NN". These paradigms can be combined, in top-down architectures to form hybrids as shown in Figure 7, resulting in GMDH/Neuro-Evolutionary Computing systems, GMDH/Neuro-Swarm systems, GMDH/Neuro-Immune systems, GMDH/Neuro-Fuzzy systems, etc. This means that it is feasible to form hybrids of GMDH-GA, GMDH-GP, GMDH-DE, GMDH-SS, GMDH-ACO, GMDH-PSO, GMDH-AIS, and GMDH-FS.

Other lower level hybrids which are not relevant to this book include Fuzzy-PSO systems, Fuzzy-GA systems, etc.

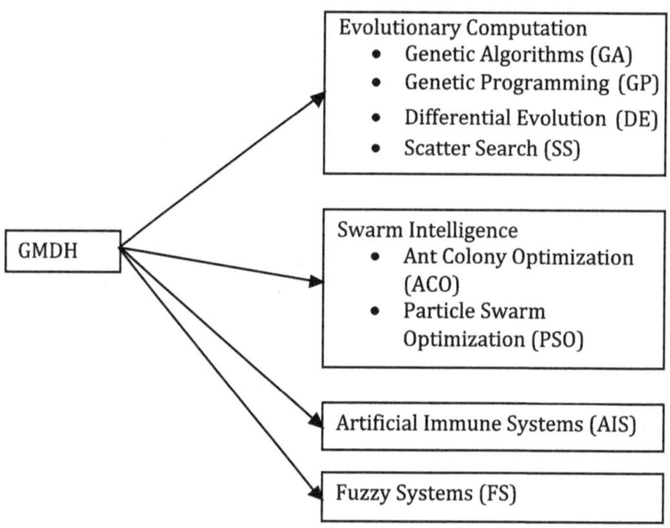

Fig. 7 Framework for hybridizing GMDH with other components of CI as presented in this book

Figure 7 Framework for hybridizing GMDH with other components of CI as presented in this book.

The other chapters of this book are organized as follows. Hitoshi Iba, the originator of Hybrid Genetic Programming and GMDH System presents STROGANOFF in chapter 2. Nader Nariman-zadeh and Ali Jamali, the originators of Hybrid Genetic Algorithm and GMDH System present chapter 3. Godfrey Onwubolu, the originator of Hybrid Differential Evolution and GMDH System presents chapter 4, which is the kernel of Knowledge Management & Mining (KMM) software that he has developed. Anuraganand Sharma and Godfrey Onwubolu, the originators of Hybrid Particle Swarm Optimization and GMDH System present chapter 5. Pavel Kordik the originator of GAME-Hybrid Self-Organizing Modeling System based on GMDH presents chapter 6.

6 Conclusion

GMDH based algorithms and self-organization can be used to automate almost the whole knowledge discovery process, i.e. models have been created adaptively and data preparation will be self-organized in special missing values are estimated and dimensionality is reduced. Automated solutions are more or less based on techniques developed in a discipline named "machine learning" as an important part of artificial intelligence. These are various techniques by which computerized algorithms can learn which patterns actually do exist in data sets. They may not be as intelligent as humans but are error-free, consistent, formidable fast, and tireless compared to humans. Experimental studies revealed that the multilayer GMDH often underperforms on non-parametric regression tasks; moreover, time series modeling GMDH exhibits a tendency to find very complex polynomials that cannot model well future, unseen oscillations of the series.

In order to alleviate the problems associated with standard GMDH approach, a number of researchers have attempted to hybridize GMDH with some evolutionary optimization techniques. This is the central theme of this book. It is hoped that researchers by sieving the contents of this book, will become active in investigating how standard GMDH could become more robust and flexible in solving complex, real-world problems which currently cannot be solved using standard GMDH approach.

References

1. Ivakhnenko, A.G.: Polynomial theory of complex systems. IEEE Trans. on Systems, Man and Cybernetics SMC-1, 364–378 (1971)
2. Madala, H.R., Ivakhnenko, A.G.: Inductive Learning Algorithms for Complex Systems Modelling. CRC Press Inc., Boca Raton (1994)
3. Myers, R.H.: Classical and modern regression with applications. PWS-KENT, Boston, Ma, vol. 4, pp. 1048–1055 (1994)

4. Bishop, C.: Neural Networks for Pattern Recognition. Oxford University Press, Oxford (1995)
5. Hertz, J., Krogh, A., Palmer, R.G.: Introduction to the Theory of Neural Computation. Addison Wesley, Reading (1991)
6. Hastie, T., Tibshirani, R., Friedman, J.: The Elements of Statistical Learning: Data Mining, Inference, and Prediction. Springer, Heidelberg (2001)
7. Nikolaev, N.Y., Iba, H.: Polynomial harmonic GMDH learning networks for time series modeling. Neural Networks 16, 1527–1540 (2003)
8. Anastasakis, L., Mort, N.: The Development of Self-Organization Technique. In: Modelling: A Review of The Group Method of Data Handling (GMDH), Research Report No. 813, Department of Automatic Control & Systems Engineering, The University of Sheffield, Mappin St, Sheffield, S1 3JD, United Kingdom (October 2001)
9. Yurachkovskiy, Y.P.: Improved GMDH algorithms for process prediction. Soviet Automatic Control c/c of Avtomatika 10(5), 61–71 (1977)
10. Duffy, J.J., Franklin, M.A.: A learning identification algorithm and its application to an environmental system. IEEE Transactions on Systems, Man and Cybernetics SMC-5(2), 226–240 (1975)
11. Sarychev, A.P.: Stable estimation of the coefficients in multilayer GMDH algorithms. Soviet Automatic Control c/c of Avtomatika 17(5), 1–5 (1984)
12. Nishikawa, T., Shimizu, S.: Identification and forecasting in management systems using the GMDH method. Applied Mathematical Modelling 6(1), 7–15 (1982)
13. Ivakhnenko, A.G.: The group method of data handling in prediction problems. Soviet Automatic Control c/c of Avtomatika 9(6), 21–30 (1976)
14. Triseyev, Y.P.: GMDH algorithm with variable freedom of choice in selection layers based on criterion of diversity of variables. Soviet Automatic Control c/c of Avtomatika 10(4), 30–33 (1977)
15. Parker, R.G.J., Tummala, M.: Identification of volterra systems with a polynomial neural network. In: Proceedings of the 1992 IEEE International Conference on Acoustics, Speech and Signal Processing - ICASSP 1992, vol. 4, pp. 561–564 (1992)
16. Styblinski, M.A., Aftab, S.: Combination of interpolation and self-organizing approximation techniques-a new approach to circuit performance modeling. IEEE Transactions on Computer Aided Design of Integrated Circuits and Systems 12(11), 1775–1785 (1993)
17. Ivakhnenko, A.G., Krotov, G.I.: A multiplicative-additive non-linear GMDH with optimization of the power of factors. Soviet Automatic Control c/c of Avtomatika 17(3), 10–13 (1984)
18. Ivakhnenko, A.G.: The group method of data handling - a rival of the method of stochastic approximation. Soviet Automatic Control c/c of Avtomatika 1(3), 43–55 (1968)
19. Muller, J.A., Ivakhnenko, A.G.: Self-organizing modelling in analysis and prediction of stock market. In: Proceedings of the Second International Conference on Application of Fuzzy Systems and Soft Computing-ICAFS 1996, Siegen, Germany, pp. 491–500 (1996)
20. Ivakhnenko, A.G.: Heuristic self-organization in problems of engineering cybernetics. Automatica 6, 207–219 (1970)
21. Ikeda, S., Fujishige, S., Sawaragi, Y.: Non-linear prediction model of river flow by self-organization method. International Journal of Systems Science 7(2), 165–176 (1976)
22. Park, H.S., Oh, S.K., Ahn, T.C., Pedrycz, W.: A study on multi-layer fuzzy polynomial inference system based on extended GMDH algorithm. In: Proceedings of the 1999 IEEE International Conference on Fuzzy Systems - FUZZ-IEEE 1999, vol. 1, pp. 354–359 (1999)

23. Mehra, R.K.: Group method of data handling (GMDH): review and experience. In: Proceedings of the IEEE Conference on Decision and Control, pp. 29–34 (1977)
24. Tumanov, N.V.: A GMDH algorithm with mutually orthogonal partial descriptions for synthesis of polynomial models of complex objects. Soviet Automatic Control c/c of Avtomatika 11(3), 82–84 (1978)
25. Ivakhnenko, A.G.: Development and application of the group method of data handling for modelling and long-range prediction. Soviet Journal of Automation and Information Sciences c/c of Avtomatika 18(3), 26–38 (1985)
26. Ivakhnenko, A.G., Zholnarskiy, A.A., Muller, J.A.: An algorithm of harmonic rebinarization of a data sample. Journal of Automation and information Sciences c/c of Avtomatika 25(6), 34–38 (1992)
27. Ivakhnenko, A.G., Ivakhnenko, G.A.: A comparison of discrete and continuous recognition systems. Pattern Recognition and Image Analysis 6(3), 445–447 (1996)
28. Muller, J.A.: Self-organization of models - present state (1996), `http://www.inf.kiev.ua/GMDH-home/articles/`
29. Krotov, G.I., Kozubovskiy, S.F.: Verification of dendroscale forecasting by a multiplicative GMDH algorithm. Soviet Journal of Automation and Information Sciences c/c of Avtomatika 20(3), 1–7 (1987)
30. Green, D.G., Reichelt, R.E., Bradbury, R.H.: Statistical behavior of the GMDH algorithm. Biometrics 44, 49–69 (1998)
31. Iba, H., de Garis, H., Sato, T.: Genetic programming using a minimum description length priniciple. In: Kinnear Jr., K.E. (ed.) Advances in Genetic Programming, pp. 265–284. MIT Press, Cambridge (1994)
32. Robinson, C.: Multi-objective optimization of polynomial models for time series prediction using genetic algorithms and neural networks, PhD Thesis in the Dept. of Automatic Control & Systems Engineering, University of Sheffield, UK (1998)
33. Nariman-Zadeh, N., Darvizeh, A., Ahmad-Zadeh, G.R.: Hybrid genetic design of GMDH-type neural networks using singular value decomposition for modeling and predicting of the explosive cutting process. In: Proc. Instn. Mech. Engrs., vol. 217, Part B, pp. 779–790 (2003)
34. Onwubolu, G.C.: Design of hybrid differential evolution and group method in data handling networks for modeling and prediction. Information Sciences 178, 3618–3634 (2008)
35. Onwubolu, G.C., Sharma, S., Dayal, A., Bhartu, D., Shankar, A., Katafono, K.: Hybrid particle swarm optimization and group method of data handling for inductive modeling. In: Proceedings of International Conference on Inductive Modeling, Kyiv, Ukraine, September 15-19 (2008)
36. Kordik, P.: Fully Automated Knowledge Extraction using Group of Adaptive Models Evolution. PhD Thesis, Dept. of Comp. Sci. and Computers, FEE, CTU Prague, Czech Republic (September 2006)
37. Turing, A.M.: Computing Machinery and Intelligence. Mind 59, 433–460 (1950)
38. Engelbrecht, A.P.: Computation Intelligence: An Introduction, 2nd edn. Wiley, Chichester (2001)
39. McCullon, W.S., Pitts, W.A.: A logical calculus of the ideas imminent in nervous activity. Bulletin of Mathematics and Biophysics 5, 115–133 (1943)
40. Huang, S.H., Zhang, H.C.: Application of neural networks in manufacturing a state-of-the-art survey. International Journal of Production Research 33, 705–728 (1995)

41. Onwubolu, G.C.: Emerging Optimization Techniques in Production Planning & Control. Imperial College Press, London (2002)
42. Theraulaz, G., Goss, S., Gervet, J., Deneubourg, J.L.: Task differentiation in polistes wasp colonies: a model for self-organising groups of robots. In: Simulation of Adaptive Behaviour: From animals to Animats, pp. 346–355. MIT Press/ Bradford Books, Cambridge, Mass (1991)
43. Dorigo, M.: Optimisation, Learning and Natural Algorithms, PhD. Dissertation, Dipartimento Elettronica e Informazione, Politecnico di Milano, Italy (1992)
44. Kennedy, J., Eberhart, R.C.: A discrete binary version of the particle swarm algorithm. In: International Conference on Systems, Man, and Cybernetics (1997)
45. Onwubolu, G.C., Babu, B.V. (eds.): New Optimization Techniques in Engineering. Springer, Heidelberg (2004)

Hybrid Genetic Programming and GMDH System: STROGANOFF

Iba Hitoshi

Abstract. This chapter introduces a new approach to Genetic Programming (GP), based on GMDH-based technique, which integrates a GP-based adaptive search of tree structures, and a local parameter tuning mechanism employing statistical search. The GP is supplemented with a local hill climbing search, using a parameter tuning procedure. More precisely, we integrate the structural search of traditional GP with a multiple regression analysis method and establish our adaptive program called .STROGANOFF' (i.e. STructured Representation On Genetic Algorithms for NOnlinear Function Fitting). The fitness evaluation is based on a Minimum Description Length (MDL) criterion, which effectively controls the tree growth in GP. Its effectiveness is demonstrated by solving several system identification (numerical) problems and comparinf the performance of STROGANOFF with traditional GP and another standard technique. The effectiveness of this numerical approach to GP is demonstrated by successful application to computational finances.

1 Introduction

This chapter introduces a new approach to Genetic Programming (GP), based on a numerical, i.e., GMDH-based, technique, which integrates a GP-based adaptive search of tree structures, and a local parameter tuning mechanism employing statistical search (i.e. a system identification technique). In traditional GP, recombination can cause frequent disruption of building blocks, or mutation can cause abrupt changes in the semantics. To overcome these difficulties, we supplement traditional GP with a local hill climbing search, using a parameter tuning procedure. More precisely, we integrate the structural search of traditional GP with a multiple regression analysis method and establish our adaptive program called "STROGANOFF"

Iba Hitoshi
Department of Information and Communication Engineering, Faculty of Engineering, University of Tokyo, 7-3-1 Hongo, Bunkyo-ku, Tokyo, 113-8656, Japan
e-mail: iba@iba.k.u-tokyo.ac.jp

G.C. Onwubolu (Ed.): Hybrid Self-Organizing Modeling Systems, SCI 211, pp. 27–98.
springerlink.com

(i.e. STructured Representation On Genetic Algorithms for NOnlinear Function Fitting). The fitness evaluation is based on a "Minimum Description Length (MDL)" criterion, which effectively controls the tree growth in GP. We demonstrate its effectiveness by solving several system identification (numerical) problems and compare the performance of STROGANOFF with traditional GP and another standard technique (i.e. "radial basis functions"). The effectiveness of this numerical approach to GP is demonstrated by successful application to computational finances.

2 Background

The target problem we solve is "system identification". Attempts have been made to apply traditional GP to the system identification problems, but difficulties have arisen due to the fact that GP recombination can cause frequent disruption of building blocks[1], or that mutation can cause abrupt changes in the semantics. We convert a symbolic (discrete) search problem into a numeric (continuous) search space problem (and vice versa).

2.1 *System Identification Problems*

A system identification problem is defined in the following way. Assume that a single valued output y, of an unknown system, behaves as a function of m input values, i.e.

$$y = f(x_1, x_2, \cdots, x_m). \tag{1}$$

Given N observations of these input-output data pairs, i.e.

INPUT	OUTPUT
$x_{11}\ x_{12}\ \cdots\ x_{1m}$	y_1
$x_{21}\ x_{22}\ \cdots\ x_{2m}$	y_2
$\cdots$	$\cdots$
$x_{N1}\ x_{N2}\ \cdots\ x_{Nm}$	y_N

the system identification task is to approximate the function f with an approximate function $\overline{f}$ called the "complete form".

System identification can be applied to a wide area of application. An example of system identification is time-series prediction, i.e. predicting future values of a variable from its previous values (see Fig.5). Expressed in system identification terms, the output $x(t)$ at time t is to be predicted from its values at earlier times $(x(t-1), x(t-2), \cdots)$, i.e.

$$x(t) = f(x(t-1), x(t-2), x(t-3), x(t-4), \cdots) \tag{2}$$

[1] In this section, a building block (i.e. schema) for GP is defined as a subtree which is a part of a solution tree.

Another example is a type of pattern recognition (or classification) problem, in which the task is to classify objects having m features $x_1, \cdots, x_m$ into one of two possible classes, i.e. "C" and "not C". If an object belongs to class C, it is said to be a positive example of that class, otherwise it is a negative example. In system identification terms, the task is to find a (binary) function f of the m features of objects such that

$$y = f(x_1, x_2, \cdots, x_m) = \begin{cases} 0, & \text{negative example} \\ 1, & \text{positive example} \end{cases} \tag{3}$$

The output y is 1 if the object is a positive example (i.e. belongs to class C), and y is 0 if the object is a negative example.

Most system identification techniques are based on parameter and function estimates. Unfortunately these earlier approaches suffered from combinatorial explosion as the number of training data, parameters, and constrained assumptions increased. One of these approaches was a heuristic algorithm called GMDH (Group Method of Data Handling) [Ivakhnenko71]. It too had its weaknesses, due to its heuristic nature, e.g. it suffered from local extrema problems, which limited its application [Tenorio *et al.*90]. However, this chapter shows that the weakness of the GMDH approach can be largely overcome by wedding it to a (structured) GP-based approach.

2.2 Difficulties with Traditional GP

GP searches for desired tree structures by applying genetic operators such as crossover and mutation. However, standard GP is faced with the following difficulties in terms of efficiency.

1. A lack of tools to guide the effective use of genetic operators.
2. Representational problems in designing node variables.
3. Performance evaluation of tree structures.

Firstly, traditional GP blindly combines subtrees, by applying crossover operations. This blind replacement, in general, can often disrupt beneficial building blocks in tree structures. Randomly chosen crossover points ignore the semantics of the parent trees. For instance, in order to construct the Pythagoras relation (i.e. $a^2 + b^2 = c^2$) from two parent trees (i.e. $c^2 = 3$ and $(a - c) \times (a^2 + b^2)$), only one pair of crossover points is valid (see Fig.1(b)). Thus crossover operations seem almost hopeless as a means to construct higher-order building blocks. [Koza90, ch.4.10.2] used a constrained crossover operator when applied to Neural Network learning, in which two types of node (weight and threshold functions) always appeared alternately in trees to represent feasible neurons. A constrained crossover operator was applied so that it would preserve the order constraint. It worked well for Neural Network learning but its applicability is limited. [Schaffer & Morishima87] discussed adaptive crossover operations for usual string-based genetic algorithms. Although the quality

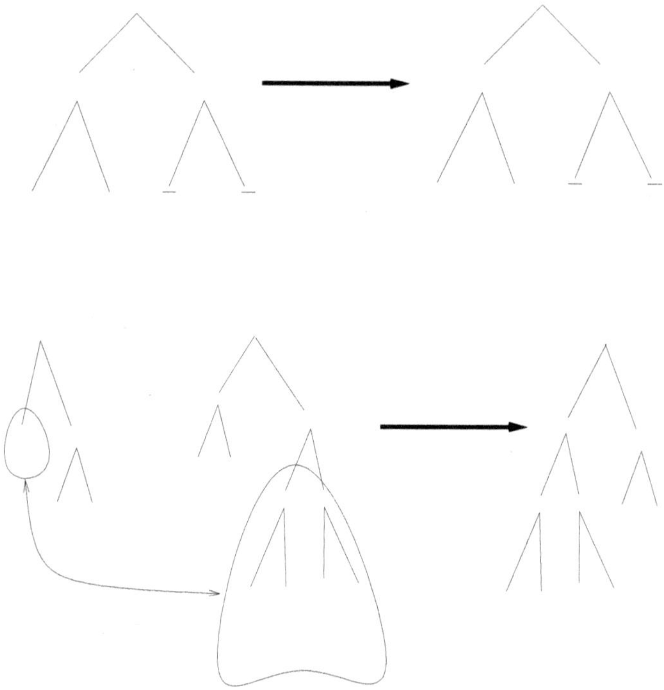

Fig. 1 Genetic Operators for GP

of adaptation is also desirable for effective search in GP, it is difficult to implement within the usual GP framework.

Secondly, choosing a good representation (i.e. designing the terminal set $\{T\}$ and functional set $\{F\}$) is essential for GP search. Recombination operations (such as swapping subtrees or nodes) often cause radical changes in the semantics of the trees. For instance, the mutation of the root node in Fig.1(a) converts a Boolean function to a totally different function, i.e. from false (i.e. $(x \wedge y) \wedge (\overline{x} \vee \overline{y}) \equiv 0$) to true (i.e. $(x \wedge y) \vee (\overline{x} \vee \overline{y}) \equiv 1$). We call this phenomenon "semantic disruption", which is due to the "context-sensitive" representation of GP trees. As a result, useful building blocks may not be able to contribute to higher fitness values of the whole tree, and the accumulation of schemata may be disturbed. To avoid this, [Koza94] proposed a strategy called ADF (Automatic Defining Function) for maintenance of useful building blocks.

Thirdly, the fitness definitions used in traditional GP do not include evaluations of the tree descriptions. Therefore without the necessary control mechanisms, trees may grow exponentially large or become so small that they degrade search efficiency. Usually the maximum depth of trees is set as a user-defined parameter in order to control tree sizes, but an appropriate depth is not always known beforehand.

2.3 Numerical Approach to GP

To overcome the above difficulties, this chapter introduces a new GP-based approach to solving system identification problems, by establishing an adaptive system we call "STROGANOFF' (i.e. STructured Representation On Genetic Algorithms for NOn-linear Function Fitting). STROGANOFF integrates a multiple regression analysis method and a GP-based search strategy. Its fitness definition is based upon a "Minimum Description Length (MDL)" criterion. The theoretical basis for this work is derived from a system identification technique due to Ivakhnenko [Ivakhnenko71].

The advantages of STROGANOFF are summarized as follows:

1. GP search is effectively supplemented with the tuning of node coefficients by multiple regression.
2. Analog (i.e. polynomial) expressions complemented the digital (symbolic) semantics. Therefore the representational problem of standard GP does not arise for STROGANOFF.
3. MDL-based fitness evaluation works well for tree structures in STROGANOFF, which controls GP-based tree search.

The effectiveness of this numerical approach is demonstrated both by successful application to numeric and symbolic problems, and by comparing STROGANOFF's performance with a traditional GP system, applied to the same problems.

We will see later how STROGANOFF overcomes the GP difficulties mentioned in section 2.2.

3 Principles of STROGANOFF

STROGANOFF consists of two adaptive processes; a) The evolution of structured representations, using a traditional genetic algorithm, b) The fitting of parameters of the nodes with a multiple regression analysis. The latter part is called a GMDH (Group Method of Data Handling) process, which is a statistical method, used to solve system identification problems [Ivakhnenko71] (see Appendix A for details of the multiple regression analysis).

3.1 STROGANOFF Algorithm

In summary, STROGANOFF algorithm is described below:

Step1 Initialize a population of tree expressions.
Step2 Evaluate each expression in the population so as to derive the MDL-based fitness (section 3.5, equation (25)).
Step3 Create new expressions (children) by mating current expressions. With a given probability, apply mutation and crossover (Figs.1 and 4) to generate the child tree expressions (sections 3.3, 3.4 and 3.7).
Step4 Replace the members of the population with the child trees.

Step5 Execute the GMDH process, so as to compute the coefficients of the intermediate nodes of the child trees (section 3.2, equation (11)).
Step6 If the termination criterion is satisfied, then halt; else go to **Step2**.

In **Step5**, the coefficients of the child trees are re-calculated using the GMDH process. However, this re-calculation is performed only on intermediate nodes, upon whose descendants crossover or mutation operators were applied. Therefore, the computational burden of **Step5** is expected to be reduced as the generations proceed. As can be seen, **Steps1$\sim$4** and **Step6** follow traditional GP, whereas **Step5** is the new local hill climbing procedure, which will be discussed in section 8.

3.2 GMDH Process in STROGANOFF

STROGANOFF constructs a feedforward network, as it estimates the output function $\overline{f}$. The node transfer functions are simple (e.g. quadratic) polynomials of the two input variables, whose parameters are obtained using regression techniques.

An example of a binary tree generated by STROGANOFF is shown in Fig.2. For instance, the upper left parent tree (P_1) can be written as a (Lisp) S-expression,

```
(NODE1
    (NODE2
        (NODE3 (x1) (x2))
        (x3)
    (x4)))
```

where x_1, x_2, x_3, x_4 are the input variables. Intermediate nodes represent simple polynomial relationships between two descendant (lower) nodes. This tree expresses a "complete form" $\overline{y}$ given by the GMDH process as follows:

1. Select two variables x_1 and x_2 and form an expression $G_{x1,x2}$ which approximates the output y (in terms of x_1 and x_2) with the least error using the multiple regression technique. Regard this function as a new variable z_1 (i.e. the new intermediate node NODE3),
$$z_1 = G_{x1,x2}(x_1, x_2). \tag{4}$$
2. Select two variables z_1 and x_3 and form an approximating expression $G_{z1,x3}$ in the same way. Regard this function as a new variable z_2 (i.e. the new intermediate node NODE2),
$$z_2 = G_{z1,x3}(z_1, x_3). \tag{5}$$
3. Select two variables z_2 and x_4 and form an approximating expression $G_{z2,x4}$. Regard this function as a "complete form" $\overline{y}$, (i.e. the root node NODE1),
$$\overline{y} = G_{z2,x4}(z_2, x_4). \tag{6}$$

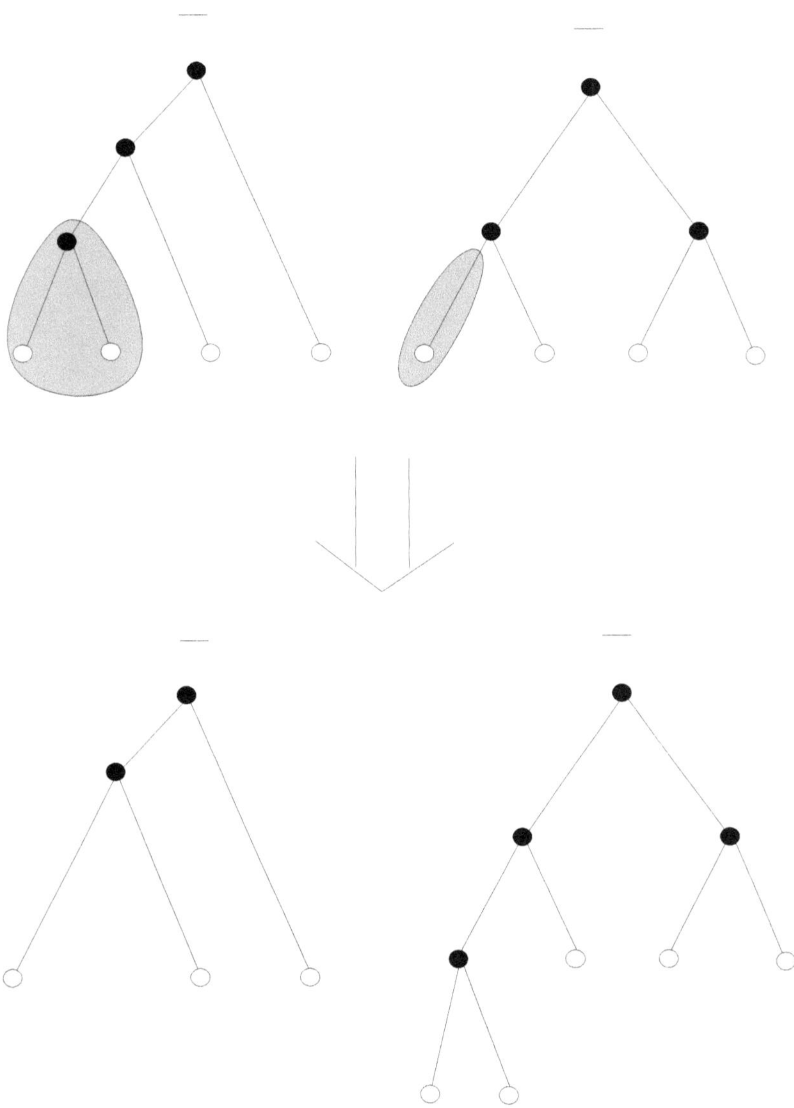

Fig. 2 Crossover Operation in STROGANOFF

For the sake of simplicity, this section assumes quadratic expressions for the intermediate nodes. Thus each node records the information derived by the following equations:

$$NODE3\ :\ z_1 = a_0 + a_1x_1 + a_2x_2 + a_3x_1x_2 + a_4x_1^2 + a_5x_2^2, \tag{7}$$

$$NODE2\ :\ z_2 = b_0 + b_1 z_1 + b_2 x_3 + b_3 z_1 x_3 + b_4 z_1^2 + b_5 x_3^2, \tag{8}$$

$$NODE1\ :\ \overline{y_1} = c_0 + c_1 z_2 + c_2 x_4 + c_3 z_2 x_4 + c_4 z_2^2 + c_5 x_4^2. \tag{9}$$

where z_1 and z_2 are intermediate variables, and $\overline{y_1}$ is an approximation of the output, i.e. the complete form. These equations are called "subexpressions". All coefficients $(a_0, a_1, \cdots, c_5\ )$ are derived from multiple regression analysis using a given set of observations (See Appendix A for details). For instance, the coefficients a_i in the equation (7) are calculated using the following least mean square method. Suppose that N data triples (x_1, x_2, y) are supplied from observation, e.g.:

$$\begin{array}{ccc} x_{11} & x_{21} & y_1 \\ x_{12} & x_{22} & y_2 \\ & \cdots & \\ x_{1N} & x_{2N} & y_N \end{array}$$

From these triples, an X matrix is constructed,

$$X = \begin{pmatrix} 1 & x_{11} & x_{21} & x_{11}x_{21} & x_{11}^2 & x_{21}^2 \\ 1 & x_{12} & x_{22} & x_{12}x_{22} & x_{12}^2 & x_{22}^2 \\ & & \cdots & & & \\ 1 & x_{1N} & x_{2N} & x_{1N}x_{2N} & x_{1N}^2 & x_{2N}^2 \end{pmatrix} \tag{10}$$

which is used to define a coefficient vector **a**, given by

$$\mathbf{a} = (X'X)^{-1}X'\mathbf{y} \tag{11}$$

where

$$\mathbf{a} = (a_0, a_1, a_2, a_3, a_4, a_5)' \tag{12}$$

and

$$\mathbf{y} = (y_1, y_2, \cdots, y_N)', \tag{13}$$

X' is the transposed matrix of X. All coefficients a_i are so calculated that the output variable z_1 approximates the desired output y. The other coefficients are derived in the same way.

Note that all node coefficients are derived locally. For instance, consider b_i's of NODE2. When applying the multiple-regression analysis to the equation (8), these b_i's are calculated from the values of z_1 and x_3 (i.e. the two lower nodes), not from x_4 or $\overline{y_1}$ (i.e. the upper node). Therefore, the GMDH process in STROGANOFF can be regarded as a local-hill climbing search, in the sense that the coefficients of a node are dependent only on its two descendent (lower) nodes.

3.3 Crossover in STROGANOFF

We now consider the recombination of binary trees in STROGANOFF. Suppose two parent trees P_1 and P_2 are selected for recombination (Fig.2). Besides the above equations, internal nodes record polynomial relationships as listed below:

$$NODE5 \ : \ z_3 = d_0 + d_1x_1 + d_2x_4 + d_3x_1x_4 + d_4x_1^2 + d_5x_4^2, \tag{14}$$

$$NODE6 \ : \ z_4 = e_0 + e_1x_3 + e_2x_1 + e_3x_3x_1 + e_4x_3^2 + e_5x_1^2, \tag{15}$$

$$NODE4 \ : \ \overline{y_2} = f_0 + f_1z_3 + f_2z_4 + f_3z_3z_4 + f_4z_3^2 + f_5z_4^2. \tag{16}$$

Suppose z_1 in P_1 and x_1 in P_2 (shaded portions in Fig.2) are selected as crossover points in the respective parent trees. This gives rises to the two child trees C_1 and C_2 (lower part of Fig.2). The internal nodes represent the following relations:

$$NODE8 \ : \ z_1' = a_0' + a_1'x_1 + a_2'x_3 + a_3'x_1x_3 + a_4'x_1^2 + a_5'x_3^2, \tag{17}$$

$$NODE7 \ : \ \overline{y_1}' = b_0' + b_1'z_1' + b_2'x_4 + b_3'z_1'x_4 + b_4'z_1'^2 + b_5'x_4^2, \tag{18}$$

$$NODE12 \ : \ z_2' = c_0' + c_1'x_1 + c_2'x_2 + c_3'x_1x_2 + c_4'x_1^2 + c_5'x_2^2, \tag{19}$$

$$NODE10 \ : \ z_3' = d_0' + d_1'z_2' + d_2'x_4 + d_3'z_2'x_4 + d_4'z_2'^2 + d_5'x_4^2, \tag{20}$$

$$NODE11 \ : \ z_4' = e_0' + e_1'x_3 + e_2'x_1 + e_3'x_3x_1 + e_4'x_3^2 + e_5'x_1^2, \tag{21}$$

$$NODE9 \ : \ \overline{y_2}' = f_0' + f_1'z_3' + f_2'z_4' + f_3'z_3'z_4' + f_4'z_3^{2'} + f_5'z_4'^2. \tag{22}$$

Since these expressions are derived from multiple regression analysis, we have the following equations:

$$z_2' = z_1, \tag{23}$$

$$z_4' = z_4. \tag{24}$$

Thus, when applying crossover operations, we need only derive polynomial relations for $z_1', z_3', \overline{y_1}', \overline{y_2}'$. In other words, recalculation of the node coefficients for the replaced subtree (z_2') and non-replaced subtree (z_4') is not required, which reduces much of the computational burden in STROGANOFF.

3.4 Mutation in STROGANOFF

When applying mutation operations, we consider the following cases:

1. A terminal node (i.e. an input variable) is mutated to another terminal node (i.e. another input variable).
2. A terminal node (i.e. an input variable) is mutated to a nonterminal node (i.e. a subexpression).
3. A nonterminal node (i.e. a subexpression) is mutated to a terminal node (i.e. an input variable).
4. A nonterminal node (i.e. a subexpression) is mutated to another nonterminal node (i.e. another subexpression).

3.5 Fitness Evaluation in STROGANOFF

STROGANOFF uses a Minimum Description Length (MDL)-based fitness function for evaluating the tree structures. This fitness definition involves a tradeoff between certain structural details of the tree, and its fitting (or classification) errors.

$$\text{MDL fitness} = (Tree\_Coding\_Length) + (Exception\_Coding\_Length). \tag{25}$$

The MDL fitness definition for our binary tree is defined as follows [Tenorio *et al.*90]:

$$Tree\_Coding\_Length = 0.5k \log N, \tag{26}$$

$$Exception\_Coding\_Length = 0.5N \log S_N^2, \tag{27}$$

where N is the number of input-output data pairs, S_N^2 is the mean square error, i.e.

$$S_N^2 = \frac{1}{N} \sum_{i=1}^{N} \mid \overline{y_i} - y_i, \mid^2 \tag{28}$$

and k is the number of parameters of the tree, e.g. the k-value for the tree P_1 in Fig.2 is $6+6+6=18$ because each internal node has six parameters ($a_0, \cdots, a_5$ for $NODE3$ etc).

An example of this MDL calculation is given in section 4.1.

3.6 Overall Flow of STROGANOFF

The STROGANOFF algorithm is described below:

Input: t_{max}, I, Pop_size
Output: x, the best individual ever found.

1 $t \leftarrow 0$;
{I is a set of input variables (see eq.(1)). $NODE\_2$ is a nonterminal node of 2-arity.}
2 $P(t) \leftarrow$ initialize($Pop\_size, I, \{NODE\_2\}$);
3 $F(t) \leftarrow$ evaluate($P(t), Pop\_size$);
4 $x \leftarrow a_j(t)$ **and** $Best\_so\_far \leftarrow MDL(a_j(t))$, where $MDL(a_j(t)) = \min(F(t))$;
{the main loop of selection, recombination, mutation.}
5 **while** ($\iota(P(t), F(t), t_{max}) \neq$ true) **do**
6 **for** $i \leftarrow 1$ **to** $\frac{Pop\_size}{2}$ **do**
{select parent candidates according to the MDL values.}
$Parent_1 \leftarrow select(P(t), F(t), Pop\_size)$;
$Parent_2 \leftarrow select(P(t), F(t), Pop\_size)$;
{apply GP crossover operation, i.e. swapping subtrees (Fig.2).}
$a'_{2i-1}(t), a'_{2i}(t) \leftarrow$ GP_recombine($Parent_1, Parent_2$);
{apply GP mutation operation,
i.e. changing a node label and deleting/inserting a subtree.}

```
                a''_2i(t) ← GP_mutate(a'_2i(t));
                a''_2i-1(t) ← GP_mutate(a'_2i-1(t));
        od
        P''(t) ← (a''_1(t), ···, a''_Pop_size(t));
        F(t) ← evaluate(P''(t), Pop_size);
        tmp ← a''_k(t), where MDL(a''_k(t)) = min(F(t));
        if (Best_so_far > MDL(a''_k(t)))
                then x ← tmp and Best_so_far ← MDL(a''_k(t));
        P(t+1) ← P''(t);
        t ← t+1;
  od
  return (x);
```

```
  {terminate if more than t_max generations are over.}
  ι(P(t), F(t), t_max) :
          if (t > t_max)
                  then return true;
                  else return false;
```

```
  {initialize the population randomly.}
  initialize(Pop_size, T, F):
      for i ← 1 to Pop_size do
              generate a tree a_i randomly,
                  where the terminal and nonterminal sets are T and F.
      od
      return (a_1, ···, a_Pop_size);
```

```
  {evaluate of a population of size Pop_size.}
  evaluate(P(t), Pop_size):
      for i ← 1 to Pop_size do
              {calculate eq.(28).}
              GMDH_Process(a_i);
              S_N^2(a_i) ← the mean square error of a_i;
              {calculate eqs.(25),(26) and (27).}
              MDL(a_i) ← Tree_Coding_Length(a_i) + Exception_Coding_Length(a_i);
      od
      return (MDL(a_1), ···, MDL(a_Pop_size));
```

```
  {execute the GMDH process.}
  GMDH_Process(a):
          nd ← the root node of a;
          if (nd is a terminal node)
                  then return;
          {if the node coefficients of nd are already derived, then return.}
```

```
4        if (Coeff(nd) ≠ NULL)
             then return;
5        nl ← left_child(nd);
6        nr ← right_child(nd);
7        GMDH_Process(nl);
8        GMDH_Process(nr);
9        Coeff(nd) ← Mult_Reg(nl,nr);
         return;
```

```
   {execute the multiple-regression analysis.}
1  Mult_Reg(n1,n2):
       Assume n1 is the first variable and n2 is the second variable.
       For instance, x1 ← n1, x2 ← n2 for eq.(7)
       Derive and return the fitting coefficients, i.e. eq.(12)
       return;
```

In the GMDH_Process called by the evaluate routine, the coefficients of the child trees are recalculated using the multiple regressions. However, this recalculation is performed only on intermediate nodes, upon whose descendants crossover or mutation operators were applied (see the fourth lines in GMDH_Process). Therefore, the computational burden of the GMDH process is expected to be reduced as the generations proceed. As can be seen, lines from 6 to 7 in the STROGANOFF algorithm follow traditional GP, whereas GMDH_Process is the new local hill climbing procedure, which will be discussed later.

3.7 Recombination Guidance in STROGANOFF

Multiple-regressions in STROGANOFF tune the node coefficients so as to guide GP recombination effectively with MDL values. By this mechanism, STROGANOFF can avoid the disruption problem caused by the traditional GP crossover or mutation (Fig.2). This section explains the recombination guidance of STROGANOFF.

Fig.3 illustrates an exemplar STROGANOFF tree for the time series prediction (see section 4.1 for details), in which the error of fitness ratios (i.e. mean square error, MSE) and MDL values are shown for all subtrees. As can be seen from the figure, the MSE values monotonically decrease towards the root node in a given tree. Thus the root node has the lowest (i.e. best) MSE value. However, the MDL values do not monotonically change. The subtree whose MDL value is lowest is expected to give the best performance of all subtrees. Therefore, it can work as a building-block for crossover operations.

We realize a type of adaptive recombination based on MDL values. For this purpose, in applying crossover or mutation operators, we follow the rules described below:

1. Apply a mutation operator to a subtree whose MDL value is larger.
2. Apply a crossover operator to a subtree whose MDL value is larger, and get a subtree whose MDL value is smaller from another parent.

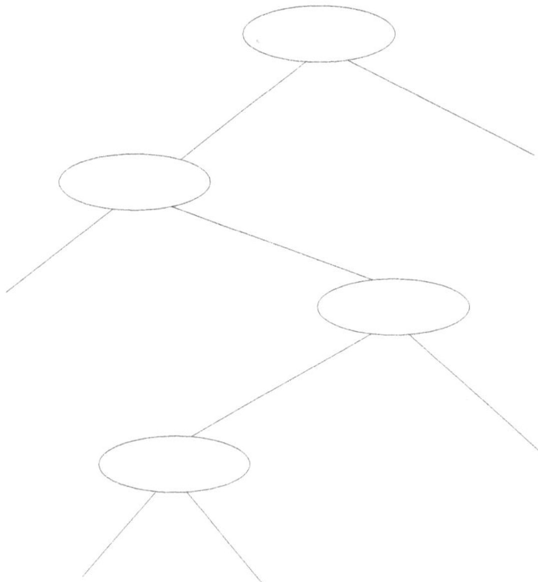

Fig. 3 An Exemplar STROGANOFF Tree

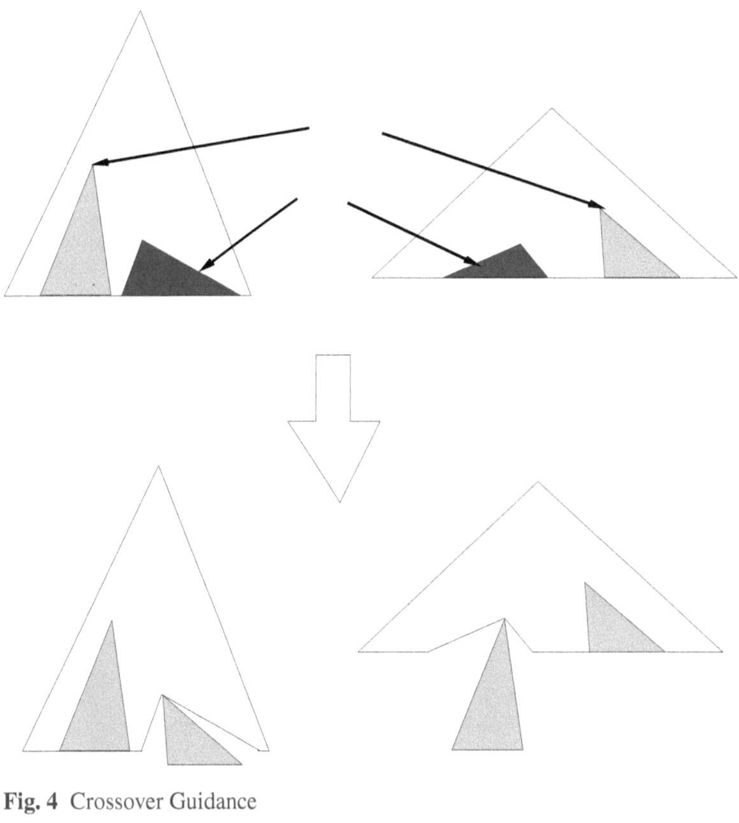

Fig. 4 Crossover Guidance

When the second operator is applied to two parents P_1 and P_2, execute the following steps (see Fig.4).

1. Let W_1 and W_2 be the subtrees with the largest MDL values of P_1 and P_2.
2. Let B_1 and B_2 be the subtrees with the smallest MDL values of P_1 and P_2.
3. A new child C_1 is a copy of P_1, in which W_1 is replaced by B_2.
4. A new child C_2 is a copy of P_2, in which W_2 is replaced by B_1.

The above mechanism exploits already built structures (i.e. useful building-blocks) with adaptive recombination guided by MDL values.

We have confirmed the effectiveness of this guidance by experiments (see [Iba *et al.*96b] for details). Therefore, we believe STROGANOFF can guide GP recombination effectively in the sense that the recombination operation is guided using MDL values.

4 Numerical Problems with STROGANOFF

We applied STROGANOFF to several problems such as time series prediction, pattern recognition, and 0-1 optimization [Iba *et al.*93, Iba *et al.*94b]. The results obtained were satisfactory. This section describes the experiments with time series predictions and compare the performance of STROGANOFF with other techniques.

4.1 Time Series Prediction with STROGANOFF

The Mackey-Glass differential equation

$$\frac{dx(t)}{dt} = \frac{ax(t-\tau)}{1+x^{10}(t-\tau)} - bx(t), \tag{29}$$

is used for time series prediction problems, where a=0.2, b= 0.1 and τ=17 (Fig.5(a)). This is a chaotic time series with a strange attractor of fractal dimension of approximately 3.5 [Tenorio *et al.*90].

In order to predict this series, the first 100 points (i.e. the values of $x(1), \cdots x(100)$) were given to STROGANOFF as training data. The aim was to obtain a prediction of $x(t)$ in terms of M past data, i.e.

$$x(t) = f(x(t-1), x(t-2), \cdots, x(t-M)). \tag{30}$$

The parameters for STROGANOFF were as follows:
$N_{popsize}$: 60
P_{cross} : 0.6
P_{mut} : 0.0333
T : $\{x(t-1), x(t-2), \cdots, x(t-10)\}$
We used 10 past data for simplicity.

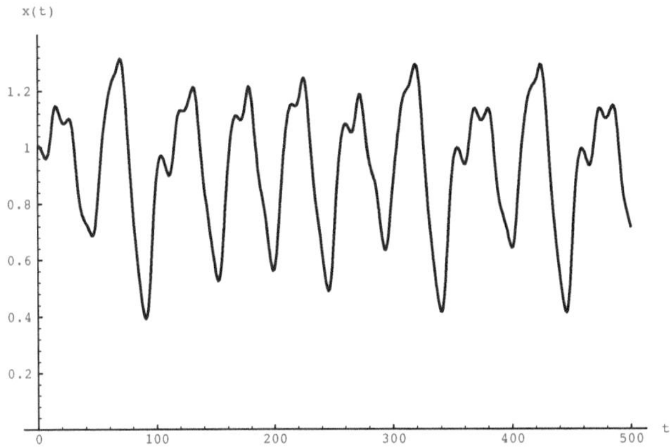

(a) Chaotic Time Series

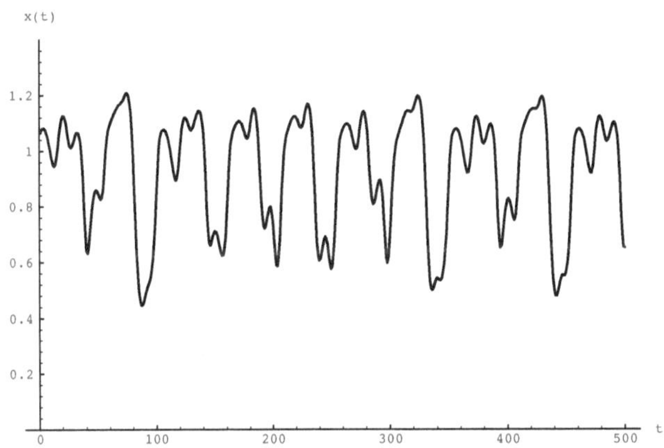

(b) Prediction at 233rd Generation

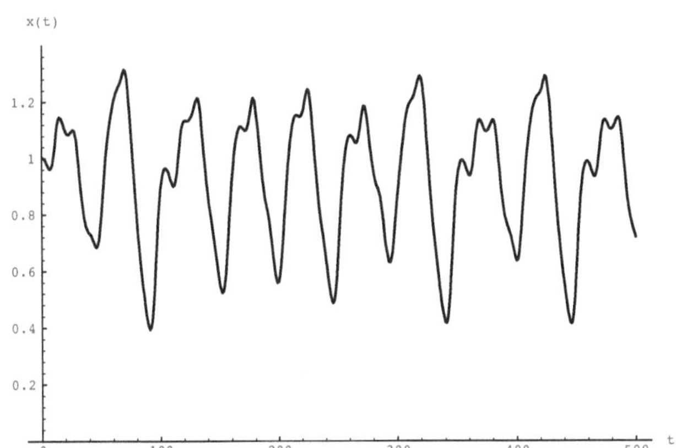

(c) Prediction at 1740th Generation

Fig. 5 Predicting the Mackey–Glass Equation

Fig.6 shows the results of this experiment, namely the mean square error (S_N^2) and the MDL value as a function of the number of generations. Figs.5(b) and (c) are the time series predicted by STROGANOFF (generations 233 and 1740 respectively). The MDL fitness values did not decrease monotonically, because the MDL values were plotted only when the minimum error-of-fit ratios improved. Note that the selection process of STROGANOFF is based on the MDL-value, and not on the raw fitness (i.e. the error-of-fit ratio). The resulting structure of Fig.5(c) was as follows:

```
(NODE95239 (7)
  (NODE95240
    (NODE95241
      (NODE95242
        (NODE95243
          (NODE95244
            (8)
            (NODE95245
              (8)
              (NODE95130 (2) (3))))
          (NODE95173
            (10)
            (NODE95174
              (NODE95175 (4) (1))
              (5))))
        (5))
      (6))
    (NODE95178 (NODE95179 (8) (3)) (10))))
```

Where (i) represents $x(t-i)$. Some of the node coefficients were in Table 1. The mean square errors (i.e. MSEs) for this period are summarized in Table 2. The MDL value (i.e. fitness) of this tree is given as follows:

$$\text{MDL fitness} = 0.5k \log N + 0.5N \log S_N^2 \quad (31)$$

$$= 0.5 \times (6 \times 13) \times \log 100 + 0.5 \times 100 \times \log(4.70 \times 10^{-6}) \quad (32)$$

$$= -433.79. \quad (33)$$

Where the number of training data (i.e. N) is 100, and the MSE (i.e. S_N^2) is 4.70×10^{-6}. Since the number of intermediate nodes is 13, the k-value is roughly estimated as 6×13, because each internal node has six parameters.

Note that in Fig.5(c) the prediction at the 1740th generation fit the training data almost perfectly. We then compared the predicted time series with the testing time series (i.e. $x(t)$ for $t > 100$). This also produced good results (compare Fig.5(a) and Fig.5(c)).

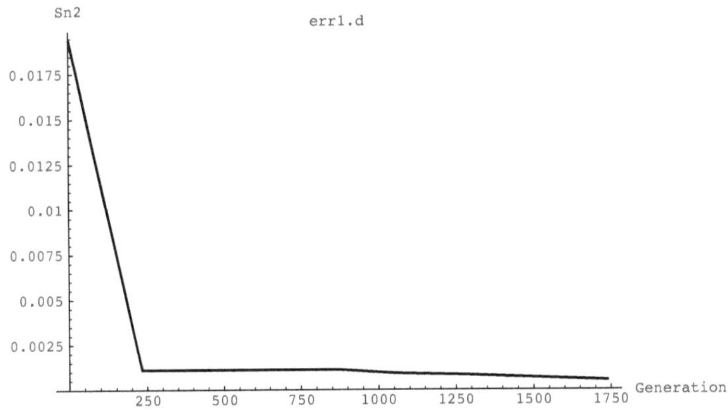

(a) Test Data.

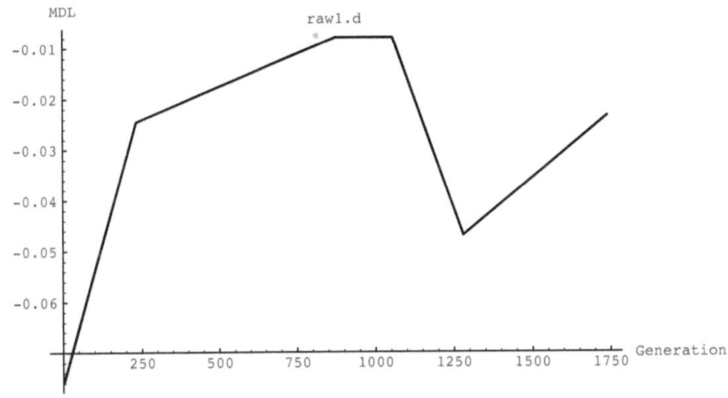

(b) Prediction Result.

Fig. 6 Time Series Prediction

Table 1 Node Coefficients

Node	NODE95239	NODE95240	NODE95179
a_0	0.093	-0.090	0.286
a_1	0133	1.069	-0.892
a_2	0.939	-0.051	1.558
a_3	-0.029	1.000	1.428
a_4	0.002	-0.515	-0.536
a_5	-0.009	-0.421	-0.844

4.2 Comparison with a Traditional GP

Traditional GP has also been applied to the prediction task. In order to compare the performance of STROGANOFF, we applied a traditional GP system "sgpc1.1"

Table 2 Mean Square Errors (STROGANOFF)

Generation	Training data	Testing data	MDL
233	0.01215	0.01261	-192.86
1740	4.70×10^{-6}	5.06×10^{-6}	-433.79

Table 3 GP Parameters (Predicting the Mackey–Glass equation)

Objective:	Predict next data $X(t)$ in Mackey–Glass mapping series.
Terminal set:	Time–embedded data series from $t = 1, 2, \cdots, 10$. i.e. $\{X(t-1), X(t-2), \cdots, X(t-10)\}$, with a random constant.
Function set:	$\{+, -, \times, \%, SIN, COS, EXP10\}$.
Fitness cases:	Actual members of the Mackey–Glass mapping ($t = 1, 2, \cdots, 500$).
Raw fitness:	Sum over the fitness cases of squared error between predicted and actual points.
Standardized fitness:	Same as raw fitness.
Parameters:	$M = 5000$. $G = 101$.
Max. depth of new individuals:	6
Max. depth of mutant subtrees:	4
Max. depth of individuals after crossover:	17
Fitness–proportionate reproduction fraction:	0.1
Crossover at any point fraction:	0.2
Crossover at function points fraction:	0.7
Selection method:	fitness–proportionate
Generation method:	ramped half-and-half

Table 4 Mean Square Errors (GP vs. STROGANOFF)

System	Gen.	#Pop×Gen.	Training data	Testing data
STROGANOFF	233	13,980	0.01215	0.01261
	1740	104,440	4.70×10^{-6}	5.06×10^{-6}
sgpc1.1(GP)	67	325,000	9.62×10^{-4}	2.08×10^{-3}
	87	435,000	6.50×10^{-6}	1.50×10^{-4}

(a Simple Genetic Programming in C written by Walter Alden Tackett) to the same chaotic time series (i.e. Mackey–Glass equation). For the sake of comparison, all the arameters chosen were the same as those used in the previous study [Oakley94, p.380,Table17.3], except that the terminal set consisted of ten past data for the short-term prediction (see Table 3).

Table 4 gives the results of the experiments, which show the mean square error of the best performance over 20 runs. For the sake of comparison, we also list the results given by STROGANOFF. The numbers of individuals to be processed are shown in the third column (i.e. #Pop×Gen.).

The trees resulting from traditional GP are as follows:

```
<< Generation 67>>
(%
  (SIN
    (+
      (%
        X(t-8)
        X(t-3))
      X(t-9)))
  (%
    (%
      X(t-5)
      X(t-1))
    (%
      X(t-4)
      (EXP10
        X(t-7)))))

<< Generation 87>> (-
  (+
    X(t-1)
    X(t-1))
  X(t-2))
```

The experimental results show that the traditional GP suffers from overgeneralization, in the sense that the mean square error of the test data (i.e. 1.50×10^{-4}) is much worse than that of the training data (i.e. 6.50×10^{-6}). This may be caused by the fact that traditional GP has no appropriate criterion (such as MDL for STROGANOFF) for evaluating the trade-off between the errors and the model complexities (i.e. the description length of S-expressions).

Another disadvantage of traditional GP is due to the mechanisms used to generate constants. In traditional GP, constants are generated randomly by initialization and mutation. However, there is no tuning mechanism for the generated constants. This may degrade the search efficiency, especially in the case of time series prediction tasks, which require a fine–tuning of the fitting coefficients, so that the number of processed individuals for the same quality of solution is much greater than STROGANOFF (i.e. the third column in Table 4).

4.3 Statistical Comparison of STROGANOFF and a Traditional GP

In order to clarify these performance differences more statistically, we compare our approach with other prediction methods. More precisely, the predictor errors of the following techniques are compared using a variety of dynamic systems as case studies.

1. STROGANOFF
2. Traditional GP ("sgpc1.1" based on [Oakley94])
3. Radial basis functions [Franke82, Poggio & Girosi90]

4.3.1 Comparative Method

Given an m-dimensional chaotic time series $\{x_1,\cdots,x_n,x_{n+1} \mid x_i \in \Re^m,\ x_{n+1} = f(x_n)\}$, a predictor for x_{n+1} is described as follows:

$$x_{n+1} = \widetilde{f_N}(\{x_1,\cdots,x_n\}), \tag{34}$$

where N is the number of training data. Note that $\widetilde{f_N}$ is a m-dimensional vector function, i.e. $\widetilde{f_N} = (\widetilde{f_N^1},\cdots,\widetilde{f_N^m})$. In order to quantify how well $\widetilde{f_N}$ performs as a predictor for f, the predictor error $\sigma^2(\widetilde{f_N})$ of $\widetilde{f_N}$ is defined by

$$\sigma^2(\widetilde{f_N}) = \lim_{M\to\infty} \frac{1}{M} \times \sum_{n=N}^{N+M-1} \|x_{n+1} - \widetilde{f_N}(\{x_n\})\|^2/Var, \tag{35}$$

where

$$Var = \lim_{M\to\infty} M^{-1} \sum_{m=1}^{M} \|x_m - \lim_{M\to\infty} \sum_{m=1}^{M} x_m\|^2. \tag{36}$$

Var is a normalizing factor. $\|\cdot\|$ denotes the Euclidean norm on $\Re^m$. M is the number of test data and is set to be 10^3 in the following discussions.

In order to overcome certain higher dimensional problems faced by STROGANOFF and traditional GP, we modify the previous predictor as follows. In the learning phase, we train the predictor $\widetilde{f_N^i}$ using the equation

$$x_t^i = \widetilde{f_N^i}(N(x_t)), \tag{37}$$

where $N(x_t)$ is the neighborhood of x_t. In the testing phase, we predict the future data x_{t+1} using the equation

$$x_{t+1}^i = \widetilde{f_N^i}(\{x_{j+1} \mid x_j \in N(x_t)\}). \tag{38}$$

x_{t+1} is derived using its neighborhood $N(x_{t+1})$ in the same way as the training phase. However, because $N(x_{t+1})$ is not known before x_{t+1} is derived, the predicted neighborhood $\{x_{j+1} \mid x_j \in N(x_t)\}$ is used as its substitute. The parameters used were the same as in the previous example, except that the terminal set included 10 past data in the neighborhood (i.e. $N(x_i)$). For instance, we used as the terminal set $\{x_1(t), \cdots, x_{10}(t), y_1(t), \cdots, y_{10}(t)\}$ for a two-dimensional problem, where $x_i(t)$ and $y_i(t)$ are the x- and y-coordinates of the i-th nearest past data to $x(t)$. We compare the performances of STROGANOFF and traditional GP (i.e. "sgpc1.1") with data estimated from "radial basis function predictors".

4.3.2 Comparison Tests

The following dynamic systems were chosen for computing predictor errors using the above techniques.

1. Ikeda map [Ikeda79]
   The Ikeda map is a two dimensional dynamic system described below:

$$f(x,y) = (1 + \mu(x\, cos\, t - y\, sin\, t),\ \ \mu(x\, sin\, t + y\, cos\, t)), \tag{39}$$

   where

$$t = 0.4 - \frac{6.0}{1 + x^2 + y^2}, \tag{40}$$

$$\mu = 0.7. \tag{41}$$

   We consider the fourth iterate of the above Ikeda map (see Fig.7):

$$(x_{n+1}, y_{n+1}) = f^4(x_n, y_n). \tag{42}$$

   This map was chosen because it has a complicated functional form which is not of a type used in any of the above approximation techniques, but is well behaved and slowly varying.
2. Lorenz equation [Lorenz63]
   Lorenz's equations consist of three simultaneous differential equations:

$$\frac{dx(t)}{dt} = -3(x(t) - y(t)), \tag{43}$$

$$\frac{dy(t)}{dt} = -x(t)z(t) + 26.5x(t) - y(t), \tag{44}$$

$$\frac{dz(t)}{dt} = x(t)y(t) - z(t), \tag{45}$$

   where

$$x(0) = z(0) = 0,\ y(0) = 1. \tag{46}$$

   We use sampling rates $\tau = 0.20$ (see Fig.8).

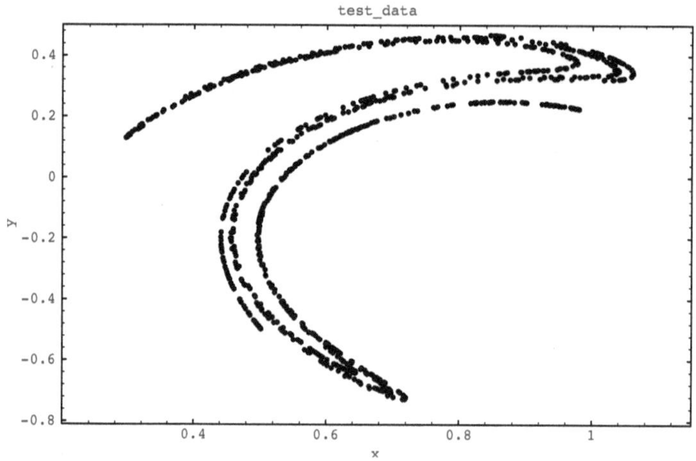

(a) Test Data.

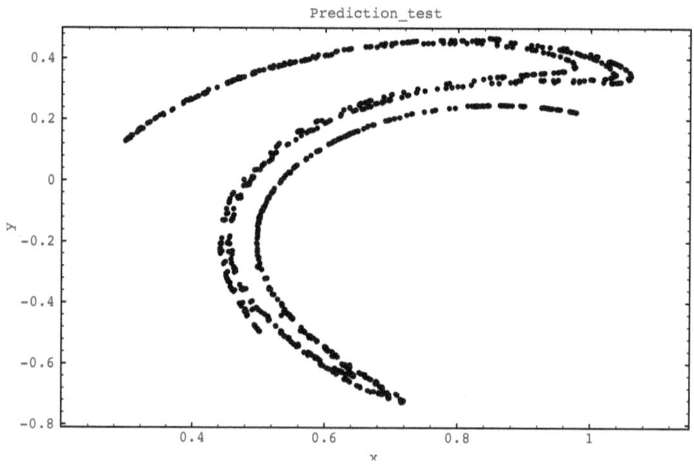

(b) Prediction Result.

Fig. 7 Ikeda Map

3. Mackey-Glass equation [Mackey & Glass77]
   This delay differential equation was presented before, i.e.

$$\frac{dx(t)}{dt} = \frac{ax(t-\tau)}{1+x^{10}(t-\tau)} - bx(t), \tag{47}$$

   The parameters used are

$$a = 0.2,\ b = 0.1,\ \tau = 17 \tag{48}$$

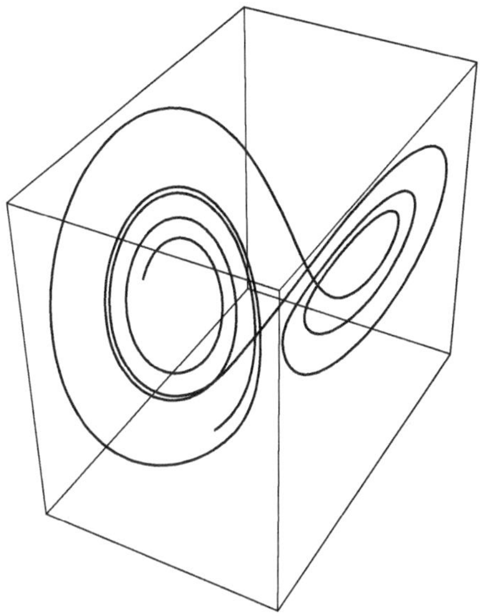

(a) Test Data.

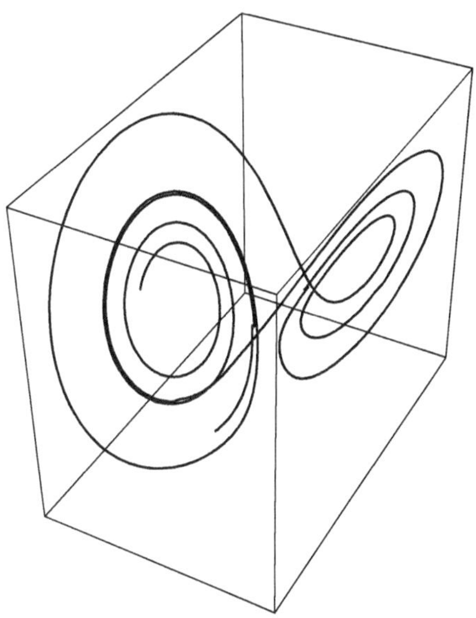

(b) Prediction Result.

Fig. 8 Lorenz Attractor

4.3.3 Results

Figs.7 and 8 show the test data and prediction results by STROGANOFF, for the Ikeda map and the Lorenz equation respectively. Table 5 shows the estimated values

Table 5 Estimated Values of $\log_{10}\sigma(\widetilde{f_N})$

	D	N	Radial	GP	STROGANOFF
Ikeda	1.32	500	-2.10	-0.99	-1.23
Lorenz	2.0	500	-1.35	-0.55	-1.20
Mackey – Glass	2.1	500	-1.97	-1.43	-2.00

of $\log_{10}\sigma(\widetilde{f_N})$ for predictors $\widetilde{f_N}$ using three techniques. Also tabulated is the information dimension D. We conducted experiments for STROGANOFF and traditional GP.

From the table, we see that the performance of traditional GP is very poor, especially in higher dimensional problems. We often observed that the best individuals acquired for the Ikeda map by traditional GP were simple expressions, shown below:

```
X(t) = (-
  X1(t)
  (%
    X8(t)
    (EXP10
      (*
        (*
          X8(t)
          Y1(t))
        X7(t))))))
```

```
X(t) = (+
  X1(t)
  (-
    Y8(t)
    Y8(t)))
```

```
X(t) = X1(t)
```

```
Y(t) = Y1(t)
```

Xi(t) and Yi(t) are the x- and y-coordinates of the i-th nearest past data to $X(t)$. Note that the second expression is identical to X1(t), because $X1(t) + (Y8(t) - Y8(t)) = X1(t) + 0 = X1(t)$. The first expression is also nearly equal to X1(t), because the second term is close to zero (i.e. $\frac{X8(t)}{EXP10(X8(t) \times Y1(t) \times X7(t))} \approx 0$). $X1(t)$ and $Y1(t)$ are considered as very rough approximations of $X(t)$ in the training phase, because they are the closest points to $X(t)$ (see equation (37)). However, these are not effective predictors in the testing phase, in the sense that $X1(t)$ and $Y1(t)$ do not necessarily belong to the appropriate neighborhood of $X(t)$ (equation (38)). Therefore, the fact that the monomial expressions (i.e. X1(t) and Y1(t)) often appeared in the resultant trees shows that traditional GP lacks generalization mechanisms, which in term, results in the poor performance on the testing data.

The performance of STROGANOFF is by no means inferior to other techniques and gives acceptable results even in the case of the Ikeda map, which has complex dynamics (see Fig.7).

Radial basis predictors seem superior to the other techniques. This technique is a global interpolation with good localization properties. It provides a smooth interpolation of scattered data in arbitrary dimensions and has proven useful in practice. However, with radial basis functions, the calculation of coefficients can be very costly for large N.

On the other hand, STROGANOFF has the following advantages. Firstly, the calculation of the inverse matrix (equation (11)) consumes much less computation time, because STROGANOFF requires the coefficients for only two variables at each nonterminal node. Fig.9 plots the computation costs with the degrees of fitting polynomials. The costs are estimated as the numbers of the loop iterations of the inverse calculations. GLMS(i) represents the general least mean square method for the fitting equation of i input variables. The vertical axis is translated (i.e. divided by $O(6^3)$) for the sake of convenience. As can be seen in the figure, the advantage of STROGANOFF comes about when dealing with large, complex systems, i.e. when fitting a higher-order polynomial of multiple input variables (see Appendix B for details).

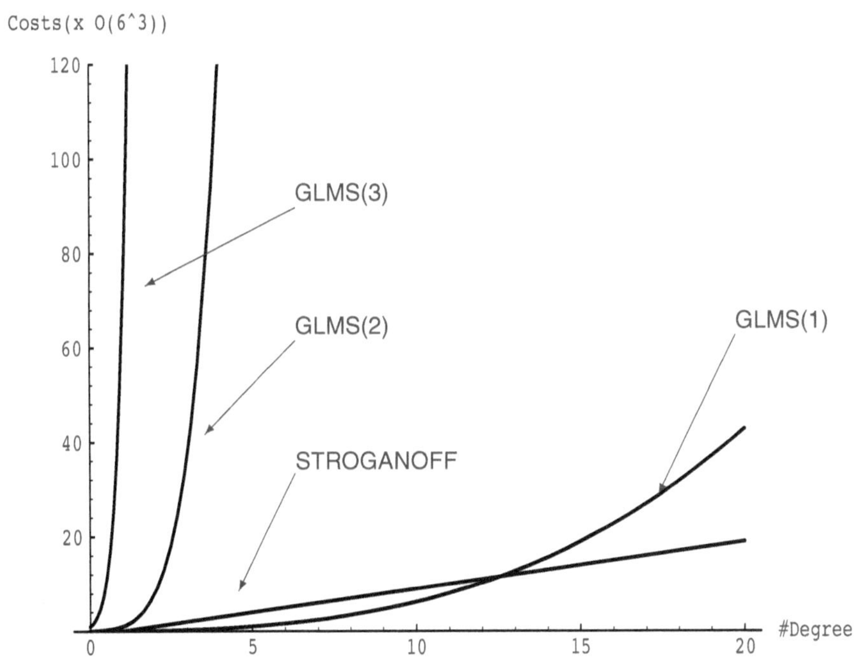

Fig. 9 Computational Costs

Secondly, the degree of the polynomial (the depth of the STROGANOFF tree) is adaptively tuned during the evolution of the trees. Therefore, we can conclude STROGANOFF offers an effective technique, which integrates a GP-based adaptive search of tree structures, and a local parameter tuning mechanism employing statistical search.

5 Symbolic Problems with STROGANOFF

5.1 Extension of STROGANOFF

Symbolic (non-numeric) reasoning problems, such as Boolean concept formation or symbolic regression, differ from the above-mentioned system identification (numerically based) problems, in the sense that the "complete form" needs to be constructed from a set of symbols, and should be as simple as possible. We therefore extend STROGANOFF as follows:

1. Introducing new kinds of functional nodes (see Tables 6 and 9 for example). In order to interpret resulting trees symbolically, divide functional nodes into two types.

   a. Digital (Symbolic)-type nodes, which correspond to the functional nodes of traditional GP, e.g. logical functions such as AND ($\wedge$), OR ($\vee$), arithmetic operations (+,-,*,/), and transcendental functions such as SQRT, SIN or COS.
   b. Analog-type nodes, which perform multiple regression analysis. We use a variety of subexpressions for internal nodes. These are simple polynomials of linear or quadratic expressions, e.g. $\alpha x_1 x_2$, $\alpha x_1 + \beta x_2$, $\alpha x_1^2 + \beta x_2^2$, $\alpha(x_1^2 - x_2^2)$, $\alpha x_1 + \beta x_2 + \gamma x_1 x_2 + \delta x_1^2 + \varepsilon x_2^2$. The choice of subexpressions is decided either randomly or optimally. Since the above subexpressions include no constant terms, we introduce a virtual node (x_0) for constant fitting. The value of x_0 is always 1.

2. Modifying the MDL-based fitness definition. The MDL-based fitness definition is modified by weighing the tree description part. This suppresses the divergence of search, by reducing the exploration of polynomials of higher degrees when minimizing errors. More precisely, we modify the previous equation (25) as follows:

$$\text{MDL fitness} = (Tree\_Coding\_Length) + C_E \times (Exception\_Coding\_Length). \tag{49}$$

   where C_E is a weighting coefficient. In the previous experiments, we set $C_E = 1.0$. For symbolic problems, we choose $C_E > 1.0$ (e.g. $C_E = 3.0$) which favors the search for simple expressions, but at the expense of generating greater errors.

3. Pruning redundant nodes.
   Redundant nodes are those which do not contribute statistically to the overall fitness. We perform prunings of nodes whose coefficients of the subexpression in its parent node are nearly zero. For instance, if $z_1 = a_0 + a_1x_1 + a_2x_2 + a_3x_1x_2 + a_4x_1^2 + a_5x_2^2$, $a_1 \approx 0$, $a_3 \approx 0$, and $a_4 \approx 0$, then prune the node x_1. This pruning reduces the size of nodes and hence improves the efficiency of STROGANOFF.
4. Symbolic interpretation of analog-type nodes (see equations (51) and (56) for example).
   Construct a polynomial for the final "complete form" and translate it into a desired symbolic form.

STROGANOFF was applied successfully to the problems of Boolean concept formation [Iba *et al.*94c]. The next subsection presents an experiment in symbolic regression.

5.2 *Symbolic Regression*

The goal of symbolic regression is to discover a set of numerical coefficients for a combination of independent variable(s), which minimizes some measure of error. In other words, the problem is both the discovery of the correct functional form that fits the data and the discovery of the appropriate numeric coefficients [Koza90, Ch.4.3.2]. This problem is closely related to the discovery of various scientific laws from empirical data, such as the well-known BACON system [Langley and Zytkow89].

When applying traditional GP to symbolic regression, it is usual to assign the terminal set $\{T\}$ to the independent variables and to devise some appropriate function set $\{F\}$. However the discovery of the appropriate numerical coefficients is very difficult, because there is no coefficient modification mechanism other than random creation of constants. In addition, there is no explicit control mechanism over GP search for symbolic manipulation.

The following two experiments show how STROGANOFF can be applied to polynomial symbolic regression.

Exp.1 Two-box Problem [Koza94, ch.3.1]
The two-box problem is to find a polynomial relationship of six independent variables (a,b,c,d,e,f), where the relationship among these variables is the difference y in volumes of the first box, whose length, width, and height are a,b,c, and the second box, whose length, width, and height are d,e,f. Thus

$$y = a \times b \times c - d \times e \times f \tag{50}$$

The goal of this symbolic regression is to derive the above equation as a "complete form", when given a set of N observations $\{(a_1,b_1,c_1,d_1,e_1,f_1,y_1),\cdots,(a_N,b_N,c_N,d_N,e_N,f_N,y_N)\}$. We used the parameters shown in Table 6, where x_0 is a virtual variable, for the purpose of representing a constant, i.e. the value of x_0 is always 1.

Table 6 STROGANOFF Parameters for Symbolic Regression

Population Size	120
Probability of Crossover	60%
Probability of Mutation	3.3%
Terminal Nodes	$\{x_0, a, b, c, d, e, f\}$
Functional Nodes	Digital-type: None Analog-type: $\{\alpha x_1 x_2,\ \alpha x_1 + \beta x_2\}$
# of Training Data (N)	10

Table 7 Subexpressions for Two-Box Problem

Node	subexpression
NODE216612	$1.17682x_1 + 1.30238x_2$
NODE216609	$0.18895x_1 x_2$
NODE216610	$-4.49710x_1 x_2$
NODE216232	$0.163891x_1 x_2$
NODE209664	$4.684920x_1 x_2$

Fig.10 shows the results of this experiment, namely the mean square error (S_N^2) and MDL value as a function of the number of generations. The acquired structure with subexpressions at generation 440 is shown below. This is a typical acquired tree from several runs. The tree gives 100% correct answers to all 10 data.

```
(NODE216612
    (NODE216609 f (NODE216610 e d))
    (NODE216232 (NODE209664 c b) a))
```

This "complete form" expresses the exact equation as follows:

$$1.17682\,(0.18895\,f\,(-4.49710\,e\,d)) + 1.30238\,(0.163891(4.684920\,c\,b)\,a)$$

$$= 0.99998\,a\,b\,c - 0.99997\,d\,e\,f$$

$$\approx abc - def \tag{51}$$

Rounding the coefficients reduces the above expression to equation (50). Thus we regard the complete form as a desired symbolic regression.

In order to compare the performance of STROGANOFF with the traditional GP, we derived the computational effort E, required to yield a solution to a problem with a satisfactorily high probability. Following [Koza94, Ch.4], for each generation i, we can compute an estimate of the "cumulative probability of success", $P(M,i)$, that a particular run with a population size M yields a solution by generation i. If we want to satisfy the success predicate with a certain specified probability z, then the number of independent runs, $R(M,i,z)$, required to satisfy the success predicate by generation i with a probability of z, depends upon z and $P(M,i)$ as follows:

$$z = 1 - [1 - P(M,i)]^{R(M,i,z)}. \tag{52}$$

The total number of "individuals that must be processed", $I(M,i,z)$, in order to yield a solution to the problem with the probability of z for a population size M, by generation i, is given as follows:

$$I(M,i,z) = M(i+1)R(z). \tag{53}$$

Thus, the computational effort E is given as

$$E = \max_i I(M,i,z) = I(M,i^*,z) = M(i^*+1)R(z). \tag{54}$$

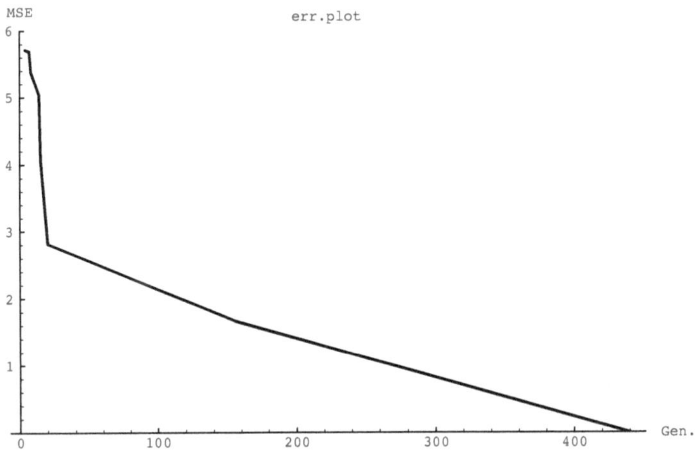

(a) Error-of-fit vs. Generations.

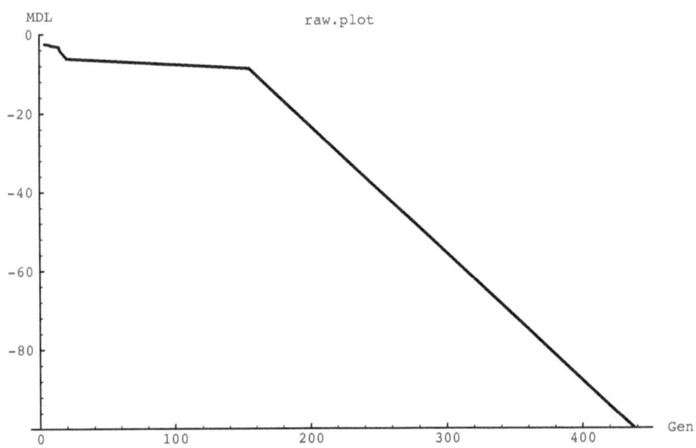

(b) MDL vs. Generations.

Fig. 10 Experimental Results (Two-Box Problem)

Table 8 Computational Effort

Method	E	i^*	M	z
STROGANOFF	43,725	29	120	99%
GP (with ADF)	2,220,000	14	4,000	99%
GP (without ADF)	1,176,000	5	4,000	99%

Table 9 STROGANOFF Parameters for Symbolic Regression (2)

Population Size	120
Probability of Crossover	60%
Probability of Mutation	3.3%
Terminal Nodes	$\{(0),(1),(2),(3),(sqr1),(sqr2),(sqr3)\}$
Functional Nodes	Digital-type: $\{+, -\}$ Analog-type: $\{\alpha x_1 x_2,\ \alpha x_1 + \beta x_2\}$
# of Training Data (N)	10

Fig.5.2 shows the performance curves for STROGANOFF based on 20 runs, with $M = 120$ and $z = 99\%$. The figure shows the computational effort E at the 29th generation. The performance comparison is given in Table 8. The GP data was extracted from [Koza94, p.120,p.104]. The table shows the computational effort for STROGANOFF is about 27 times less than that of traditional GP without ADF and about 50 times less than GP with ADF.

Exp.2 Heron formula
Next we experimented with a more complex symbolic regression problem. We tried to find the Heron formula for the area S of a triangle when given the lengths of its three sides (a,b,c):

$$S = \sqrt{\frac{(a+b+c)(a+b-c)(a+c-b)(b+c-a)}{16}}. \tag{55}$$

The function discovery of this formula from a set of observations has been studied by [Barzdins and Barzdins91], in which a heuristic enumeration method was used as a traditional machine learning technique. The study showed the difficulty of this problem due to complicated terms in the Heron formula. In order to use polynomial symbolic regression, we experimented in acquiring a formula of the square of the area (S^2), using the three lengths (a,b,c). We used the parameters shown in Table 9. Where (1),(2), and (3) indicate the variable a,b, and c respectively. (0) is a virtual variable x_0 for the purpose of representing a constant, i.e. the value of x_0 is always 1. The terminal nodes $sqr1$, $sqr2$, and $sqr3$ are square values of a, b, and c. As we will see later, these input variables are not essential for our system and used only for

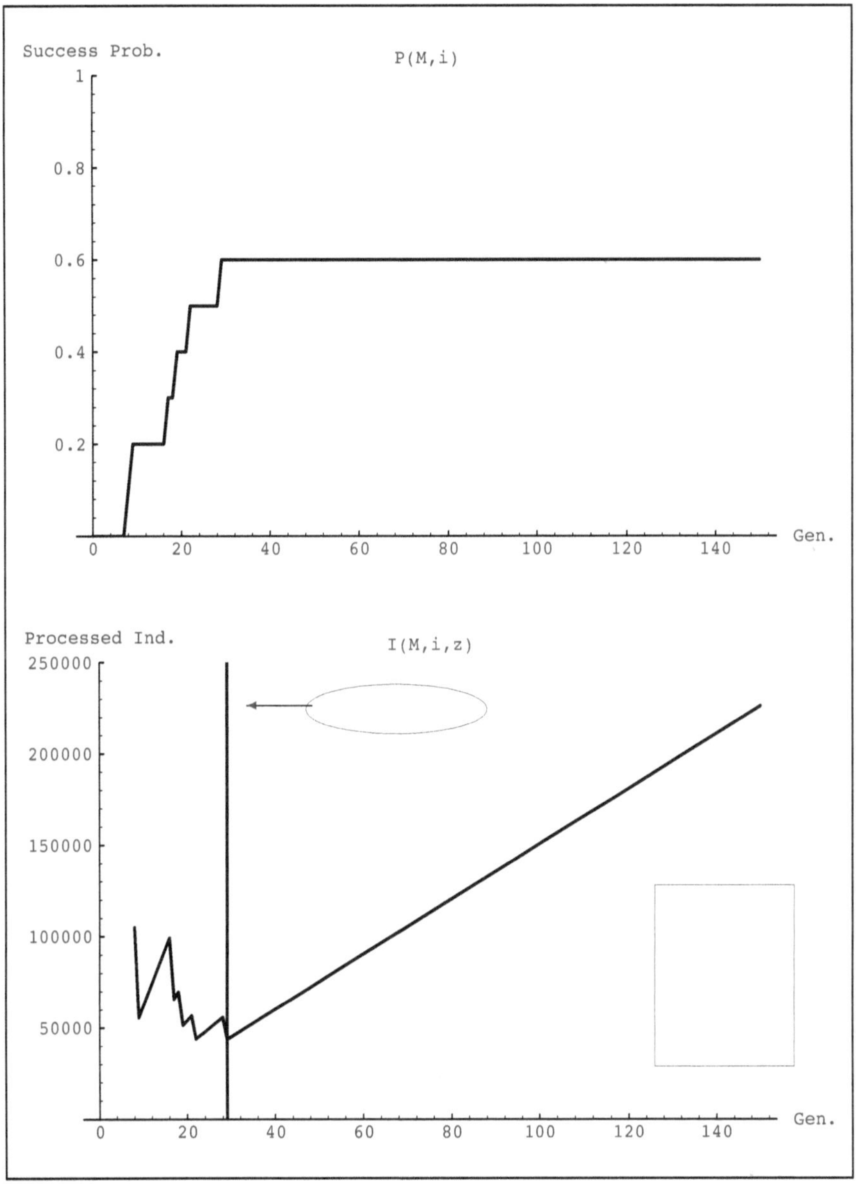

Fig. 11 Performance Curves (Two-Box Problem)

the sake of simplification. The acquired structure with subexpressions at generation 1105 is shown below (see also Fig.12). This is a typical acquired tree from several runs. The tree gives 100% correct answers to all 10 data.

```
(NODE376268 (NODE370522 (NODE370523 (2) (2)) (sqr1))
 (NODE376269 (NODE375704 (NODE375705 (sqr2) (sqr1)) (sqr3))
  (NODE376270 (sqr3) (NODE376271 (sqr2) (sqr1)))))
```

This "complete form" expresses the exact formula as follows:

$$1.26976(0.41361(0.47600 \times b \times b) \times a^2) + (-0.51711(-0.12086((b^2+a^2) - c^2) \times (c^2 - (a^2+b^2))))$$

$$= 0.24998a^2b^2 - 0.06249(b^2+a^2-c^2)^2$$

$$\approx 0.0625\{4a^2b^2 - (b^2+a^2-c^2)^2\}$$

$$= \frac{1}{16}(a+b+c)(a+b-c)(a+c-b)(b+c-a) \tag{56}$$

Thus we regard the complete form as a desired Heron formula (i.e. the square of equation (55)). In NODE370523, *sqrt*2 is expanded to $0.47600 \times b \times b$. Therefore, the introduction of square values (i.e. *sqrt*1, *sqrt*2, and *sqrt*3) is not necessary for this experiment. They are used in order to improve efficiency.

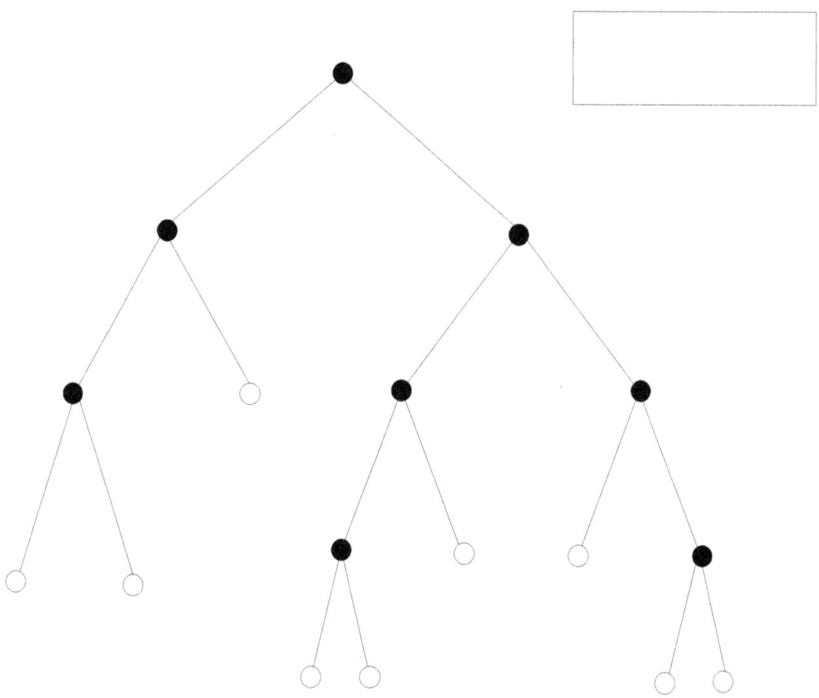

Fig. 12 Acquired Structure (Heron Formula)

Table 10 Subexpressions for Heron Formula

Node	subexpression
NODE376268	$1.26976x_1 + (-0.51711x_2)$
NODE370522	$0.41361x_1x_2$
NODE370523	$0.47600x_1x_2$
NODE376269	$-0.12086x_1x_2$
NODE375704	$-$
NODE375705	$+$
NODE376270	$-$
NODE376271	$+$

Since STROGANOFF can only handle polynomial relations for subexpressions at the moment, general symbolic regressions (including transcendental functions, e.g. square root function in this example) are beyond its scope. However, by increasing the types of subexpressions, and changing the regression procedures, we expect to be able to cope with more general cases. We are currently working on this topic.

6 Applying STROGANOFF to Computational Finances

We present the application of STROGANOFF to predicting a real-world time series, i.e., the prediction of the price data in the Japanese stock market. Our goal is to make an effective decision rule as to when and how many stocks to deal, i.e., sell or buy.

Evolutionary algorithms have been applied to the time series prediction, such as sun spot data [Angeline96] or the time sequence generated from the Mackey-Glass equation (section 4.1). Among them, the financial data prediction provides a challenging topic. This is because the stock market data are quite different from other time series data for the following reasons:

1. The ultimate goal is not to minimize the prediction error, but to maximize the profit gain.
2. Stock market data are highly time-variant, i.e., changeable every minute.
3. The stock market data are given in an event-driven way. They are highly influenced by the indeterminate dealing.

There have been several applications of GA or GP to the financial tasks, such as portfolio optimization, bankruptcy prediction, financial forecasting, fraud detection and scheduling.

We show how successfully the decision rule derived by STROGANOFF predicts the stock pricing so as to gain high profits from the market simulation. The comparative experiments are conducted with standard GP and neural networks to show the effectiveness of our approach.

6.1 *Predicting Stock Market Data*

This chapter utilizes our method to predict the price data in Japanese stock market. The financial data we use is the stock price average of Tokyo Stock Exchange, which is called Nikkei225.

6.1.1 Target Financial Data

The Nikkei225 average is computed by the Nihon Keizai Shimbun-Sha, a well-known financial newspaper publishing firm. The derivation is based upon the Dow formula. As of Feb.,12th, 2008, the Nikkei average stood at 13,021.96 Japanese yen (JPY) However, this average is a theoretical number and should be rigidly distinguished from the real average price in the market place. The computation formula for the Nikkei average is as follows:

$$\text{Nikkei Average} = \frac{\sum_{x \in 225\ \text{stocks}} \text{Price}_x}{D} \tag{57}$$

The sum of the stock price Price_x is over 225 representative stocks in Tokyo Stock Exchange market. Originally, the divisor D was 225, i.e., the number of component stocks. However, the divisor is adjusted whenever price changes resulting from factors other than those of market activity take place. The Nikkei averages are usually given every minute from 9:00am to 12:00pm and from 1:00pm to 3:00pm. The data we use in the following experiments span over a period from April 1st 1993 to September 30th 1993. Fig.13 shows the example tendency of the Nikkei225 average

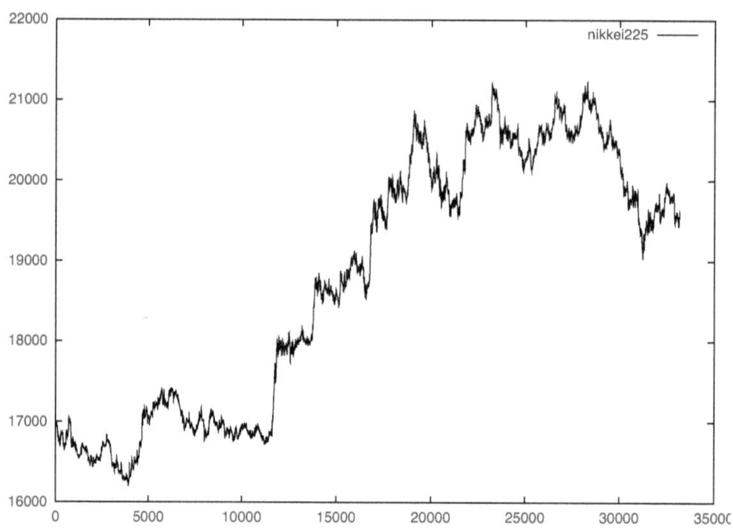

Fig. 13 Nikkei225 Data

during the above period. All data are normalized between 0.0 and 1.0 as the input value. The total number of data is 33,177. We use the first 3,000 time steps for the training data and the rest for the testing data.

6.1.2 STROGANOFF Parameters and Experimental Conditions

We have applied STROGANOFF to predicting the Nikkei225 stock price average. The used parameters are shown in Table 11. For the sake of comparison, STROGANOFF was run using a variety of terminal sets described below.

- Condition A: The terminal set is $\{\mathrm{y1},\cdots,\mathrm{y10},\Re\}$, in which $\mathrm{y}i$ is the Nikkei225 price average observed i minutes before the predicted time. That is, if $x(t)$ is the Nikkei225 price average at time t, then $\mathrm{y}i = x(t-i)$. $\Re$ is a constant generated randomly.
- Condition B: The terminal set is $\{\mathrm{ave1},\cdots,\mathrm{ave10},\Re\}$. The $\mathrm{ave}i$ terminal is the average of the Nikkei225 value every 10 minutes, i.e.,

$$\mathrm{ave}i = \frac{\sum_{k=1}^{10} x(t-10*(i-1)-k)}{10}.$$

- Condition C: The terminal set is $\{\mathrm{m1},\cdots,\mathrm{m10},\Re\}$. The $\mathrm{m}i$ terminal is the variance of the Nikkei225 value every 10 minutes, i.e.,

$$\mathrm{m}i = \frac{\sum_{k=1}^{10} (x(t-10*(i-1)-k)-\mathrm{ave}i)^2}{10}.$$

- Condition D: The terminal set is $\{\mathrm{m1},\cdots,\mathrm{m10},\mathrm{ave1},\cdots,\mathrm{ave10},\Re\}$.
- Condition E: The terminal set is $\{\mathrm{v1},\cdots,\mathrm{v10},\mathrm{r1},\cdots,\mathrm{r10},\Re\}$, where the terminals $\mathrm{v}i$ and $\mathrm{r}i$ are defined as follows:

$$\mathrm{v}i = |x(t-i)-x(t-i-1)|$$

$$\mathrm{r}i = \frac{x(t-i)-x(t-i-1)}{x(t-i-1)}$$

The predicted value, i.e., the target output of a STROGANOFF tree, is the current Nikkei225 price average for the conditions from A to D. On the other hand, for the condition E, the target is the difference between the current Nikkei225 price average and the price observed one minute before. The mean square error is derived from the predicted value and the target data. Then, the fitness value is calculated as follows:

$$\text{MDL fitness} = 0.5kW\log N + 0.5N\log S_N^2, \tag{58}$$

where where N is the number of input-output data pairs, S_N^2 is the mean square error. In this equation, we modified the previous definition of MDL (eq.(25)) so as to use the weight value W.

Table 11 STROGANOFF Parameters

max_generation	100	max_depth_after_crossover	17
population_size	100	max_depth_for_new_trees	6
steady_state	0	max_mutant_depth	4
grow_method	GROW	crossover_any_pt_fraction	0.2
tournament_K	6	crossover_func_pt_fraction	0.7
selection_method	TOURNAMENT	fitness_prop_repro_fraction	0.1
Weigh value w	$w \in \{0.2, 0.1, 0.01, 0.001, 0.0001, 0.0, -0.01\}$		

6.1.3 GP Parameters and Experimental Conditions

For the sake of comparison, standard GP was also applied to the same data. We chose sgpc1.1, a simple GP system in C language, for predicting the Nikkei225 stock price average. The used parameters are shown in Table 12. GP was run using the same terminal sets as those used by STROGANOFF (see section 6.1.2).

The GP fitness value is defined to be the mean square error of the predicted value and the target data. The smaller fitness value, the better.

Table 12 GP Parameters for sgpc1.1

max_generation	100	max_depth_after_crossover	17
population_size	1000	max_depth_for_new_trees	6
steady_state	0	max_mutant_depth	4
grow_method	GROW	crossover_any_pt_fraction	0.2
tournament_K	6	crossover_func_pt_fraction	0.7
selection_method	TOURNAMENT	fitness_prop_repro_fraction	0.1
function set	$\{+, -, *, \%, \sin, \cos, \exp\}$		

6.1.4 Validation Method

In order to confirm the validness of the predictor acquired by STROGANOFF and GP, we examine the best evolved tree with the stock market simulation during the testing period. Remember that the output prediction of a tree is the current Nikkei225 price average for conditions from A to D. Thus, we use the following rule to choose the dealing, i.e., to decide whether to buy or sell a stock. Let $Pr(t)$ be the observed Nikkei225 average at the time step of t.

Step1 Initially, the total budget BG is set to be 1,000,000 JPY. Let the time step t be 3000, i.e., the beginning of the testing period. The stock flag ST is set to be 0.

Step2 Derive the output, i.e., the predicted Nikkei225 average, of the GP tree. Let $\widetilde{Pr}(t)$ be the predicted value.

Step3 If $Pr(t-1) < \widetilde{Pr}(t)$ and $ST = 0$, then buy the stock. That is, set ST to be 1.

Step4 Else, if $Pr(t-1) > \widetilde{Pr}(t)$ and $ST = 1$, then sell the stock. That is, set ST to be 0.

Step5 If $ST = 1$, let $BG := BG + Pr(t) - Pr(t-1)$.
Step6 If $BG < 0$, then return 0 and stop.
Step7 If $t < 33,177$, i.e., the end of the testing period, then $t := t + 1$ and go to Step2. Else return the total profit, i.e., $BG - 1,000,000$ yen.

The stock flag ST indicates the state of holding stock, i.e., if $ST = 0$, then no stock is shared at present, whereas if $ST = 1$, then a stock is shared. In **Step**5, the total property is derived according to the newly observed stock price. The satisfaction of the **Step**6 condition means that the system has gone into bankruptcy.

For the condition E, the tree outputs the difference between the current Nikkei225 price average and the price observed one minute before. Let the predicted output be $\widetilde{Pr'}(t)$. Then the dealing condition depends on the output value itself. More precisely, the above steps are revised as follows:

Step3 If $0 < \widetilde{Pr'}(t)$ and $ST = 0$, then buy the stock. That is, set ST to be 1.
Step4 Else, if $0 > \widetilde{Pr'}(t)$ and $ST = 1$, then sell the stock. That is, set ST to be 0.

We use the above dealing rules for the validation of the acquired STROGANOFF or GP tree. For the sake of simplicity, we put the following assumptions on the market simulation:

1. At most one stock is shared at any time.
2. The dealing stock is imaginary, in the sense that its price behaves exactly the same as the Nikkei225 average price.

The optimal profit according to the above dealing rule is 80,106.63 yen. This profit is ideally gained when the prediction is perfectly accurate during the testing period.

6.1.5 Experimental Results

STROGANOFF and GP runs were repeated under each condition 10 times. The training and the validation performance is shown in Tables 14 and 13. The MSE values are the average of mean square errors given by the best evolved tree for the training data. The hit percentage means how accurately the GP tree made an estimate of the qualitative behavior of the price. That is, the hit percentage is calculated as follows:

$$\text{hit} = \frac{N_{\text{up\_up}} + N_{\text{down\_down}}}{N_{\text{up\_up}} + N_{\text{up\_down}} + N_{\text{down\_up}} + N_{\text{down\_down}}} = \frac{N_{\text{up\_up}} + N_{\text{down\_down}}}{30,177}, \quad (59)$$

where $N_{\text{up\_up}}$ means the number of times when the tree makes an upward tendency while the observed price rises, and $N_{\text{down\_up}}$ means the number of times when the tree makes a downward tendency while the observed price falls, and so on. The total number of the predictions is 30,177, which equals the number of testing data.

All experimental results show that there seems to be a strong relationship between the MSE value, the hit percentage, and the profit gain. The lower the MSE value is,

Table 13 Experimental Results (STROGANOFF)

		Training	Testing			
			Hit(%)		Profit gain(yen)	
Condition	Weight	MSE	Average	Best	Average	Best
A	0.2	9.40E-06	62.3	62.4	30712	30762
	0.1	9.38E-06	62.3	62.4	30744	30762
	0.01	9.37E-06	62.2	62.3	30516	30823
	0.001	9.37E-06	62.2	62.4	30651	30804
	0.0001	9.37E-06	61.7	62.4	27511	30769
	0.0	9.38E-06	62.3	62.4	30654	30762
B	0.2	1.25E-05	57.5	57.7	18636	19194
	0.1	1.25E-05	57.3	57.7	18594	19194
	0.01	1.24E-05	55.3	57.7	13266	19194
C	0.2	6.57E-04	50.0	50.3	1599	3156
	0.1	6.57E-04	50.0	50.3	1517	3156
	0.01	6.57E-04	50.0	58.2	841	4044
	0.001	6.57E-04	49.9	50.1	890	1921
	0.0001	6.57E-04	50.0	50.8	1092	4044
	0.0	6.57E-04	50.0	50.2	471	2577
D	0.2	1.26E-05	57.6	57.7	18995	19194
	0.1	1.25E-05	57.2	57.7	18390	19194
	0.01	1.25E-05	54.9	57.7	13569	19194
E	0.2	7.25E-04	51.2	51.3	5785	6071
	0.1	7.24E-04	51.6	51.7	5381	5443
	0.01	7.24E-04	51.7	51.7	5443	5443
	0.001	7.24E-04	51.1	51.7	5381	5443
	0.0001	7.24E-04	51.7	51.7	5443	5443
	0.0	7.24E-04	51.7	51.7	5443	5443
	-0.01	7.24E-04	51.6	51.7	5381	5443

Table 14 Experimental Results (GP)

	Training	Testing			
		Hit(%)		Profit gain(yen)	
Condition	MSE	Average	Best	Average	Best
A	1.79e-06	55.02	62.78	12411.01	31256.06
B	1.22e-05	47.47	48.17	-4093.22	-2341.50
C	5.82e-04	50.42	51.00	127.03	305.13
D	1.28e-05	41.09	51.64	-19727.52	-3811.19
E	1.80e-06	61.38	62.56	28942.03	30896.56

the higher both the hit percentage and the profit gain are. However, this is not necessarily a matter of course, because achieving the high profit requires more accurate prediction for the critical tendency change, i.e., when the stock price suddenly falls (rises) reversely after the price rises (falls) before.

Table 14 shows that different weight values, i.e., w, resulted in different performance by STROGANOFF. We can observe that STROGANOFF gave relatively better performance under the condition A. The example acquired tree, i.e., the best evolved STROGANOFF predictors, under the condition A is shown in Fig.14. The average and best hit percentages were well over 50% under the conditions A, B, and D. Especially, STROGANOFF runs under the condition A resulted in the average hit percentage of 60% and over, which led to the high and positive profit gain. Using small weight values often gave rise to relatively long STROGANOFF trees so that the execution was aborted due to memory extinction. Fig.17 shows the prediction of the normalized Nikkei225 price by the best evolved tree under the conditions A and E. The predicted value of Nikkei225 price for the first 100 minutes is shown for condition A. The predicted difference between the current Nikkei225 price and the price one minute before is plotted for condition E. Fig.18 illustrates the optimal profit and the profits gained by the predicted trees. These results provide the evidence that the predicted difference under the condition E corresponds to the observed qualitative behavior, i.e., the upward or downward tendency, of the Nikkei225 price. This causes the high profit gain shown in Fig.17.

Table 13 presents that the average and best hit percentages were below 50% by standard GP under the conditions B, C and D, which resulted in the low profit and the negative returns except the condition C. On the other hand, under the conditions A and E, the average hit percentage was over 50% and the best one was over 60%, which led to the high and positive profit gain. Especially, GP runs under the condition E resulted in the average hit percentage of 60% and over. Fig.15 shows the prediction of the normalized Nikkei225 price by the best evolved tree under condition A. The predicted value (cond.A) of Nikkei225 price for the first 100 minutes is shown for condition A. The target Nikkei price (cash93A) is also shown in the figure. Fig.15 illustrates the optimal profit and the profits gained by the predicted trees.

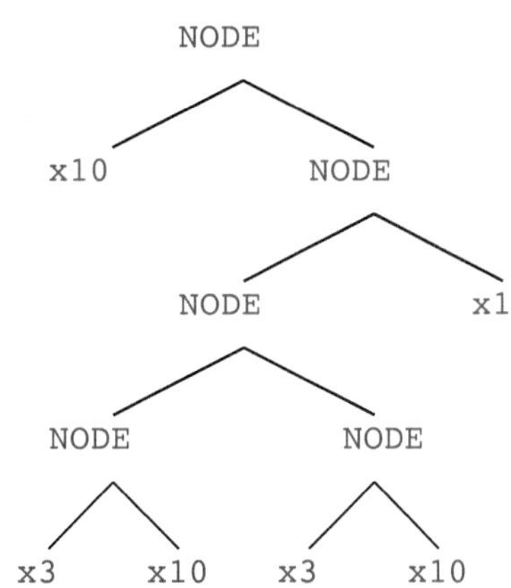

Fig. 14 The best evolved tree by STROGANOFF under condition A

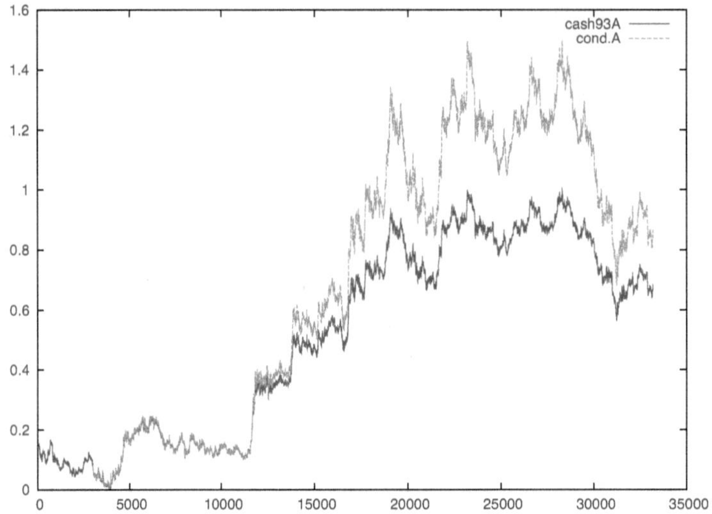

Fig. 15 Time series predicted by STROGANOFF under condition A

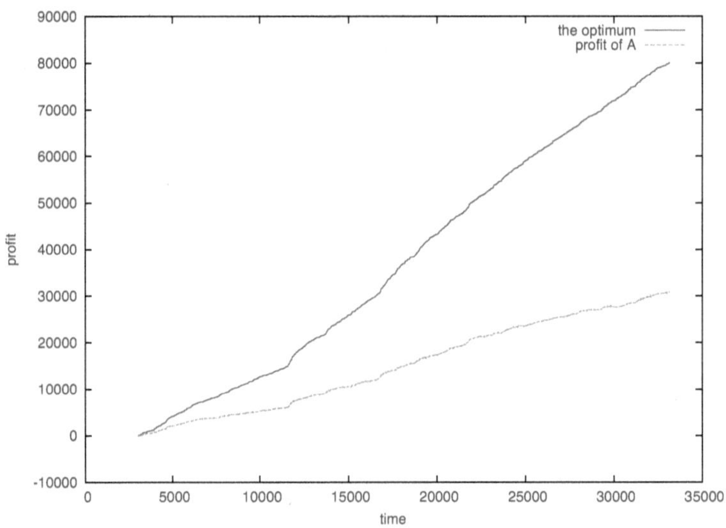

Fig. 16 Profit gained by STROGANOFF under condition A

To summarize the above GP experimental results, we can confirm the following points:

1. The average or variance terminals were not effective for the prediction (conditions B and C).
2. Using only past data or difference values led to the unstable prediction (condition A).

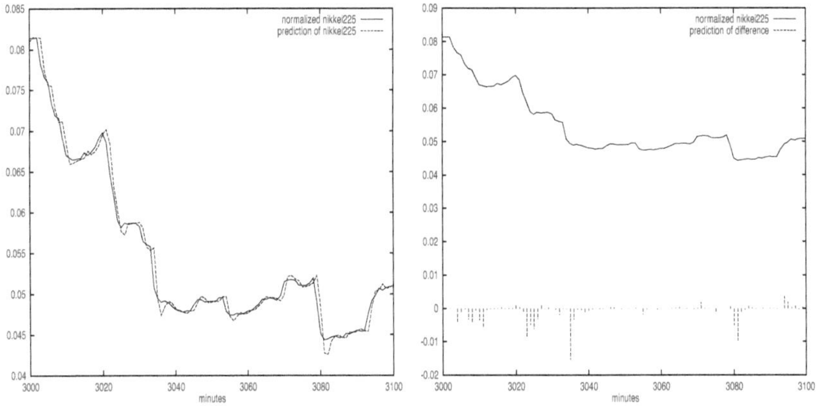

(a) Condition A. (b) Condition E.

Fig. 17 Prediction Results by GP

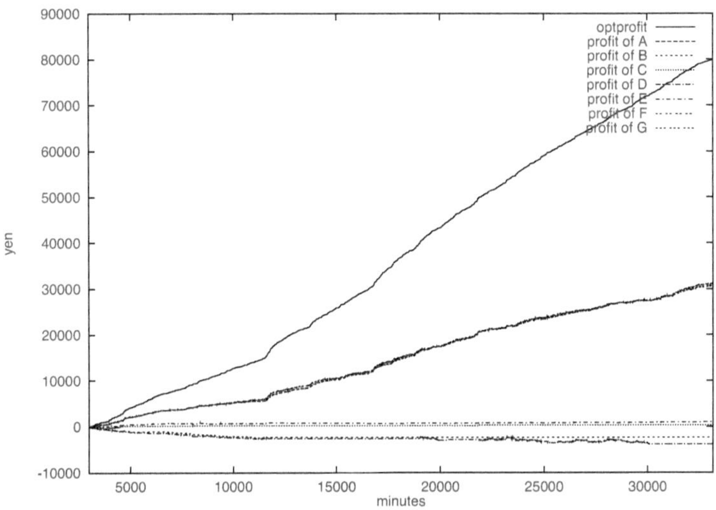

Fig. 18 Optimal Profit and Profits Gained by GP

3. The most effective terminal set included the absolute values and the directional values of the difference between the current Nikkei225 price and the past one (condition E).

Although the best profit is obtained by GP under condition A, the average profit is not necessarily high under the same condition. As can be seen in these results, GP performance is extremely dependent upon the terminal choice. However, there is not much theoretical background for the best choice. In general, the terminal and function sets play an essential role in GP search, but they are problem-dependent and not easy to choose. On the other hand, STROGANOFF's performance is relatively stable independently from the terminal choice.

6.1.6 Comparative Experiment with Neural Networks

For the sake of comparison, we apply Neural Network (NN) to the same prediction task and examine the performance difference. We used the program available at "Neural Networks at Your Fingertips" [Kutza96]. This NN program implements the classical multi-layer backpropagation network with bias terms and momentum. It is used to detect structure in time-series, which is presented to the network using a simple tapped delay-line memory. The program originally learned to predict future sunspot activity from historical data collected over the past three centuries. To avoid overfitting, the termination of the learning procedure is controlled by the so-called stopped training method.

The NN parameters used are shown in Table 15. The network was trained under the previous condition A. That is, the input variables of the network was set to be $\{y1,\cdots,y10\}$. The random constant $\Re$ is omitted. Table 16 shows the experimental results. The data are averaged over 10 runs with different numbers of hidden units. Comparing these results with the ones in Tables 14 and 13, we can confirm that NN gave much worse results than STROGANOFF. The reason seems to be that the neural network suffers from the overfitting, as can be seen in the table. Moreover, the computational time is much longer for the convergence for the neural network. Thus, we can conclude the superiority of STROGANOFF over NN.

Table 15 Neural Network Parameters

#. of Layers	3	#. of hidden nodes	5, 10, 15
α	0.5	BIAS	1
η	0.05	EPOCHS	1000
Gain	1	#. of LOOP	100

Table 16 Experimental Results (NN)

	Training	Testing			
		Hit(%)		Profit gain(yen)	
#.hidden units	MSE	Average	Best	Average	Best
5	2.92e-06	58.2	60.2	23682	27586
10	2.70e-06	58.7	59.4	24725	26427
15	2.73e-06	58.3	59.5	23990	26245

6.2 Developping Day-Trading Rules

As trading systems become more common, we see a number of different algorithms being used both by stock brokers and individual investors. However most of the algorithms that come bundled with these trading systems are closed rule-based decision systems, which either depend on many different parameters, or were developed for a given reality of the market place, and cannot adapt as this reality changes. GP has been successfully applied in the generation of algorithmic trading systems [Potvin *et al.*04, Aranha *et al.*07]. In this section, we present a new system of generating trading rules by means of STROGANOFF, which is able to face the challenges of day-trading described above.

We test this system in a simulation using historical data of the Japanese stock market, and compare our results with default rule based methodologies.

6.2.1 Day-Trading

We denominate as Day-Traders those investors who will, as part of their strategy, open and close all their positions in the same day. The goal of the day-trader is short-term profit from the daily fluctuations of an asset. Day-trading strategy consists of first determining the overall trade tendency for the asset. If it is a rising tendency, the day-trader will buy the asset in the beginning of the day, and then wait for the optimal opportunity to sell it. Similarly, if it is believed that the market will show a downwards tendency, the trader will open a short position by selling stock, and then waiting for the opportunity of buying that stock back.

In order to operate, the day trader must then decide the key times when it will sell high and buy low. Ideally, these points should match the high price and low price of the day.

6.2.2 Predicting High and Low Price Values

Our system generates a function that determines the High Price and the Low Price (HP and LP), which can be used to derive the Buying Price and Selling Prince (BP and SP). The inputs of the function are the Opening Price, Closing Price, High Price, Low Price and Volume of the 6th day before the transaction (OP, CP, HP, LP, respectively). We also include techinical analysis indicators such as RSI (Relative Strength Index) and EMA (Exponential Moving Average). RSI was developed by J. Welles Wilder in 1978 and is defined as follows:

$$RSI = 100 - \frac{100}{(1+RS)} \quad (60)$$

$$RS = \frac{\text{Average Gain over RSI Period}}{\text{Average Loss over RSI Period}} \quad (61)$$

The RSI ranges from 0 to 100. An asset is overbought when the RSI approaches the 70 level, meaning that it may be getting overvalued and is a good candidate for a pullback. On the contrary, if the RSI approaches 30, it indicates that the asset may be getting oversold and therefore likely to become undervalued. The 80 and 20 levels are sometimes preferred by traders.

EMA is a moving average that gives extra weight to more recent price data, whichi is defined as follows:

$$\text{Today's EMA} = (\text{current day's CP}) \times k + (\text{previous EMA}) \times (1-k) \tag{62}$$

k is called the exponential percentage and more commonly known as the smoothing constant described below:

$$k = \frac{2}{n+1}, \tag{63}$$

where n is the number of periods to average.

The output of the function is a price that will determine HP and LP. The input variables are $\{\mathrm{OP}n, \mathrm{CP}n, \mathrm{HP}n, \mathrm{LP}n, \mathrm{EMA}n, \mathrm{RSI}n \ (n = 1, \ldots, 6)\}$, i.e., the Opening Price, Closing Price, High Price, Low Price and Volume of the 6th day before the transaction (OP, CP, HP, LP, respectively). An exmaple STROGANOFF tree is given as follows:

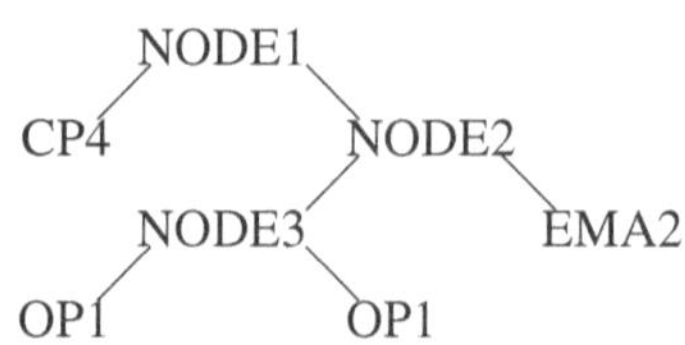

Where CPn and LPn mean CP and LP values of the n-th day before.

In case of GP, an exmaple tree is given as follows:

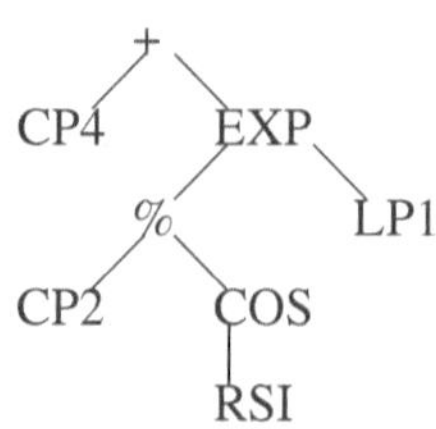

For the sake of predicting HP and LP values, the fitness function is based on the MSE value between the true values and the predicted ones. In case of GP, MSE value is directly used to derive the raw fitness, where as the MDL criterion is derived in addition to the MSE values (see eq.(25)).

6.2.3 Generating Dealing Rules

Given the above two values, i.e., HP and LP, by means of STROGANOFF or GP, we set the buying point (BP) and selling point (SP) as follows:

- If $HP - OP > OP - LP$, then
- set $BP := OP$ and $SP := HP \cdot k + OP \cdot (1-k)$,
- else set $SP := OP$ and $BP := LP \cdot k + OP \cdot (1-k)$.

With these two valuses, the system trader executes the trade in the following manner. If the asset price reaches BP or SP, it buys or sells (shorts) the asset, respectively, opening the position. Once the position is open, the system trader waits for the asset price to become the corresponding value (SP, if bought, and BP if sold), then it closes the position. If the asset does not reach the desired price by the end of the day, the system closes the position at that day's closing price. Finally, if the asset reaches neither BP nor SP during the day, the system does not execute any trades for that day. (In that case, it suffers a 500JPY penalty to its fitness value, in order to discourage passive behavior).

6.2.4 Experimental Results

To test and validate our proposal, we have conducted a series of simulation experiments based on historical data of the Japanese financial market. We ran a simulation on 218 of the 225 stocks listed in the NIKKEI index that operated continuously in the period from 2005 August (80 days). The former half, i.e., 40 days, is used for the training and the latter, i.e., 40 days, is used for the testing.

STROGANOFF and GP runs were repeated under the conditions shown in Tables 17 and 18. Remember that the input variables are $\{\mathrm{OP}n, \mathrm{CP}n, \mathrm{HP}n, \mathrm{LP}n, \mathrm{EMA}n, \mathrm{RSI}n\ \ (n = 1,\ldots,6)\}$, i.e., the Opening Price, Closing Price, High Price, Low Price and Volume of the 6th day before the transaction (OP, CP, HP, LP, respectively). The MSE value for a specific stock by STROGANOFF is shown in Table 19. The MSE values are the average of mean square errors between the true HP or LP values and the predicted ones, which are given by the best evolved tree for the training data. The training and the validation performance, i.e., the profit rate according to the above dealing policy is shown in Table 20.

Table 17 STROGANOFF Parameters for HP and LP prediction

max_generation	100	population_size	100
selection_method	TOURNAMENT	Weigh value w	1.0
terminal set	$\{\mathrm{OP}n, \mathrm{CP}n, \mathrm{HP}n, \mathrm{LP}n, \mathrm{EMA}n, \mathrm{RSI}n\ \ (n = 1,\ldots,6)\}$		

Table 18 GP Parameters for HP and LP prediction

max_generation	200	max_depth_after_crossover	8
population_size	500	max_depth_for_new_trees	8
steady_state	0	max_mutant_depth	4
grow_method	GROW	crossover_any_pt_fraction	0.2
tournament_K	6	crossover_func_pt_fraction	0.7
selection_method	TOURNAMENT	fitness_prop_repro_fraction	0.1
function set	$\{+, -, *, \sin, \mathrm{IF}\}$		
terminal set	$\{\mathrm{OP}n, \mathrm{CP}n, \mathrm{HP}n, \mathrm{LP}n, \mathrm{EMA}n, \mathrm{RSI}n\ \ (n = 1,\ldots,6)\}$		

Table 19 MES values

	Stock #	101	8001	9531	9501	5002	5401	8801	Avg.	Std.
STROGANOFF	HP	6.5E-04	3.3E-04	9.0E-04	4.5E-04	5.6E-04	7.0E-04	1.6E-04	5.4E-04	2.5E-04
	LP	6.8E-04	3.3E-04	4.2E-04	2.7E-04	2.3E-04	5.3E-04	2.7E-04	3.9E-04	1.6E-04
Simple GP	HP	2.26E-05								
	LP	4.73E-05								

Table 20 Profit Rates

	Stock #	101	8001	9531	9501	5002	5401	8801	Avg.	Std.
STROGANOFF	Training	0.132	0.366	0.337	0.331	0.638	0.730	0.486	0.431	0.203
	Testing	-0.004	-0.199	-0.063	-0.002	0.008	0.079	-0.086	-0.038	0.089
Simple GP	Training	0.0734	-0.112	0.136	0.215	0.177	0.294	0.169	0.136	0.129
	Testing	-0.055	0.0122	0.011	0.00188	-0.184	-0.116	-0.029	-0.051	0.074
Buy and Hold	Testing	0.143	0.268	0.105	-0.00374	-0.00816	0.0125	0.127		

Table 21 Transaction Details

Stock #	101	8001	9531	9501	5002	5401	8801
Profit (JPY)	-745.72	9.40	5.00	5.41	-268.88	-47.88	-52.52
#. positive transactions	17	25	20	19	11	19	22
#. negative transactions	18	15	19	19	28	17	17
Profit rate	-0.055	0.0122	0.011	0.00188	-0.184	-0.116	-0.029
Profitiw/o comissionj(JPY)	253.50	75.86	41.72	224.82	-160.04	-18.07	96.91
Commision	999.22	66.46	36.72	219.41	108.84	29.81	149.43
Profit:Test (JPY)	899.53	-74.05	56.04	573.20	223.66	88.46	244.77
Profit rate:Test	0.0734	-0.112	0.136	0.215	0.177	0.294	0.169

Table 22 Optimal Transactions

Stock #	101	8001	9531	9501	5002	5401	8801
Profit (JPY)	5,040.95	740.26	302.70	889.11	889.11	410.59	1,822.18
Profit rate	0.375	0.963	0.676	0.309	0.609	0.992	1.01

7 Inductive Genetic Programming

Nikolaev and Iba have proposed Inductive Genetic Programming (IGP) for the sake of extending STROGANOFF. This section describes the basics of IGP and its applications[2].

Inductive Genetic Programming is a specialization of the Genetic programming (GP) paradigm for inductive learning. The reasons for using this specialized term are: 1) *inductive* learning is a search problem and GP is a versatile framework for

[2] This section is mainly based on Nikolaev and Iba's recent works on the extension of STROGANOFF. The readers should refer to [Nikolaev and Iba06] for the details of IGP and other applications.

exploration of large multidimensional search spaces; 2) GP provides *genetic* learning operators for hypothetical model sampling that can be tailored to the data; and 3) GP manipulates *program*-like representations which adaptively satisfy the constraints of the task. An advantage of inductive GP is that it discovers not only the parameters but also the structure and size of the models.

The basic computational mechanisms of a GP system are inspired by those from natural evolution. GP conducts a search with a population of models using mutation, crossover and reproduction operators. Like in the nature these operators have a probabilistic character. The mutation and crossover operators choose at random the model elements that will undergo changes, while the reproduction selects random good models among the population elite. Another characteristic of GP is its flexibility in the sense that it allows us easily to adjust its ingredients for the particular task. It enables us to change the representation, to tune the genetic operators, to synthesize proper fitness functions, and to apply different reproduction schemes.

7.1 Polynomial Neural Networks

Polynomial neural networks (PNN) are a class of feedforward neural networks [3]. They are developed with the intention of overcoming the computational limitations of the traditional statistical and numerical optimization tools for polynomial identification, which practically can only identify the coefficients of relatively low order terms. The adaptive PNN algorithms are able to learn the weights of highly nonlinear models.

A PNN consists of nodes, or neurons, linked by connections associated with numeric weights. Each node has a set of incoming connections from other nodes, and one (or more) outgoing connections to other nodes. All nonterminal nodes, including the fringe nodes connected to the inputs, are called hidden nodes. The input vector is propagated forward through the network. During the forward pass it is weighted by the connection strengths and filtered by the activation functions in the nodes, producing an output signal at the root. Thus, the PNN generates a nonlinear real valued mapping $P : \mathscr{R}^d \to \mathscr{R}$, which taken from the network representation is a *high-order polynomial model*:

$$P(\mathbf{x}) = a_0 + \sum_{i=1}^{L} a_i \prod_{j=1}^{d} x_j^{r_{ji}} \tag{64}$$

where a_i are the term coefficients, i ranges up to a pre-selected maximum number of terms L: $i \leq L$; x_j are the values of the independent variables arranged in an input vector $\mathbf{x}$, i.e. $j \leq d$ numbers; and $r_{ji} = 0, 1, ...$ are the powers with which the j-th element x_j participates in the i-th term. It is assumed that r_{ji} is bounded by a maximum polynomial order (degree) s: $\sum_{j=1}^{d} r_{ji} \leq s$ for every i. The above

[3] This section is basically based on our previous works. Refer to [Nikolaev and Iba06] for more details.

polynomial is linear in the coefficients a_i, $1 \leq i \leq L$, and non-linear in the variables x_j, $1 \leq j \leq d$.

Strictly speaking, a power series contains an infinite number of terms that can exactly represent a function. In practice a finite number of them is used for achieving the predefined sufficient accuracy. The polynomial size is manually fixed by a design decision.

7.2 PNN Approaches

The differences between the above PNN are in the representational and operational aspects of their search mechanisms for identification of the relevant terms from the power series expansion, including their weights and underlying structure. The main differences concern: 1) what is the polynomial network topology and especially what is its connectivity; 2) which activation polynomials are allocated in the network nodes for expressing the model, are they linear, quadratic, or highly-nonlinear mappings in one or several variables; 3) what is the weight learning technique; 4) whether there are designed algorithms that search for the adequate polynomial network structure; 5) what criteria for evaluation of the data fitting are taken for search control.

The models evolved by IGP are genetic programs. IGP breeds a population $\mathscr{P}$ of genetic programs $\mathscr{G} \in \mathscr{P}$. The notion of a *genetic program* means that this is a sequence of instructions for computing an input-output mapping. The main approaches to encoding genetic programs are: 1) tree structures [Koza94]; 2) linear arrays [Banzhaf *et al.*98]; and 3) graphs [Teller and Veloso1996]. The tree-like genetic programs originate from the expressions in functional programming languages, where an expression is arranged as a tree of elementary functions in its nodes and variables in its leaves. The linear genetic programs are linear arrays of instructions, which can be written in terms of a programming language or written in machine code. The graph-based programs are made as directed graphs with stacks for their processing and memory for the variables. The edges in the graph determine the sequence for execution of the programs. Each node contains the function to be performed, and a pointer to the next instruction.

Tree-like genetic programs are suitable for IGP as they offer two advantages: 1) they have parsimonious topology with sparse connectivity between the nodes, and 2) they enable efficient processing with classical algorithms. Subjects of particular interest here are the linear genetic program trees that are genotypic encodings of PNN phenotypes which exhibit certain input-output behaviors.

A genetic program has a *tree* structure. In it a node is below another node if the other node lies on the path from the root to this node. The nodes below a particular node are a subtree. Every node has a parent above it and children nodes under it. Nodes without children are leaves or terminals. The nodes that have children are nonterminals or functional nodes.

PNN are represented with *binary trees* in which every internal functional node has a left child and a right child. A binary tree with Z functional nodes has $Z+1$

terminals. The nodes are arranged in multiple levels, called also *layers*. The level of a particular node is one plus the level of its parent, assuming that the root level is zero. The *depth*, or height of a tree, is the maximal level among the levels of its nodes. A tree may be limited by a maximum tree depth, or by a maximum tree size which is the number of all nodes and leaves.

Trees are now described formally to facilitate their understanding. Let $\mathscr{V}$ be a vertex set from two kinds of components: *functional nodes* $\mathscr{F}$ and *terminal leaves* $\mathscr{T}$ ($\mathscr{V} = \mathscr{F} \cup \mathscr{T}$). A *genetic program* $\mathscr{G}$ is an ordered tree $s_0 \equiv \mathscr{G}$, in which the sons of each node $\mathscr{V}$ are ordered, with properties:

- it has a distinguishing parent $\rho(s_0) = \mathscr{V}_0$ called the *root* node;
- its nodes are labelled $\nu : \mathscr{V} \to \mathscr{N}$ from left to right and $\nu(\mathscr{V}_i) = i$;
- any functional node has a number of children, called arity $\kappa : \mathscr{V} \to \mathscr{N}$, and a terminal leaf $\rho(s_i) = \mathscr{T}_i$ has zero arity $\kappa(\mathscr{T}_i) = 0$;
- the children of a node $\mathscr{V}_i$, with arity $k = \kappa(\mathscr{V}_i)$, are roots of disjoint subtrees $s_{i1}, s_{i2}, ..., s_{ik}$. A subtree s_i has a root $\rho(s_i) = \mathscr{V}_i$, and subtrees $s_{i1}, ..., s_{ik}$ at its k children: $s_i = \{(\mathscr{V}_i, s_{i1}, s_{i2}, ..., s_{ik}) \mid k = \kappa(\mathscr{V}_i)\}$.

This vertex labeling suggests that the subtrees below a node $\mathscr{V}_i$ are ordered from left to right as the leftmost child s_{i1} has smallest label $\nu(s_{i1}) < \nu(s_{i2}) < ... < \nu(s_{ik})$. This ordering of the nodes is necessary for making efficient tree implementations, as well as for the design of proper genetic learning operators for manipulation of tree structures.

The construction of binary tree-like PNN requires us to instantiate its parameters. The terminal set includes the explanatory input variables $\mathscr{T} = \{x_1, x_2, ..., x_d\}$, where d is the input dimension. The function set contains the activation polynomials in the tree nodes $\mathscr{F} = \{p_1, p_2, ..., p_m\}$, where the number m of distinct functional nodes is given in advance. A reasonable choice are the incomplete bivariate polynomials up to second-order that can be derived from the complete one (1.2) assuming that some of its coefficients are zero. The total number of such incomplete polynomials is 25 from all $2^5 - 1$ possible combinations of monomials $w_i h_i(x_i, x_j)$, $1 \leq i \leq 5$, having always the leading constant w_0, and two different variables. A subset $p_i \in \mathscr{F}$, $1 \leq i \leq 16$ of them is taken after elimination of the symmetric polynomials (Table 23).

The notion of *activation polynomials* is considered in the context of PNN instead of transfer polynomials to emphasize that they are used to derive backpropagation network training algorithms.

The motivations for using all distinctive *complete* and *incomplete* (first-order and second-order) bivariate activation polynomials in the network nodes are: 1) having a set of polynomials enables better identification of the interactions between the input variables; 2) when composed higher-order polynomials rapidly increase the order of the overall model, which causes overfitting even with small trees; 3) first-order and second-order polynomials are fast to process; and 4) they define a search space of reasonable dimensionality for the GP to explore. The problem of using only the complete second-order bivariate polynomial is that the weights of the superfluous terms do not become zero after least squares fitting, which is an obstacle for achieving good generalization.

Table 23 Activation polynomials for genetic programming of PNN

1.	$p_1(x_i,x_j) = w_0 + w_1x_1 + w_2x_2 + w_3x_1x_2$
2.	$p_2(x_i,x_j) = w_0 + w_1x_1 + w_2x_2$
3.	$p_3(x_i,x_j) = w_0 + w_1x_1 + w_2x_1x_2$
4.	$p_4(x_i,x_j) = w_0 + w_1x_1 + w_2x_1x_2 + w_3x_1^2$
5.	$p_5(x_i,x_j) = w_0 + w_1x_1 + w_2x_2^2$
6.	$p_6(x_i,x_j) = w_0 + w_1x_1 + w_2x_2 + w_3x_1^2$
7.	$p_7(x_i,x_j) = w_0 + w_1x_1 + w_2x_1^2 + w_3x_2^2$
8.	$p_8(x_i,x_j) = w_0 + w_1x_1^2 + w_2x_2^2$
9.	$p_9(x_i,x_j) = w_0 + w_1x_1 + w_2x_2 + w_3x_1x_2 + w_4x_1^2 + w_5x_2^2$
10.	$p_{10}(x_i,x_j) = w_0 + w_1x_1 + w_2x_2 + w_3x_1x_2 + w_4x_1^2$
11.	$p_{11}(x_i,x_j) = w_0 + w_1x_1 + w_2x_1x_2 + w_3x_1^2 + w_4x_2^2$
12.	$p_{12}(x_i,x_j) = w_0 + w_1x_1x_2 + w_2x_1^2 + w_3x_2^2$
13.	$p_{13}(x_i,x_j) = w_0 + w_1x_1 + w_2x_1x_2 + w_3x_2^2$
14.	$p_{14}(x_i,x_j) = w_0 + w_1x_1 + w_2x_2 + w_3x_1^2 + w_4x_2^2$
15.	$p_{15}(x_i,x_j) = w_0 + w_1x_1x_2$
16.	$p_{16}(x_i,x_j) = w_0 + w_1x_1x_2 + w_2x_1^2$

The following hierarchically composed polynomial illustrates a hierarchically composed polynomial extracted from the PNN in Fig.19 to demonstrate the transparency and easy interpretability of the obtained model.

```
(( w0 + w1 * z7^2 + w2 * z4^2 )
        z7=( w0 + w1 * x2 + w2 * x2^2 + w3 * x3^2 )
               x2
               x3 )
        z4=( w0 + w1 * z2 + w2 * z2 * x1 + w3 * z2^2 )
               z2=( w0 + w1 * x7 + w2 * x5 )
                      x7
                      x5 )
               x1 ))
```

The accommodation of a set of complete and incomplete activation polynomials in the network nodes makes the models versatile for adaptive search, while keeping the neural network architecture relatively compact. Using a set of activation polynomials does not increase the computational demands for performing genetic programming. The benefit of having a set of activation polynomials is of enhancing the expressive power of this kind of PNN representation.

An example of a tree-structured polynomial using some of these activation polynomials is illustrated in Fig. 19. The computed polynomial $P(\mathbf{x})$ at the output tree root is the multivariate composition: $P(x_1,x_2,x_3,x_5,x_7) = p_8(p_7(x_2,x_3), p_4(p_2(x_7,x_5),x_1))$.

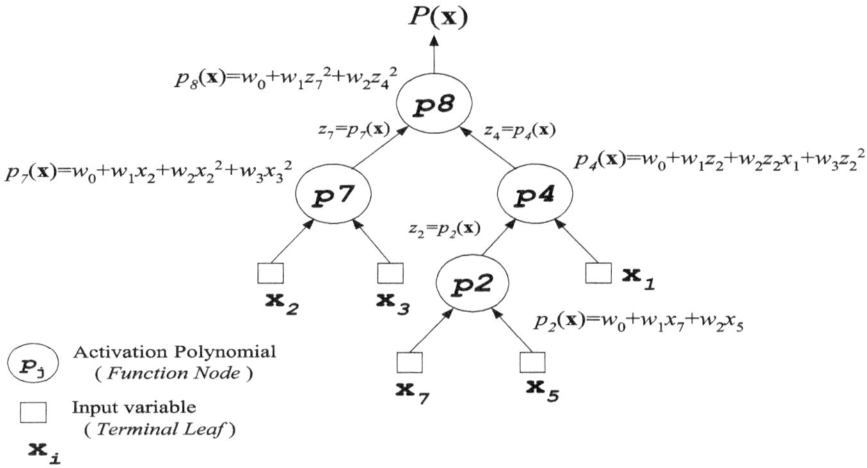

Fig. 19 Tree-structured representation of a PNN

7.3 Basic IGP Framework

The IGP paradigm can be used for the automatic programming of polynomials. It provides a problem independent framework for discovering the polynomial structure, in the sense of shape and size, as well as the weights. The IGP learning cycle involves five substeps: 1) ranking of the individuals according to their fitness; 2) selection of some elite individuals to mate and produce offspring; 3) processing of the chosen parent individuals by the crossover and mutation operators; 4) evaluation of the fitnesses of the offspring; and 5) replacement of predetermined individuals in the population by the newly born offspring. Table 24 presents the basic IGP algorithmic framework.

The formalization of the basic framework, which can be used for implementing an IGP system, requires some preliminary definitions. The IGP mechanisms operate at the genotype level, that is they manipulate linearly implemented genetic program trees g. The basic control loop breeds a population $\mathcal{P}$ of genetic programs g during a number of cycles τ called generations. Let n denote the size of the population vector, that is the population includes $g_i, 1 \leq i \leq n$ individuals. Each individual g is restricted by a predefined tree depth S and size L in order to limit the search space to within reasonable bounds. The initial population $\mathcal{P}(0)$ is randomly created.

The function *Evaluate* estimates the fitness of the genetic programs using the fitness function f to map genotypes $g \in \Gamma$ into real values f: $\Gamma \rightarrow R$. The fitness function f takes a genetic program tree g, decodes a phenotypic PNN model from it, and measures its accuracy with respect to the given data. All the fitnesses of the genetic programs from the population are kept in an array of fitnesses F of size n. The selection mechanism *Select*: $\Gamma^n \rightarrow \Gamma^{n/2}$ operates according to a predefined scheme

Table 24 Basic framework for IGP

Inductive Genetic Programming

step	*Algorithmic sequence*
1. Initialisation	Let the generation index be $\tau = 0$,and the pop size be n
	Let the initial population be: $\mathscr{P}(\tau) = [g_1(\tau), g_2(\tau), ..., g_n(\tau)]$ where g_i, $1 \leq i \leq n$, are genetic programs of depth up to S
	Let μ be a mutation parameter, κ be a crossover parameter
	Create a random initial population: $\mathscr{P}(\tau) = RandomTrees(n)$, such that $\forall g, Depth(g) < S$
	Evaluate the fitnesses of the individuals: $F(\tau) = Evaluate(\mathscr{P}(\tau), \lambda)$ and order the population according to $F(\tau)$
2. Evolutionary Learning	a) Select randomly $n/2$ elite parents from $\mathscr{P}(\tau)$ $\mathscr{P}'(\tau) = Select(\mathscr{P}(\tau), F(\tau), n/2)$
	b) Perform recombination of $\mathscr{P}'(\tau)$ to produce $n/4$ offspring $\mathscr{P}''(\tau) = CrossTrees(\mathscr{P}'(\tau), \kappa)$
	c) Perform mutation of $\mathscr{P}'(\tau)$ to produce $n/4$ offspring $\mathscr{P}''(\tau) = MutateTrees(\mathscr{P}'(\tau), \mu)$
	d) Compute the offspring fitnesses $F''(\tau) = Evaluate(\mathscr{P}''(\tau), \lambda)$
	e) Exchange the worst $n/2$ from $\mathscr{P}(\tau)$ with offspring $\mathscr{P}''(\tau)$ $\mathscr{P}(\tau+1) = Replace(\mathscr{P}(\tau), \mathscr{P}''(\tau), n/2)$
	f) Rank the population according to $F(\tau+1)$ $g_0(\tau+1) \leq g_1(\tau+1) \leq ... \leq g_n(\tau+1)$
	g) Repeat the Evolutionary Learning (*step* 2) with another cycle $\tau = \tau + 1$ until the termination condition is satisfied

for picking randomly $n/2$ elite individuals which are going to be transformed by crossover and/or mutation.

The recombination function *CrossTrees*: $\Gamma^{n/4} \times R \rightarrow \Gamma^{n/4}$ takes the half $n/4$ from the selected $n/2$ elite genetic programs, and produces the same number of offspring using size-biased crossover using parameter κ. The mutation function *MutateTrees*: $\Gamma \times R \rightarrow \Gamma$ processes half $n/4$ from the selected $n/2$ elite genetic programs, using size-biased context-preserving mutation using parameter μ.

The resulted offspring are evaluated, and replace inferior individuals in the population *Replace*: $\Gamma^{n/2} \times \Gamma^{n/2} \times N \rightarrow \Gamma^n$. The steady-state reproduction scheme is used to replace the genetic programs having worst fitness with the offspring so as to maintain a proper balance of promising individuals. Next, all the individuals in the updated population are ordered according to their fitnesses.

7.4 PNN vs. Linear ARMA Models

Linear models are widely used for time series modelling due to the sound theory that explains them [Box and Jenkins70]. Although nonlinear models can also produce linear models, they usually outperform the linear models in the presence of nonlinearities, and especially sustained oscillations, as well as in the presence of stochastic disturbances. Simpler linear models such as exponential smoothing and linear regressions may be used if there is no clear evidence of more complex nonlinearity in the data. The linear models often need specific manipulation with techniques for elimination of trends and seasonal patterns for example, which require additional knowledge.

A comparison of an evolved PNN model with a linear AutoRegressive Moving Average (ARMA) model was made recently [de Menezes and Nikolaev06]. The PNN resemble ARMA models in that the activation polynomials are treated as linear regressors. The weights of the PNN activation polynomials are learned by efficient least squares fitting as are the weights of the linear ARMA models. This provides the advantage of reaching the optimal weights due to the unique global minimum on the error surface in case of linear models.

The benchmark *Airline series* [Faraway and Chatfield98] popular in the statistical community is chosen here for performing experimental comparisons. The series contains 144 observations, which are monthly totals of international airline passengers. The initial 132 points are taken for training through input vectors $\mathbf{x}(t) = [x(t), x(t-1), ..., x(t-11)]$. Following the standard methodology of [Box and Jenkins70], a seasonal ARMA model is developed and it is fit to the logarithm of the observed values: $\log x_t \sim ARMA(0,1,1) \times (0,1,1)_{12}$. Next, a PNN model is evolved using IGP by performing 50 runs using: fitness proportional selection, both crossover and mutation operators, population of size 100, common regularization parameter for the all weights $\lambda = 0.001$ and selection threshold for pruning $z = 0.01$.

The *ARMA* model shows accuracy of fitting the series $MSE_{ARMA} = 90.53$ which is better than the PNN accuracy $MSE_{PNN} = 152.12$. The prediction performance of the *ARMA* model is much worse showing one-step-ahead forecasting error $MSE^f_{ARMA} = 356.75$ while the PNN shows $MSE^f_{PNN} = 185.27$. The fitting accuracy and the prediction of the examined PNN model are illustrated in Fig.20 and Fig.21.

This brief study allows us to make several observations that are indicative of the advantages of genetically programmed PNN over linear ARMA models for time series modelling: 1) the use of PNN eliminates the need to perform data transformations before learning, so the need to decide whether and how to preprocess the given data is avoided; 2) the IGP of PNN are able to find polynomials that capture the time series characteristics well and predict well in the short-term; 3) the IGP of PNN can help to discover the relevant input variables for learning, and thus they help to understand the lag dependencies in time series; and 4) the PNN structure as a hierarchical composition of simple polynomials is a factor that affects the forecasting performance.

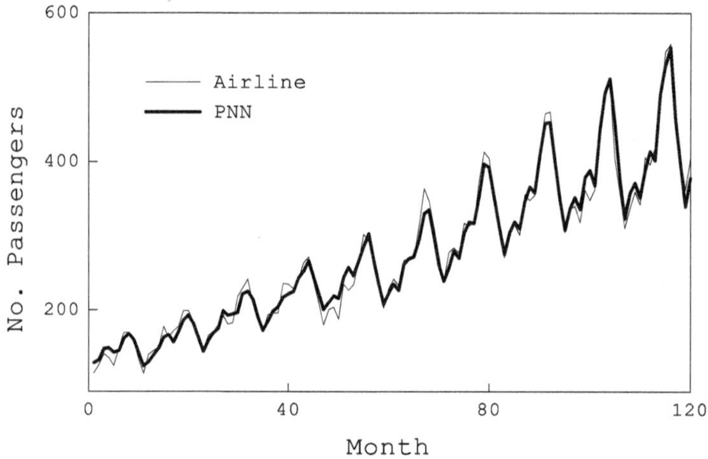

Fig. 20 Fitting of the Airline series by a PNN model evolved by IGP

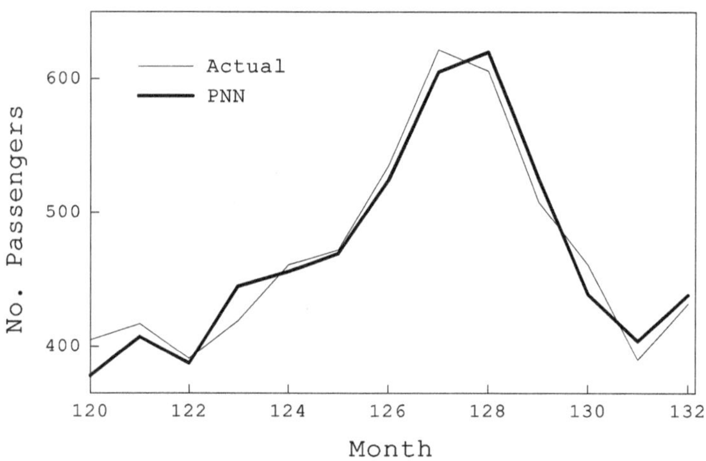

Fig. 21 Forecasting (single-step ahead prediction) of the Airline series by a PNN model evolved by IGP

7.5 PNN vs. Neural Network Models

The PNN generated by the IGP system belong to the category of feed-forward MLP (multilayer polynomial) networks [Rumelhart *et al.*86]. Both kinds of networks, MLP and PNN, implement nonlinear functions as hierarchical compositions. The practical problem of MLP is that the proper number of layers and the number of nodes must usually be found experimentally. A distinctive feature of PNN is that their model structure and variables can be found automatically using the evolutionary micromechanisms of IGP.

PNN and MLP both use adaptive learning by backpropagation (BP) techniques for gradient descent search in the weight space. In this sense PNN benefit from the efficacy, simplicity and power of the backprop techniques. At the same time both PNN and MLP suffer from the need to identify suitable values for the parameters of the algorithm such as the learning rate, the momentum, the regularization parameter, and the termination criterion. There are approaches to finding suitable parameter values that can be applied directly to PNN such as those based on the Bayesian evidence procedure [MacKay95]. PNN also assumes the strategies for improving the generalization performance developed originally for MLP such as network pruning and early stopping [Bishop95].

A PNN evolved by IGP and improved after that by BP is compared to an MLP network on the benchmark *Far-Infrared-Laser* series [Hübner *et al.*94]. This Laser series contains fluctuations of a physical laser recorded in a chaotic state during a laboratory experiment with an oscilloscope. The objective is to learn the description of a far-infrared NH_3 laser given its intensity pulsations. The initial 900 points are taken for training, and the next 100 points for testing as in the other research. The embedding dimension is $d = 10$. Approximately fifty runs are conducted with IGP using populations of size 100, $MaxTreeSize = 40$, and $MaxTreeDepth = 6$. The IGP system uses parameters: mutation probability $p_m = 0.01$, crossover probability $p_c = 1.5$, regularization $\lambda = 0.001$, and selection threshold $z = 0.01$.

The BP training algorithm is run to perform 150 epochs with parameters: learning rate $\eta = 0.001$ and momentum $\alpha = 0.01$. The MLP network is manually designed with one hidden layer of 10 sigmoidal activation functions and a summation output node. Training of the MLP by the backpropagation algorithm is made using a fixed learning rate $\eta_{MLP} = 0.01$ and momentum $\alpha_{MLP} = 0.02$.

The fitting accuracy and the prediction capacity of the best discovered PNN model are given in Fig. 22 and Fig. 23. The evolved PNN has 15 nodes with 34

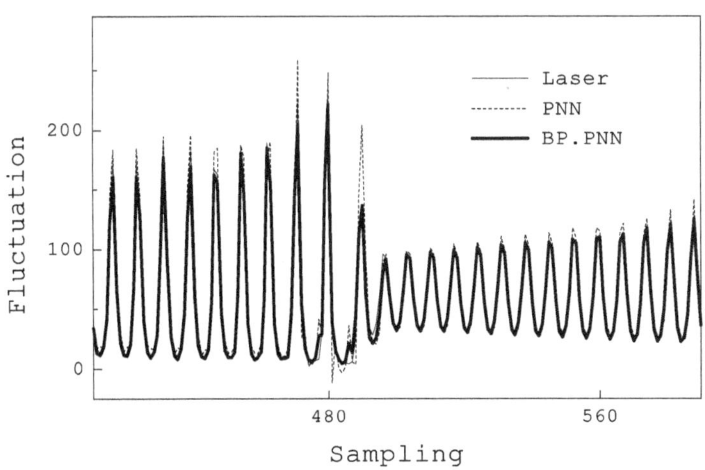

Fig. 22 Fitting the Laser series by an evolved PNN model retrained by BP

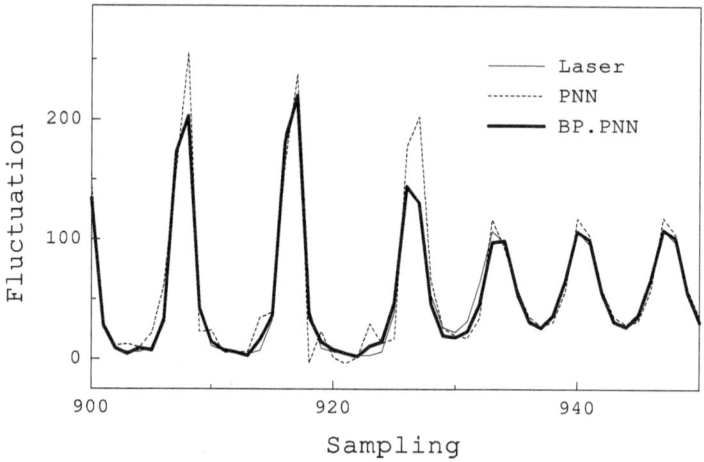

Fig. 23 Forecasting (single-step ahead prediction) of the Laser series by a PNN model evolved by IGP and re-trained by BP

coefficients, while the MLP is fully connected with 10 hidden nodes. The PNN model shows accuracy on fitting the series $MSE_{PNN} = 32.45$ which is better than accuracy of the MLP $MSE_{MLP} = 48.62$. The prediction performance of PNN is also better demonstrating one-step-ahead forecasting error $MSE^f_{PNN} = 55.67$ while the MLP shows $MSE^f_{MLP} = 80.07$.

MLP can benefit from using the input variables from the best PNN found by IGP, and this helps to achieve neural networks with improved forecasting performance. The IGP system, however, has similar computational disadvantages to the MLP: their algorithms require tuning many free parameters and there are random initializations that can affect their operation. While the MLP uses randomly initialized weights and derivatives to start the learning process, the IGP uses a random initialization of the initial population of PNN, fitness proportional randomized selection, and random selection of transformation nodes for the learning crossover and mutation operators. All these random effects require a large number of runs in order to acquire convincing results.

The benefit of evolving PNN by IGP is that polynomials of almost unlimited order could be discovered due to the hierarchical polynomial network construction inherited from the multilayer GMDH algorithm. The identification of the higher-order term weights is made efficiently by cascading low-order activation polynomials whose weights are estimated without serious computational problems. This is advantage over traditional multilayer feedforward neural networks trained by backpropagation which are limited in modelling very high order functions by the computer capacity to calculate higher-order weights [Wray and Green94]. The precision of linear polynomial networks [Wray and Green94] is also sensitive to the computational limitations of the BP training algorithm.

8 Discussion

8.1 Comparison of STROGANOFF and Traditional GP

The previous sections showed the experimental results of our STROGANOFF program. This section discusses the effectiveness of our numerical approach to GP.

Due to the difficulties mentioned in section 1.2, we have observed the following inefficiencies with traditional GP:

1. The number of individuals to be processed for a solution is much greater than with other methods.
2. Overgeneralization occurs in time series prediction tasks (Table 4).
3. Randomly generated constants do not necessarily contribute to the desired tree construction, because there is no tuning mechanism for them.

To overcome these difficulties, we have introduced a new approach to GP, based on a numerical technique, which integrates a GP-based adaptive search of tree structures, and a local parameter tuning mechanism employing statistical search. Our approach has overcome the GP difficulties mentioned in section 1.2 in the following ways:

1. GP search is effectively supplemented with the tuning of node coefficients by multiple regression. Moreover, STROGANOFF can guide GP recombination effectively in the sense that the recombination operation is guided using MDL values (section 3.7).
2. MDL-based fitness evaluation works well for tree structures in STROGANOFF, which controls GP-based tree search.
3. STROGANOFF performance is affected by the terminal choice less than GP's (Tables 13 and 14).

First, node coefficients can be tuned by our statistical method. This tuning is done "locally", in the sense that coefficients of a certain data point are derived from the data of its child nodes. Thus, STROGANOFF integrates the local search of node tuning with GP-based global search. Furthermore, as described in section 3.7, this mechanism together with MDL values leads to the recombinative guidance of STROGANOFF.

Second, MDL-based fitness is well-defined and used in our STROGANOFF trees. This is because a STROGANOFF tree has the following features:

Size-based Performance The more the tree grows, the better its performance (fitness) is. This is a basis for evaluating the tradeoff between the tree description and the error.

Decomposition The fitness of a substructure is well-defined, i.e. the fitness of a subtree (substructure) reflects that of the whole structure. If a tree has good substructures, its fitness is necessarily high.

The complexity-based fitness evaluation has already been introduced in order to control GA search strategies. We have shown that an MDL-based fitness can also be used for controlling the tree growth in STROGANOFF, i.e. an MDL-based fitness prevents overgeneralization in learning. The effectiveness of an MDL-based fitness definition for GP has also been discussed in [Iba *et al.*94b] and [Zhang & Mühlenbein95].

Third, as we have observed financial applications (Tables 13 and 14), STROGANOFF performs less dependently upon the terminal choice than GP. This feature is desirable in the sense that the best choice of terminals is not always known beforehand. Also note that although the best profit is obtained by GP under condition A, the average profit of GP is not necessarily high under the same condition. Thus, we can believe that the STROGANOFF's performance is more stable than GP, which is more suitable for the real-world applications.

8.2 Genetic Programming with Local Hill Climbing

The main feature of our work is that our approach introduces a way to modify trees, by integrating node coefficient tuning and traditional GP recombination. Our numerical approach builds a bridge from traditional GP to a more powerful search strategy. We have introduced a new approach to GP, by supplementing it with a local hill climbing approach. Local hill climbing search uses local parameter tuning (of the node functionality) of tree structures, and works by discovering useful substructures in STROGANOFF trees. Our proposed augmented GP paradigm can be considered schematically in several ways:

augmented GP = global search + local hill climbing search
= structured search + parameter tuning of node functionalities

The local hill climbing mechanism uses a type of relabeling procedure[4], which finds a locally (if not globally) optimal assignment of nodes for an arbitrary tree. Therefore, speaking generally, our new approach can be characterized as:

augmented GP = traditional GP + relabeling procedure

The augmented GP algorithm is described below:

Step1 Initialize a population of tree expressions.
Step2 Evaluate each expression in the population.
Step3 Create new expressions (children) by mating current expressions. Apply mutation and crossover to the parent tree expressions.
Step4 Replace the members of the population with the child trees.
Step5 A local-hill climbing mechanism (called "relabeling") is executed periodically, so as to relabel nodes of the trees of the population.
Step6 If the termination criterion is satisfied, then halt; else go to **Step2**.

As can be seen, **Steps1**$\sim$**5** follow traditional GP, where **Step4** is the new local hill-climbing procedure. In our augmented GP paradigm, the traditional GP representation (i.e. the terminal and non-terminal nodes of tree-expressions) is constrained so

[4] The term "label" is used to represent the information (such as a function or polynomial) at a nonterminal node.

Table 25 Properties of GP Variants

	STROGANOFF	$\Re$-STROGANOFF	BF-STROGANOFF
Problem Domain	System identification	Temporal data processing	Boolean concept formation
Tree Type	binary tree	network	binary tree
Terminal Nodes	input variables	input variables their negations	input variables
Non-terminal Nodes	polynomial relationships	polynomial relationships, memory	AND, OR, LEFT, RIGHT
Relabeling Process	GMDH	Error Propagation	ALN

that our new relabeling procedure can be applied. The sufficient condition for this applicability is that the designed representation have the property of "insensitivity" or "semantic robustness", i.e. changing a node of a tree does not affect the semantics of the tree. In other words, the GP representation is determined by the choice of the local-hill climbing mechanism.

In this chapter, we have chosen a GMDH algorithm as the relabeling procedure for system identification problems. We are currently pursuing other relabeling procedures for various kinds of problem domains. The characteristics of these resulting GP variants are summarized in Table 25.

For instance, in our previous research [Iba *et al.*95], we extended STROGANOFF to cope with temporal events and established a new system $\Re$-STROGANOFF (Recurrent STROGANOFF). $\Re$-STROGANOFF integrates a GP-based adaptive search of tree structures, and a parameter tuning mechanism employing an error-propagation method. We demonstrated the effectiveness of our new system with several experiments in learning FSA (Finite State Automata). The readers should refer to [Iba *et al.*95] for more details.

We have chosen another vehicle to perform the relabeling procedure for the sake of Boolean concept formation [Iba *et al.*94b]. Boolean concept learning is an important part of traditional machine learning. The goal is to identify the following function,

$$y = f(x_1, x_2, \cdots, x_n) = \begin{cases} 0, & \text{False value} \\ 1, & \text{True value} \end{cases} \tag{65}$$

where $x_1, x_2, \cdots x_n$ are binary values (i.e. $\{0,1\}$), from a given set of observable input and output pairs $\{(x_{i1}, x_{i2}, \cdots, x_{in}, y_i) \in \{0,1\}^{n+1} \mid i = 1, \cdots, N\}$. N is number of observations. For Boolean concept formation, we introduced the ALN (Adaptive Logic Network) algorithm [Armstrong *et al.*79, Armstrong91] as our relabeling procedure (**Step5** in the above algorithm), and used it to establish the Boolean GP variant, i.e., BF-STROGANOFF (Boolean concept Formation by STROGANOFF). BF-STROGANOFF helped overcome the problem of semantic disruption. The terminal nodes of an ALN tree are the input variables (i.e. $x_1, x_2, \cdots, x_n$) and their negations (i.e. $\overline{x_1}, \overline{x_2}, \cdots, \overline{x_n}$). The non-terminal nodes consist of the following four Boolean functions of two variables: AND, OR, LEFT (which outputs the first input),

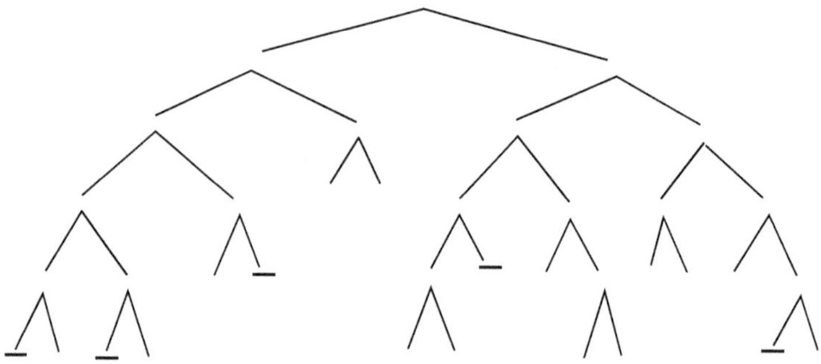

Fig. 24 An exmaple tree for 6-multiplexor

RIGHT (which outputs the second input). Fig. 24 shows an example tree for the following function (called 6-multiplexor – "mx6"):

$$y = f(x_1,x_2,x_3,x_4,x_5,x_6) = \overline{x_1}\,\overline{x_2}x_3 \vee \overline{x_1}x_2x_4 \vee x_1\overline{x_2}x_5 \vee x_1x_2x_6, \tag{66}$$

where x_1,x_2 are address variables and x_3,x_4,x_5,x_6 are data variables.

The ALN algorithm gives a good node assignment, which is sometimes globally optimal. The adaptive process of the ALN is based on the concept of "true responsibility". A node is truly responsible if changing its output would also change the whole output, all the others remaining the same. The concept can be defined recursively starting at the root by examining the node labels and the inputs to the nodes. For example, if an input to a truly responsible AND-node is 0, then the other input to the AND node will have no effect on the node's output. Hence the child node on the opposite child is not truly responsible. If the input is a 1, then the opposite child is truly responsible. If a node is truly responsible for the input vector at a given step of training, then the state of the node is enabled to change during that step (see [Armstrong91] for a formal definition).

In a BF-STROGANOFF tree, Each non-terminal node is associated with two counters C_{01},C_{10}. These counters are updated so that they determine the outputs for the (0,1) and the (1,0) input pairs respectively. Notice that all the node functions used have the property that a (0,0) input gives a 0 output, and a (1,1) input gives a 1 output. The fundamental algorithm of an ALN is described as follows:-

Step1 Randomly assign one of the four functions AND, OR, LEFT and RIGHT to the nodes of a tree. Set all counters $\{C_{01},C_{10}\}$ of these nodes to zero.
Step2 For each training set $\{(x_1,x_2,\cdots,x_n,y) \mid y = f(x_1,x_2,\cdots,x_n)\}$ do:.

1. Calculate the outputs of all nodes.
2. For each node N do:
   if N is truly responsible, then its two counters C^N_{01},C^N_{10} are updated depending upon their received input pairs and the desired output y in the following way:-

input	y	action
(0,1)	1	$C_{01}^{N} := C_{01}^{N} + 1$
(0,1)	0	$C_{01}^{N} := C_{01}^{N} - 1$
(1,0)	1	$C_{10}^{N} := C_{10}^{N} + 1$
(1,0)	0	$C_{10}^{N} := C_{10}^{N} - 1$

Step3 For each node N, set its label to:

AND	if $C_{01} < 0 \wedge C_{10} < 0$.
LEFT	if $C_{01} < 0 \wedge C_{10} > 0$.
RIGHT	if $C_{01} > 0 \wedge C_{10} < 0$.
OR	if $C_{01} > 0 \wedge C_{10} > 0$.

Therefore, in the relabeling procedure of BF-STROGANOFF, node functionalities (i.e. AND, OR, LEFT, RIGHT) at non-terminal nodes of a tree are locally tuned by using the ALN algorithm so that the tree outputs local, if not global, optima for a given tree structure of input relationships. Thus BF-STROGANOFF helps overcome the problem of semantic disruption.

To confirm the effectiveness of BF-STROGANOFF, we conducted several experiments using the parameters shown in Table 26. *ALN_Period* parameter is used to designate the period of execution of ALN (i.e., relabeling procedure). *Depth_limit* dictates the maximum depth size of individual trees.

We first experimented in learning a simple function mx6 (i.e., equation (66)). The population size is 40 and the maximum depth limit 8. All 64 ($= 2^6$) input-output pairs are given as the training data. The raw fitness value is the percentage of correct outputs of a given tree. Experiments were repeated 10 times with different *ALN_Period*'s. An example of the acquired tree is shown in Fig.24. Fig.25 plots the average number of individuals required to yield a solution (black dots) and their standard deviations (vertical bars) with different *ALN_Period*'s. As can be seen in the figure, the smaller the ALN period is, the fewer individuals are required. Since *Depth_limit* was set to be 8, the maximum number of nodes for this experiment was 128 ($= 2^{8-1}$), which is much smaller than the number required by the original ALN described above.

To compare the performance of BF-STROGANOFF with traditional GP, we experimented with the learning of more complex functions such as "even 3 parity", "even 4

Table 26 Parameters for BF-STROGANOFF

Variable	Meaning
Popsize	Population Size
P_{Cross}	Probability of Crossover (usually 60%)
P_{Mut}	Probability of Mutation (usually 3.3%)
T	Terminal Nodes $\{x_1, x_2, \cdots, x_n, \overline{x_1}, \overline{x_2}, \cdots, \overline{x_n}\}$
F	Functional Nodes $\{AND,\ OR,\ RIGHT,\ LEFT\}$
ALN_Period	Period of ALN process
Depth_limit	Maximum Depth Limit

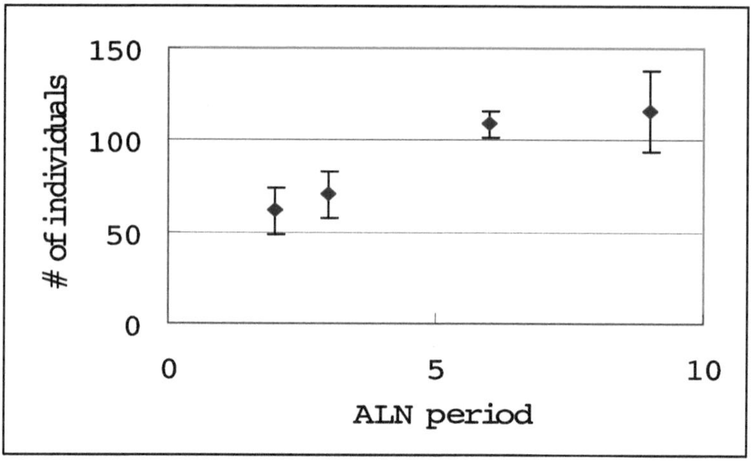

Fig. 25 The average number of individuals required to yield a solution (6-multiplexor problem)

parity", "even 5 parity", 11-multiplexor [Koza92, Koza94] or Emerald's robot world problem [Janikow93], and confirmed the effectiveness of BF-STROGANOFF. For example, the average number of individuals to yield a solution for this problem is about 8,000. Using traditional GP, this value is about 38,400 [Koza94][1]. Since the terminal and non-terminal nodes are different for the two methods, it is not possible to make a direct comparison. However it should be noted that BF-STROGANOFF required 50 times fewer evaluations compared to traditional GP.

Next we conducted an experiment in learning a nonstationary Boolean function. A given ALN cannot easily adapt to nonstationary situations, because it is necessary to construct a new tree from scratch. However, BF-STROGANOFF retains retain useful building-blocks which enables it to quickly discover non-stationary optima. To confirm this, we used a time-varying environment described below:-

1. The initial target function was mx6 (i.e., equation (66)).
2. Every 10th generation after the 40th, one of the data variables (x_3, x_4, x_5, x_6) was randomly chosen and negated. For instance, if x_4 was chosen, the new target function would be

$$y = f(x_1, x_2, x_3, x_4, x_5, x_6) = \overline{x_1}\,\overline{x_2}x_3 \vee \overline{x_1}x_2\overline{x_4} \vee x_1\overline{x_2}x_5 \vee x_1x_2x_6. \qquad (67)$$

The other experimental conditions were the same as the previous mx6. In Fig.26, the number of correct outputs (fitness) is plotted against the number of generations. Of course the fitness decreased every 10th generation. Notice that the fitness values quickly rose after these decreases and much more quickly than during the first

[1] [Koza94, ch.5] chose input variables $(x_1, x_2, \cdots, x_n)$ as terminal nodes and $\{AND, OR, NAND, NOR\}$ as non-terminal nodes.

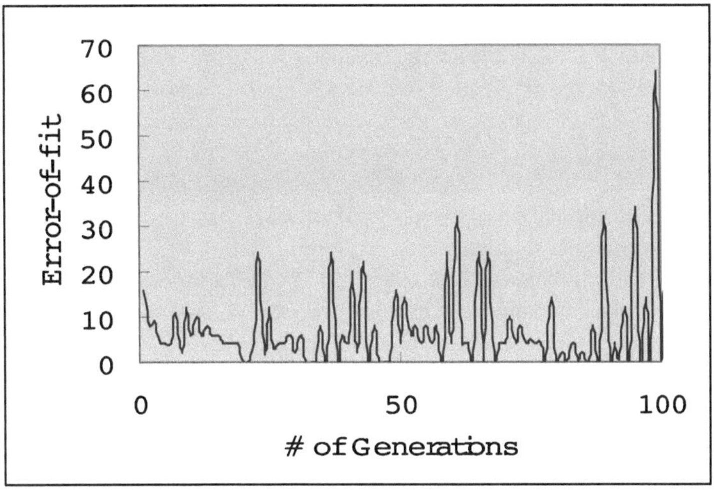

Fig. 26 The number of correct outputs with generations (6-multiplexor problem)

40 generations. Therefore BF-STROGANOFF effectively adapts itself to a time-varying environment.

8.3 *Limitations and Further Extensions of STROGANOFF*

Whereas traditional GP relies upon a large population to maintain diversity, and requires only several generations, our method can function with a small population, and can construct useful building blocks as the generations proceed. Also, the total number of evaluations of individuals is probably much less for STROGANOFF. For instance, we showed that the computational effort of STROGANOFF was 20 to 50 times less than that of traditional GP, for several symbolic regression problems [Iba *et al.*96a], and that the number of individuals to be processed by traditional GP (for the same quality of solution) in the time series prediction problem was much greater than that of STROGANOFF (Table 4). However this difference does not reflect the difference in computational complexities between the two, because a STROGANOFF evaluation involves many regression derivations. Most of the computational burden concentrates on the multiple regression analysis (i.e. the derivation of the inverse matrix, equation(11)). We have not yet studied the computational complexity of STROGANOFF theoretically. Thus it is difficult to compare the proposed algorithm with other approaches. The purpose of this chapter is to propose a numerical approach to GP and to show its feasibility through experiment. Theoretical studies, including a mathematical analysis of the computational complexities of STROGANOFF, and the improvement of its efficiency, remain important research topics.

One limitation of our approach is the memory space required for statistical calculation. In general, each intermediate node requires the storage of a set of data, whose size is equal to that of the training data. For instance, consider the P_1 tree in Fig.2. Let N be the number of training data. In order to derive the coefficients $(b_0, \cdots, b_5)$ of NODE2 (z_2), N data of (z_1, x_3) are used to deduce N equations of (8). Thus N values of z_1 should be kept in NODE3 rather than be calculated on request, for the purpose of saving the computation of the same z_1 values for later usage. Therefore a large memory space may be needed for the entire population of GMDH trees in our STROGANOFF system. Another limitation is the computational time needed to perform the multiple regression analysis, as mentioned above. However, we believe that parallelizing STROGANOFF (i.e. both the GP process and the statistical process) leads to a reduction of the computational burden. We are currently working on this topic.

8.4 Applicability to computational finances

The above experimental results have shown the effectiveness of GP-based approach for the sake of predicting financial data. However, there are several points to be improved for practical use. For instance, the following extensions should be considered:

1. The dealing simulation should be more realistic including the payment of the commission. The profit gain is offset with the fee.
2. The prediction accuracy should be improved. Especially, we should put much more emphasis on the short-term or real-time prediction, rather than the long-term prediction.
3. The problem-specific knowledge, such as economical index options or foreign exchange rates, could be introduced for the further performance improvement.

As for the third point, we are now in pursuit of the quantitative factor analysis for the purpose of choosing the significant economical features. This will have an essential impact on the prediction accuracy, especially for the short-term prediction.

We have been applying STROGANOFF to the financial problem as our main research concerns. STROGANOFF is a numerical GP system, which effectively integrates traditional GP adaptive search and statistical search [Iba *et al.*96a]. The preliminary results obtained by STROGANOFF were satisfactory and promising. However, we also observed the overfitting difficulty. This is probably because STROGANOFF used the polynomial regression, which led to finding the highly fit polynomials in terms of MSE or MDL values. But this did not necessarily give rise to the high profit gain as mentioned earlier. We believe that this difficulty will be avoided by using the discrete terminals, such as a step function or a sign function. The extension of STROGANOFF in this direction is our future research topic.

9 Conclusion

This chapter has introduced a numerical approach to Genetic Programming (GP), which integrates a GP-based adaptive search of tree structures and a statistical search technique. We have established an adaptive system called STROGANOFF, whose aim is to supplement traditional GP with a local parameter tuning mechanism. More precisely, we have augmented the traditional structural search of GP with a local hill climbing search which employs a relabeling procedure. The effectiveness of this approach to GP has been demonstrated by its successful application to numerical and symbolic problems.

In addition, we described a new GP-based approach to temporal data processing, and presented an adaptive system called $\Re$-STROGANOFF. The basic idea was derived from our previous system STROGANOFF. $\Re$-STROGANOFF integrates an error-propagation method and a GP-based search strategy. The effectiveness of our approach was confirmed by successful application to an oscillation task, to inducing languages from examples, and to extracting finite-state automata (FSA).

We have also applied STROGANOFF to such "real world" problems as predicting stock-market data or developping effective dealing rules. We presented the application of STROGANOFF to the prediction of stock price data in order to gain the high profit in the market simulation. We confirmed the following points empirically:

1. STROGANOFF was successfully applied to predicting the stock price data. That is, the MSE value for the training data was satisfactorily low, which gave rise to the high profit gain in the dealing simulation.
2. The performance under a variety of conditions, i.e., different terminal sets, was compared. Using the terminals based upon the delayed difference of the stock price were more effective than using the exact price values.
3. The STROGANOFF result was compared with those of neural networks and GP, which showed the superiority of our method.

As for the future, we intend to extend STROGANOFF by:

- parallelization.
- introducing recurrency to the GMDH network.
- performing a theoretical analysis of computational complexities.

Another important area of research concerns the extension of STROGANOFF framework to other symbolic applications, such as concept formation or program generation. We believe the results shown in this chapter are a first step toward this end.

Multiple Regression Analysis

Consider the previous unknown system,

$$y = f(x_1, x_2, \cdots, x_m). \tag{68}$$

Multiple-regression analysis gives a rough approximation by fitting the above unknown function f to a straight-line model. This method is also called a "general linear least square method".

Given N observations of these input-output data pairs, i.e.

INPUT	OUTPUT
$x_{11}\ x_{12}\ \cdots\ x_{1m}$	y_1
$x_{21}\ x_{22}\ \cdots\ x_{2m}$	y_2
$\cdots$	$\cdots$
$x_{N1}\ x_{N2}\ \cdots\ x_{Nm}$	y_N

this method fits a set of N data points to a model which is a linear combination of input variables, i.e.,

$$y_1 = \beta_0 + \beta_1 x_{11} + \beta_2 x_{12} + \cdots + \beta_m x_{1m} + e_1, \tag{69}$$

$$y_2 = \beta_0 + \beta_1 x_{21} + \beta_1 x_{22} + \cdots + \beta_m x_{2m} + e_2, \tag{70}$$

$$\cdots\cdots \tag{71}$$

$$y_N = \beta_0 + \beta_1 x_{N1} + \beta_1 x_{N2} + \cdots + \beta_m x_{Nm} + e_N. \tag{72}$$

β_i's are called partial regression coefficients, and e_i's are observational errors, i.e. residuals. With vector and matrix notations, the above linear relationships can be written as

$$\mathbf{y} = \mathbf{X}\beta + \mathbf{e}, \tag{73}$$

where

$$\mathbf{y} = \begin{bmatrix} y_1 \\ y_2 \\ \cdots \\ y_N \end{bmatrix}, \tag{74}$$

$$\mathbf{X} = \begin{pmatrix} 1 & x_{11} & x_{12} & \cdots & x_{1m} \\ 1 & x_{12} & x_{22} & \cdots & x_{2m} \\ & & \cdots & & \\ 1 & x_{N1} & x_{N2} & \cdots & x_{Nm} \end{pmatrix}, \tag{75}$$

$$\beta = \begin{bmatrix} \beta_0 \\ \beta_1 \\ \beta_2 \\ \cdots \\ \beta_N \end{bmatrix}, \tag{76}$$

and

$$\mathbf{y} = \begin{bmatrix} e_1 \\ e_2 \\ \cdots \\ e_N \end{bmatrix}. \tag{77}$$

The goal of the regression analysis is to get a solution that is the best approximation of the equation (73) in the least-squares sense. In terms of the above notations, the problem can be written as

$$\text{find } \beta \text{ which minimizes } \mid \mathbf{e} \mid = \mid \mathbf{X}\beta - \mathbf{y} \mid . \tag{78}$$

This minimization problem is equivalent to solving the following equation:

$$\mathbf{X}'\mathbf{X}\beta = \mathbf{X}'\mathbf{y}. \tag{79}$$

Where $\mathbf{X}'$ is the transposed matrix of $\mathbf{X}$. This equation is called a normal equation. If the inverse matrix of $\mathbf{X}'\mathbf{X}$ exists (i.e. $\mathbf{X}'\mathbf{X}$ is not a singular matrix), the coefficient matrix is given as

$$\beta = (\mathbf{X}'\mathbf{X})^{-1}\mathbf{X}'\mathbf{y}. \tag{80}$$

If the determinant of $\mathbf{X}'\mathbf{X}$ is zero (i.e. $\mathbf{X}'\mathbf{X}$ is a singular matrix), $(\mathbf{X}'\mathbf{X})^{-1}$ should be replaced by the Moore-Penrose generalized inverse matrix $(\mathbf{X}'\mathbf{X})^{+}$. Thus we get the following equation,

$$\beta = (\mathbf{X}'\mathbf{X})^{+}\mathbf{X}'\mathbf{y}. \tag{81}$$

A Moore-Penrose generalized inverse matrix is a coefficient matrix which gives a minimal-norm solution to a least square problem. [Spiegel75] and [Press *et al.*88] should be referred to for the details of this process and its theoretical explanation.

Computation Cost of Regressions

Because $(\mathbf{X}'\mathbf{X})$ of the equation (80) is a $(m+1) \times (m+1)$ matrix, the multiple regression analysis requires the inverse calculation of a matrix whose size is the number of terms of a fitting equation (i.e. equations (69)$\sim$ (72)).

Now let us consider the number of terms of a fitting equation for the general linear least square method. The number of terms in a complete multinomial of degree n (i.e. the sum of all homogeneous multinomials from 0-th degree through n-th degree) in m variables is given as follows [Farlow84]:

$$N_C(n,m) = \frac{(n+m)!}{n! \times m!}. \tag{82}$$

Computing an inverse of $N \times N$ matrix requires $O(N^3)$ loop executions by means of either Gaussian elimination or LU decomposition [Press *et al.*88, p.38]. Therefore, the computational cost of the general least mean square method for m input variables is given as,

$$O(N_C^3) = O(\{\frac{(n+m)!}{n! \times m!}\}^3). \tag{83}$$

These costs are plotted in 9. GLMS(i) represents the general least mean square method for a fitting equation of i input variables. The vertical axis is translated (i.e. divided by $O(6^3)$) for the sake of convenience. As can be seen in the figure, finding coefficients by this method is clearly out of question for multiple input variables.

On the other hand, the GMDH process in STROGANOFF is able to find the higher-order regression polynomial by repeatedly solving two-variable regressions of low-order. If we use the following quadratic expression,

$$z(x_1,x_2) = a_0 + a_1x_1 + a_2x_2 + a_3x_1x_2 + a_4x_1^2 + a_5x_2^2, \tag{84}$$

the computational cost for the inverse matrix is estimated as $O(6^3)$. Fig.27 shows the repetition of these multiple regressions for a GMDH tree. As you can see, it is necessary to construct a d-depth binary tree for getting a 2^d-degree expression. A d-depth binary tree contains $2^d - 1$ internal nodes. Therefore, the number of inverse

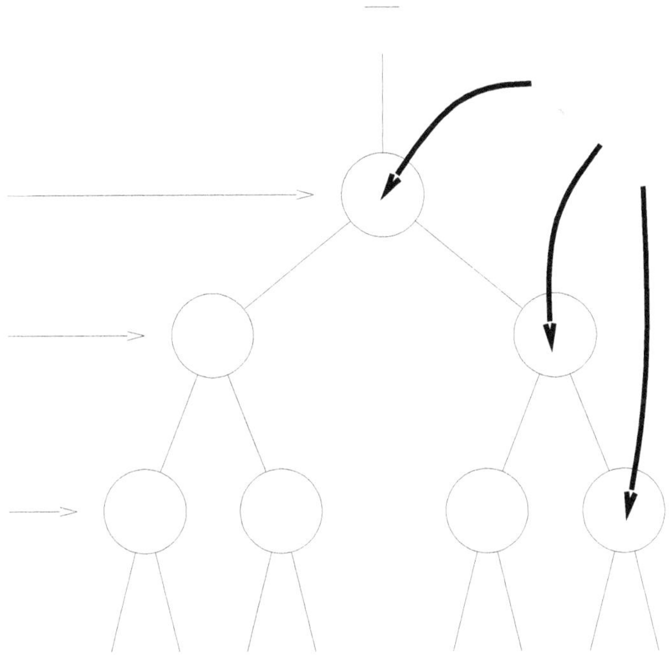

Fig. 27 A GMDH Tree

matrix calculations is $2^d - 1$ in order to obtain a multiple regression of a 2^d-degree expression with 2^d input variables. In other words, the computational cost for a GMDH tree for an N-degree regression is given as:

$$(N-1) \times O(6^3). \tag{85}$$

This computational cost is plotted in Fig.9 (i.e. STROGANOFF). The figure shows the advantage of STROGANOFF over the general least mean square method, especially in case of the regression of a multiple-input higher-order equation.

To conclude, the STROGANOFF (or its GMDH tree) is superior in terms of computational costs for large, complex systems.

References

[Angeline *et al.*94] Angeline, P.J., Saunders, G.M., Pollack, J.B.: An Evolutionary Algorithm that Constructs Recurrent Neural Networks. IEEE Tr. Neural Networks 5(1) (January 1994)

[Angeline96] Angeline, P.: Two Self-Adaptive Crossover Operators for Genetic Programming. In: Angeline, P., Kinnear, K. (eds.) Advances in Genetic Programming 2. MIT Press, Cambridge (1996)

[Aranha *et al.*07] Aranha, C., Kasai, O., Uchide, U., Iba, H.: Day-Trading Rules Development by Genetic Programming. In: Proc. 6th International Conference on Computational Intelligence in Economics & Finance (CIEF), pp. 515–521 (2007)

[Armstrong *et al.*79] Armstrong, W.W., Gecsei, J.: Adaptation Algorithms for Binary Tree Networks. IEEE TR. SMC SMC-9(5) (1979)

[Armstrong91] Armstrong, W.W.: Learning and Generalization in Adaptive Logic Networks. In: Kohonen, T. (ed.) Artificial Neural Networks, pp. 1173–1176. Elsevier Science Pub., Amsterdam (1991)

[Astrom *et al.*71] Astrom, K.J., Eykhoff, P.: System Identification, a survey. Automatica 7, 123–162 (1971)

[Banzhaf *et al.*98] Banzhaf, W., Nordin, P., Keller, R.E., Francone, F.D.: Genetic Programming: An Introduction. In: On the Automatic Evolution of Computer Programs and Its Applications. Morgan Kaufmann, San Francisco (1998)

[Barzdins and Barzdins91] Barzdins, J.M., Barzdins, G.J.: Rapid Construction of Algebraic Axioms from Samples. Theoretical Computer Science 90, 179–208 (1991)

[Belew *et al.*91] Belew, R.K., McInerney, J., Schraudolph, N.N.: Evolving Networks: Using Genetic Algorithm with Connectionist Learning. In: Langton, C.G., et al. (eds.) Artificial Life II. Addison-Wesley, Reading (1991)

[Bishop95] Bishop, C.: Neural Networks for Pattern Recognition. Oxford University Press, Oxford (1995)

[Box and Jenkins70] Box, G.E.P., Jenkins, G.M.: Time Series Analysis Forecasting and Control, Holden-Day, San Francisco, CA (1970)

[Chidambaran *et al.*98] Chidambaran, N.K., Lee, C.H.J., Trigueros, J.R.: An Adaptive Evolutionary Approach to Option Pricing via Genetic Programming. In: Proc. of the 3rd Annual Genetic Programming Conference (1998)

[de Menezes and Nikolaev06] de Menezes, L., Nikolaev, N.: Forecasting with Genetically Programmed Polynomial Neural Networks. Int. J. of Forecasting (2006)

[Farlow84] Farlow, S.J. (ed.): Self-Organizing Methods in Modeling, GMDH Type Algorithms. Marcel Dekker, Inc., New York (1984)

[Faraway and Chatfield98] Faraway, J., Chatfield, C.: Time Series Forecasting with Neural Networks: A Comparative Study using the Airline Data. Applied Statistics 47(2), 231–250 (1998)

[Fogel93] Fogel, D.B.: Evolving Behaviors in the Iterated Prisoner's Dilemma. Evolutionary Computation 1(1) (1993)

[Franke82] Franke, R.: Scattered Data Interpolation: Tests of Some Methods. Math. Comp. 38, 181–200 (1982)

[Giles *et al.*92] Giles, C.L., Miller, C.B., Chen, D., Chen, H.H., Sun, G.Z., Lee, Y.C.: Learning and Extracting Finite State Automata with Second-Order Recurrent Neural Networks. Neural Computation 4 (1992)

[Hiemstra96] Hiemstra, Y.: Applying Neural Networks and Genetic Algorithms to Tactical Asset Allocation. Neuro Ve$t Journal (May/June 1996)

[Hübner *et al.*94] Hübner, U., Weiss, C.-O., Abraham, N.B., Tang, D.: Lorenz-Like Chaos in NH3-FIR Lasers. In: Weigend, A.S., Gershenfeld, N.A. (eds.) Time Series Prediction: Forecasting the Future and Understanding the Past, pp. 73–104. Addison-Wesley, Reading (1994)

[Iba *et al.*93] Iba, H., Kurita, T., degaris, H., Sato, T.: System Identification using Structured Genetic Algorithms. in. In: Proc. of 5th International Joint Conference on Genetic Algorithms, pp. 279–286 (1993)

[Iba *et al.*94a] Iba, H., degaris, H., Sato, T.: Genetic Programming using a Minimum Description Length Principle. In: Kinnear Jr., K.E. (ed.) Advances in Genetic Programming, pp. 265–284. MIT Press, Cambridge (1994)

[Iba *et al.*94b] Iba, H., Sato, T.: Genetic Programming with Local Hill-Climbing. In: Davidor, Y., Männer, R., Schwefel, H.-P. (eds.) PPSN 1994. LNCS, vol. 866, pp. 302–411. Springer, Heidelberg (1994)

[Iba *et al.*94c] Iba, H., deGaris, H., Sato, T.: System Identification Approach to Genetic Programming. In: Proc. of IEEE World Congress on Computational Intelligence, pp. 401–406. IEEE Press, Los Alamitos (1994)

[Iba *et al.*95] Iba, H., deGaris, H., Sato, T.: Temporal Data Processing Using Genetic Programming. In: Proc. of 6th International Conference on Genetic Algorithms, pp. 279–286 (1995)

[Iba *et al.*96a] Iba, H., deGaris, H.: Numerical Approach to Genetic Programming for System Identification Evolutionary Computation 3(4), 417–452 (1996)

[Iba *et al.*96b] Iba, H., deGaris, H.: Extending Genetic Programming with Recombinative Guidance. In: Angeline, P., Kinnear, K. (eds.) Advances in Genetic Programming 2. MIT Press, Cambridge (1996)

[Ikeda79] Ikeda, K.: Multiple-valued Stationary State and its Instability of the Transmitted Light by a Ring Cavity System. Opt. Commun. 30, 257–261 (1979)

[Ivakhnenko71] Ivakhnenko, A.G.: Polynomial Theory of Complex Systems. IEEE Tr. SMC SMC-1(4) (1971)

[Janikow93] Janikow, C.Z.: A Knowledge-Intensive Genetic Algorithm for Supervised Learning. Machine Learning 13 (1993)

[Kitano90] Kitano, H.: Designing Neural Networks using Genetic Algorithms with Graph Generation System. Complex Systems 4 (1990)

[Koza90] Koza, J.: Genetic programming: A paradigm for genetically breeding populations of computer programs to solve problems, Report No. STAN-CS-90-1314, Dept. of Computer Science, Stanford Univ. (1990)

[Koza92] Koza, J.: Genetic Programming, On the Programming of Computers by means of Natural Selection. MIT Press, Cambridge (1992)

[Koza94] Koza, J.: Genetic Programming II: Automatic Discovery of Reusable Subprograms. MIT Press, Cambridge (1994)

[Kutza96] Kutza, K.: Neural Networks at Your Fingertips (1996), `http://www.geocities.com/CapeCanaveral/1624/`

[Langley and Zytkow89] Langley, P., Zytkow, J.M.: Data-driven Approaches to Empirical Discovery. Artificial Intelligence 40, 283–312 (1989)

[Lorenz63] Lorenz, E.N.: Deterministic Non-Periodic Flow. J. Atoms. Sci. 20, 130 (1963)

[Mackey & Glass77] Mackey, M.C., Glass, L.: Oscillation and Chaos in Physiological Control Systems. Science 197, 287–107 (1977)

[MacKay95] MacKay, D.J.C.: Probable Networks and Plausible Predictions- A Review of Practical Bayesian Methods for Supervised Neural Networks. Network: Computation in Neural Systems 6(3), 469–505 (1995)

[McDonnell *et al.*94] MacDonnell, J.R., Waagen, D.: Evolving Recurrent Perceptrons for Time-Series Modeling. IEEE Tr. Neural Networks 5(1) (January 1994)

[Nikolaev and Iba06] Nikolaev, N., Iba, H.: Adaptive Learning of Polynomial Networks Genetic Programming. In: Backpropagation and Bayesian Methods. Series: Genetic and Evolutionary Computation. Springer, Heidelberg (2006)

[Oakley94] Oakley, H.: Two Scientific Applications of Genetic Programming: Stack Filters and Non-Linear Equation Fitting to Chaotic Data. In: Kinnear Jr., K.E. (ed.) Advances in Genetic Programming, pp. 369–389. MIT Press, Cambridge (1994)

[Poggio & Girosi90] Poggio, T., Girosi, F.: Networks for Approximation and Learning. Proc. of the IEEE 78(9), 1481–1497 (1990)

[Potvin *et al.*04] Potvin, J.-Y., Soriano, P., Vallee, M.: Generating trading rules on the stock markets. Computer & Operations Research 31, 1033–1047 (2004)

[Press *et al.*88] Press, W.H., Flannery, B.P., Teukolsky, S.A., Vetterling, W.T.: Numerical Recipes in C, Cambridge (1988)

[Rumelhart *et al.*86] Rumelhart, D.E., Hinton, G.E., Williams, R.J.: Learning Internal Representations by Error Propagation. In: Rumelhart, D.E., et al. (eds.) Parallel Distributed Processing: Explorations in the Microstructure of Cognition, vol. 1, pp. 318–362. The MIT Press, Cambridge (1986)

[Schaffer & Morishima87] Schaffer, J.D., Morishima, A.: An Adaptive Crossover Distribution Mechanism for Genetic Algorithms. In: Proc. of 2nd International Joint Conference on Genetic Algorithms, pp. 36–40. Lawrence Erlbaum, Mahwah (1987)

[Spiegel75] Spiegel, M.R.: Theory and Problems of Statistics. McGraw-Hill, New York (1975)

[Sun90] Sun, G.Z., Chen, H.H., Giles, C.L., Lee, Y.C., Chen, D.: Connectionist Pushdown Automata that Learn Context-Free Grammars. In: IJCNN 1990 WASH D.C. Lawrence Erlbaum, Mahwah (1990)

[Teller and Veloso1996] Teller, A., Veloso, M.: PADO: A New Learning Architecture for Object Recognition. In: Ikeuchi, K., Veloso, M. (eds.) Symbolic Visual Learning, pp. 81–116. Oxford University Press, Oxford (1996)

[Tenorio *et al.*90] Tenorio, M.F., Lee, W.: Self-organizing Network for Optimum Supervised Learning. IEEE Tr. Neural Networks 1(1), 100–109 (1990)

[Tomita82] Tomata, M.: Dynamic Construction of Finite Automata from Examples using Hill-Climbing. In: Proc. 4th International Cognitive Science Conference (1982)

[Watrous *et al.*92] Watrous, R.L., Kuhn, G.M.: Induction of Finite-State Languages using Second-Order Recurrent Networks. Neural Computation 4 (1992)

[Williams *et al.*89] Williams, R.J., Zipser, D.: Experimental Analysis of the Real-Time Recurrent Learning Algorithm. Connection Science 1(1) (1989)

[Wray and Green94] Wray, J., Green, G.G.R.: Calculation of the Volterra Kernels of Nonlinear Dynamic Systems using an Artificial Neural Networks. Biological Cybernetics 71(3), 187–195 (1994)

[Zhang *et al.*93] Zhang, B.T., Mühlenbein, H.: Genetic Programming of Minimal Neural Networks using Occam's Razor. In: Proc. of 5th International Joint Conference on Genetic Algorithms (1993)

[Zhang & Mühlenbein95] Zhang, B.-T., Mühlenbein, H.: Balancing Accuracy and Parsimony in Genetic Programming. Evolutionary Computation 3(1), 17–38 (1995)

Hybrid Genetic Algorithm and GMDH System

Nader Nariman-zadeh and Jamali Ali

Abstract. This chapter presents a specific encoding scheme to genetically design GMDH-type neural networks based on using a hybrid Genetic Algorithms and SVD to design the coefficients as well as the connectivity configuration of GMDH-type neural networks used for modelling and prediction of various complex models in both single and multi-objective Pareto based optimization processes. Such generalization of network's topology provides near optimal networks in terms of hidden layers and/or number of neurons and their connectivity configuration, so that a polynomial expression for dependent variable of the process can be achieved consequently. The important conflicting objective functions of GMDH-type neural networks have been selected as Training Error (TE), Prediction Error (PE) and Number of Neurons (N) of such neural networks. Therefore, optimal Pareto fronts of such models have been obtained in each case which exhibit the trade-offs between the corresponding pair of conflicting objectives and, thus, provide different non-dominated optimal choices of GMDH-type neural networks models. Moreover, it has been shown that the Pareto front obtained by the approach of this chapter involves those that can be found by Akaike's Information Criterion.

1 Introduction

System identification and modelling of complex processes using input-output data have always attracted many research efforts. In fact, system identification techniques are applied in many fields in order to model and predict the behaviours of unknown and/or very complex systems based on given input-output data [1]. Theoretically,

Nader Nariman-zadeh
Department of Mechanical Engineering, University of Guilan, P.O. Box 3756, Rasht, Iran
e-mail: nnzadeh@guilan.ac.ir

Jamali Ali
Department of Mechanical Engineering, University of Guilan, P.O. Box 3756, Rasht, Iran

G.C. Onwubolu (Ed.): Hybrid Self-Organizing Modeling Systems, SCI 211, pp. 99–138.
springerlink.com

in order to model a system, it is required to understand the explicit mathematical input-output relationship precisely. Such explicit mathematical modelling is, however, very difficult and is not readily tractable in poorly understood systems. Alternatively, soft-computing methods [2], which concern computation in an imprecise environment, have gained significant attention. The main components of soft computing, namely, fuzzy logic, neural network, and evolutionary algorithms have shown great ability in solving complex non-linear system identification and control problems. Many research efforts have been expended to use of evolutionary methods as effective tools for system identification [3] - [8]. Among these methodologies, Group Method of Data Handling (GMDH) algorithm is a self-organizing approach by which gradually complicated models are generated based on the evaluation of their performances on a set of multi-input-single-output data pairs (X_i, y_i) (i=1, 2, ..., M). The GMDH was first developed by Ivakhnenko [9] as a multivariate analysis method for complex systems modelling and identification. In this way, GMDH was used to circumvent the difficulty of know-ing a priori knowledge of mathematical model of the process being considered. Therefore, GMDH can be used to model complex systems without having specific knowledge of the systems. The main idea of GMDH is to build an analytical function in a feedforward network based on a quadratic node transfer function [10] whose coefficients are obtained using regression technique. In fact, real GMDH algorithm in which model coefficients are estimated by means of the least square method has been classified into complete induction and incomplete induction, which represent the combinatorial (COMBI) and multilayered iterative algorithms (MIA), respectively [11]. In recent years, however, the use of such self-organizing networks leads to successful application of the GMDH-type algorithm in a broad range of areas in engineering, science, and economics [15].

The inherent complexity in the design of feedforward neural networks in terms of understanding the most appropriate topology and coefficients has a great impact on their performance. In the case of weight or coefficient training procedures, the most commonly used learning algorithm is the gradient descent algorithm, e.g., back propagation. It is believed, however, that such learning algorithms are often trapped in a local minimum and are incapable of finding a global minimum because of multi-modality and/or non-differentiability of many error functions [16]. There have been many efforts in recent years to deploy population-based stochastic search algorithms such as evolutionary methods to design artificial neural networks since such evolutionary algorithms are particularly useful for dealing with complex problems having large search spaces with many local optima [5] [14]. A very comprehensive review of using evolutionary algorithms in the design of artificial neural networks can be found in [17]. Recently, genetic algorithms have been used in a feedforward GMDH-type neural network for each neuron searching its optimal set of connection with the preceding layer [14] [18]. In the former reference, authors have proposed a hybrid genetic algorithm for a simplified GMDH-type neural network in which the connection of neurons are restricted to adjacent layers. However, such restriction has been removed by recent works of some of authors in [19] led to a generalized-structure GMDH-type neural networks (GS-GMDH) which exhibited better performance in terms of both modelling errors and network's complexity in comparisons with those of other design methods [15]. All these methods devised previously have been based on single objective optimization process in which either training error or prediction error selected to be minimized with no control of other objectives. In order to obtain more robust models, it is required to consider all the conflicting objectives, namely, training error *(TE)*, prediction error *(PE)* and number of neurons *(N)* (representing the complexity of the models) be minimized simultaneously in the sense of multi-objective Pareto op-timization process.

2 Modelling Using GMDH-Type Neural Networks

By means of GMDH algorithm a model can be represented as set of neurons in which different pairs of them in each layer are con-nected through a quadratic polynomial and thus produce new neurons in the next layer. Such representation can be used in modelling to map inputs to outputs. The formal definition of the identification problem is to find a function $\hat{f}$ so that can be approximately used instead of actual one, f in order to predict output $\hat{y}$ for a given input vector $X = (x_1, x_2, x_3, ..., x_n)$ as close as possible to its actual output y. Therefore, given M observation of multi-input-single-output data pairs so that

$$y_i = f(x_{i1}, x_{i2}, x_{i3}, ..., x_{in})(i = 1, 2, ..M) \tag{1}$$

It is now possible to train a GMDH-type neural network to predict the output values $\hat{y}_i$ for any given input vector $X = (x_{i1}, x_{i2}, x_{i3}, ..., x_{in})$, that is

$$\hat{y}_i = \hat{f}(x_{i1}, x_{i2}, x_{i3}, ..., x_{in})(i = 1, 2, .., M) \tag{2}$$

The problem is now to determine a GMDH-type neural network so that the square of difference between the actual output and the predicted one is minimized, that is

$$\sum_{i=1}^{M} [\hat{f}(x_{i1}, x_{i2}, x_{i3}, ..., x_{in}) - y_i]^2 \rightarrow \min \tag{3}$$

General connection between inputs and output variables can be expressed by a complicated discrete form of the Volterra functional series in the form of

$$y_o = a_0 + \sum_{i=1}^{n} a_i x_i + \sum_{i=1}^{n} \sum_{j=1}^{n} a_{ij} x_i x_j + \sum_{i=1}^{n} \sum_{j=1}^{n} \sum_{k=1}^{n} a_{ijk} x_i x_j x_k + ... \tag{4}$$

where is known as the Kolmogorov-Gabor polynomial [10] [11].

This full form of mathematical description can be represented by a system of partial quadratic polynomials consisting of only two vari-ables (neurons) in the form of

$$\hat{y} = G(x_i, x_j) = a_0 + a_1 x_i + a_2 x_j + a_3 x_i x_j + a_4 x_i^2 + a_5 x_j^2 \tag{5}$$

In this way, such partial quadratic description is recursively used in a network of connected neurons to build the general mathematical relation of inputs and output variables given in equation 4. The coefficients a_i in equation 5 are calculated using regression techniques [9]- [12] so that the difference between actual output, y, and the calculated one, $\hat{y}$ for each pair of (x_i, x_j) as input variables is minimized. Indeed, it can be seen that a tree of polynomials is constructed using the quadratic form given in equation 5 whose coefficients are obtained in a least-squares sense. In this way, the coefficients of each quadratic function G are obtained to optimally fit the output in the whole set of input-output data pair, that is

$$E = \frac{\sum_{i=1}^{M} (y_i - y_o)^2}{M} \to \min \tag{6}$$

In the basic form of the GMDH algorithm, all the possibilities of two independent variables out of total n input variables are taken in order to construct the regression polynomial in the form of equation 5 that best fits the dependent observations $(y_i, i = 1, 2, ...M)$ in a least-squares sense. Consequently, $\binom{n}{2} = \frac{n(n-1)}{2}$ neurons will be built up in the first hidden layer of the feed forward network from the observations $\{(y_i, x_{ip}, x_{iq}); (i = 1, 2, ..., M)\}$ for different $p, q \in \{1, 2, ..., n\}$. In other words, it is now possible to construct M data triples $\{(y_i, x_{ip}, x_{iq}); (i - 1, 2, .., M)\}$ from observation using such $p, q \in \{1, 2, ..., n\}$ in the form

$$\left[\begin{array}{cc|c} x_{1p} & x_{1q} & y_1 \\ x_{2p} & x_{2q} & y_2 \\ \overline{x_{Mp}} & \overline{x_{Mq}} & \overline{y_M} \end{array}\right]$$

Using the quadratic sub-expression in the form of equation 5 for each row of M data triples, the following matrix equation can be readily obtained as

$$Aa = Y \tag{7}$$

where a is the vector of unknown coefficients of the quadratic polynomial in equation 5

$$a = \{a_0, a_1, ..., a_5\} \tag{8}$$

and

$$Y = \{y_1, y_2, y_3, ..., y_M\}^T \tag{9}$$

is the vector of output's value from observation. It can be readily seen that

$$A = \begin{bmatrix} 1 & x_{1p} & x_{1q} & x_{1p}x_{1q} & x_{1p}^2 & x_{1q}^2 \\ 1 & x_{2p} & x_{2q} & x_{2p}x_{2q} & x_{2p}^2 & x_{2q}^2 \\ \overline{1} & \overline{x_{Mp}} & \overline{x_{Mq}} & \overline{x_{Mp}x_{Mq}} & x_{Mp}^2 & x_{Mq}^2 \end{bmatrix} \tag{10}$$

The least-squares technique from multiple-regression analysis leads to the solution of the normal equations in the form of

$$a = (A^T A)^{-1} A^T Y \tag{11}$$

which determines the vector of the best coefficients of the quadratic equation 5 for the whole set of M data triples. It should be noted that this procedure is repeated for each neuron of the next hidden layer according to the connectivity topology of the network. However, such a solution directly from normal equations is rather susceptible to round off errors and, more importantly, to the singularity of these equations.

3 Hybrid Genetic/SVD Design of GMDH-Type Neural Networks

In this section, Genetic Algorithm (GA) and Singular Value Decomposition (SVD) are deployed simultaneously for optimal design of both connectivity configuration and the values of coefficients, respectively, involved in GMDH-type neural networks which are used for modelling of complex process.

3.1 Application of SVD in the Design of GMDH-Type Networks

Singular value Decomposition (SVD) is the method for solving most linear least squares problems that some singularities may exist in the normal equations. The SVD of a matrix, $A \in \Re^{M\times 6}$ is a factorization of the matrix into the product of three matrices, column-orthogonal matrix $U \in \Re^{M\times 6}$, diagonal matrix $W \in \Re^{6\times 6}$ with non-negative elements (singular values), and orthogonal matrix $V \in \Re^{6\times 6}$ such that

$$A = U\, W\, V^T \tag{12}$$

The problem of optimal selection of vector of the coefficients in equations 7, 11 is firstly reduced to finding the modified inversion of diagonal matrix W [20] in which the reciprocals of zero or near zero singulars (according to a threshold) are set to zero. Then, such optimal a is calculated using the following relation

$$a = V\, [diag(1/w_j)]\, U^T\, Y \tag{13}$$

Such parametric identification problem is part of the general problem of modelling when structure identification is considered together with the parametric identification problem simultaneously.

3.2 Application of SVD in the Design of GMDH-Type Networks

GAs as stochastic methods are commonly used in the training of neural networks in terms of associated weights or coefficients and have successfully performed better than traditional gradient-based techniques [14]. The literature shows that a wide range of evolutionary design approaches either for architectures or for connection weights separately, in addition to efforts for them simultaneously [17]. In the most common-structure GMDH (CS-GMDH), neurons in each layer are only connected to neurons in its adjacent layer as it was the case in Methods I and II previously reported in [15]. Taking this advantage, it was possible to present a simple encoding scheme for the genotype of each individual in the population. The encoding scheme in generalized-structure GMDH (GS-GMDH) neural networks must demonstrate the ability of representing different length and size of such neural networks.

3.2.1 The Genome Representation of CS-GMDH Neural Networks

The genome or chromosome representation, which shows a topol-ogy of a GMDH-type network, simply consists of a symbolic string, composed of alphabetic representation of input variables. In this encoding scheme, each input variable is assigned an alphabetic name and a chromosome is a string of concatenated substrings of these alphabetic names of inputs. Therefore, for a given input vector $X = (x_1, x_2, x_3, ..., x_n)$, a chromosome can be represented as a string of concatenated symbols of $a_i \in \{a, b, c, d, ...\}$ in the form of $chromosome \equiv (\alpha_1, \alpha_2, \alpha_3, ..., \alpha_i, ...)$ where a, b, c .. stand for alphabetical name of inputs $x_1, x_2, x_3, ...$, respectively. It is easily seen that, for example, in the case of 4-input data samples, 4 alphabetic symbols such as a, b, c, and d can be used to construct different strings of concatenated sub-string of such symbols. Thus, it is possible to recognize that every chromosome with a length of 2^k, $k \in \{1, 2, 3, ..., (n_l + 1)\}$ where n_l is the number of hidden layers, can be readily translated to a GMDH-type neural network topology considering the fact that each neuron in such network is constructed by only two neurons in adjacent preceding layer. Therefore, for example, a chromosome such as ***abbcadbd*** represents a unique structure topology of a GMDH-type network consisting 4 input and single output which is shown in figure 1. It should be noted that there are 2 hidden layers which corresponds to length of $2^{2+1} = 8$ genes (alphabetic symbols) for this particular chromosome. It is also clear that the number of alphabetic symbols representing neurons in each layer conforms with that relation. In such representation, each $2 = 2^1$ or $4 = 2^2$ or $8 = 2^3$,.., successive number of alphabetic symbols existed in the chromosome is related to a particular neuron in a particular layer of the neural network. For example, every part of the chromosome ***abbcadbd*** with a length of $2 = 2^1$ such as $|ab|bc|ad|bd|$ (***ab, bc, ad***, or ***bd*** has a length of $2 = 2^1$) or every part of the same chromosome with a length of $4 = 2^2$ such as $|abbc|adbd|$ (***abbc*** or ***adbd*** has a length of $4 = 2^2$) or every part of the same chromosome with a length of $8 = 2^3$ such as $|abbcadbd|$ (***abbcadbd*** has a length of $8 = 2^3$) represents a particular neuron in the first layer, second layer, and output layer, respectively, as shown in figure 1. It should be noted that the chromosomes $|ab|ab|ad|bd|$ and $|aa|bc|ad|bd|$ in which two neurons to build a neuron in the next layer are the same are not valid, unlike the chromosome $|ab|bc|ad|bd|$. Therefore, it is necessary to check the validity of the constructed chromosome either in the initialisation or during the reproduction

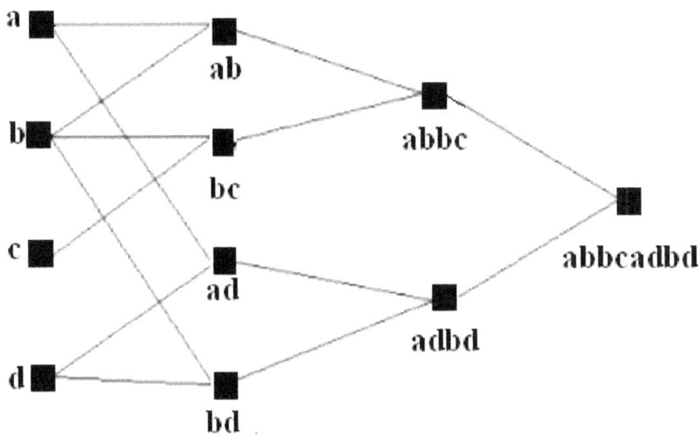

Fig. 1 A CS-GMDH-type network structure of a chromosome

processes. Such random initialisation or reproduction processes are repeated until a valid chromosome is successfully produced.

3.2.2 Genetic Operators for CS-GMDH Network Reproduction

Such genome representation of a GMDH-type neural network can now be readily used for the most two important genetic operators, namely, crossover and mutation [21]. In this work, the natural roulette wheel selection method is used for choosing two parents producing two offsprings. The crossover operator for two selected individuals is simply accomplished by exchanging the tails of two chromosomes from a randomly chosen point as shown in figure 2. It should be noted, however, such a point could only be chosen randomly from the set $\{2^1, 2^2, ..., 2^{n_l+1}\}$ where n_l is the number of hidden layers of the chromosome with the smaller length.

It is very evident from figures 2 and 3 that the crossover operation can certainly exchange the building blocks information of such GMDH-type neural networks and is therefore effective unlike some cases reported in [12]. Similarly, the mutation operation which is often given little importance in some research papers as reported

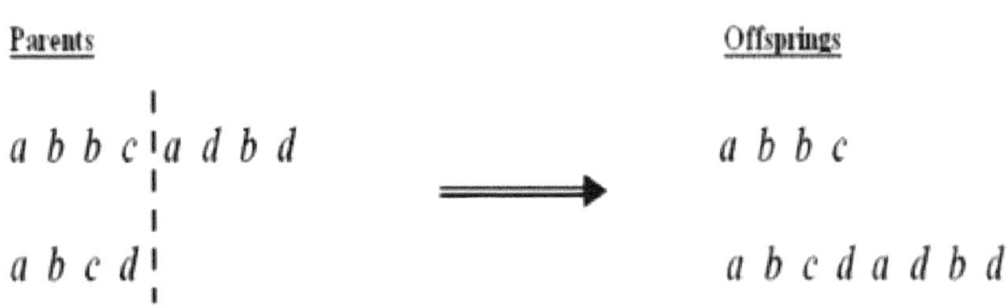

Fig. 2 Crossover operation for two individuals of CS-GMDH neural networks

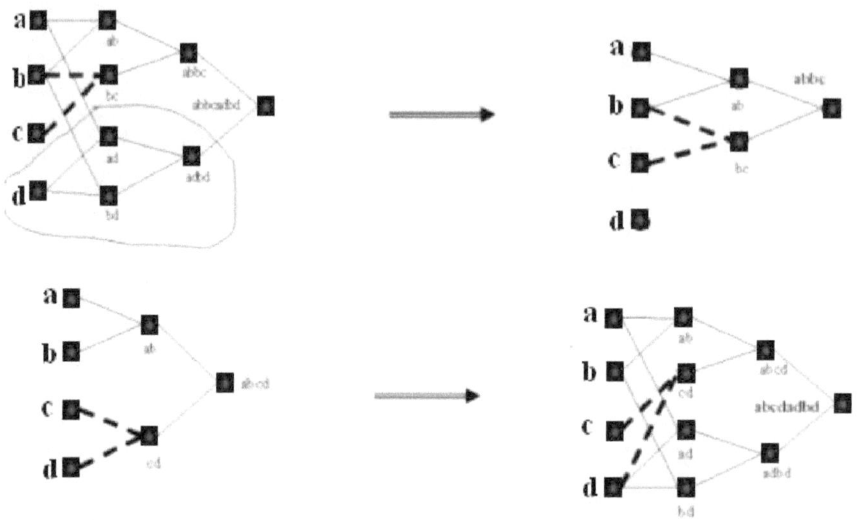

Fig. 3 Crossover operation on two CS-GMDH networks' structures

in [12], can contribute effectively to the diversity of the population. This operation is simply accomplished by charging one or more symbolic digits as genes in a chromosome to another possible symbols, for example, *abbcadbd* to *adacadbd*. It should be noted that such evolutionary operations are acceptable provided a valid chromosome is produced. Otherwise, these operations are repeated until a valid chromosome is constructed.

3.2.3 The Genome Representation of GS-GMDH Neural Networks

In figure 4, neuron *ad* in the first hidden layer is connected to the output layer by directly going through the second hidden layer. Therefore, it is now very easy to notice

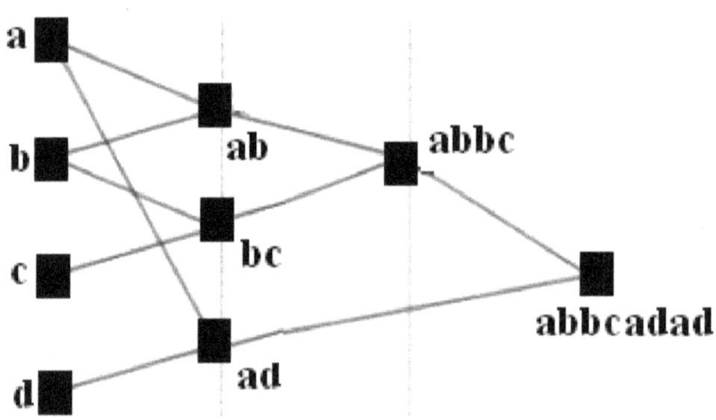

Fig. 4 A GS-GMDH network structure of a chromosome

that the name of output neuron (network's output) includes *ad* twice as *abbcadad*. In other words, a virtual neuron named *adad* has been constructed in the second hidden layer and used with *abbc* in the same layer to make the output neuron *abbcadad* as shown in the figure 4. It should be noted that such repetition occurs whenever a neuron passes some adjacent hidden layers and connects to another neuron in the next 2nd, or 3rd,or 4th,or .. following hidden layer. In this encoding scheme, the number of repetition of that neuron depends on the number of passed hidden layers, $\tilde{n}$, and is calculated as $2^{\tilde{n}}$. It is easy to realize that a chromosome such as *abab bcbc*, unlike chromosome *abab acbc* for example, is not a valid one in GS-GMDH networks and has to be simply re-written as *abbc*.

3.2.4 Genetic Operators for GS-GMDH Network Reproduction

The genetic operators of crossover and mutation can now be im-plemented to produce two offsprings from two parents. The natural roulette wheel selection method is used for choosing two parents producing two offsprings. The crossover operator for two selected individuals is simply accomplished by exchanging the tails of two chromosomes from a randomly chosen point as shown in figure 5. It should be noted, however, such a point could only be chosen randomly from the set

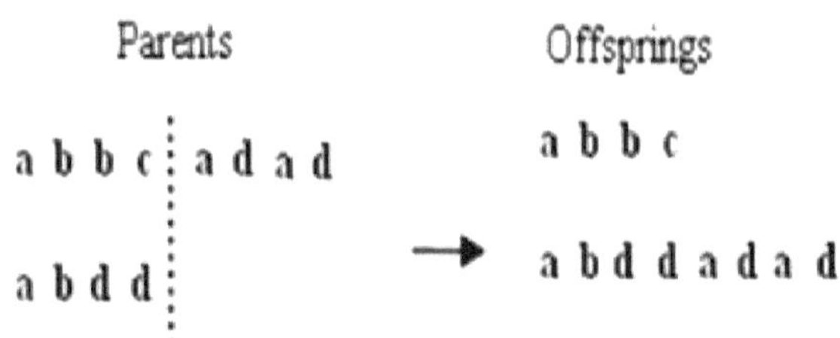

Fig. 5 Crossover operation for two individuals in GS-GMDH networks

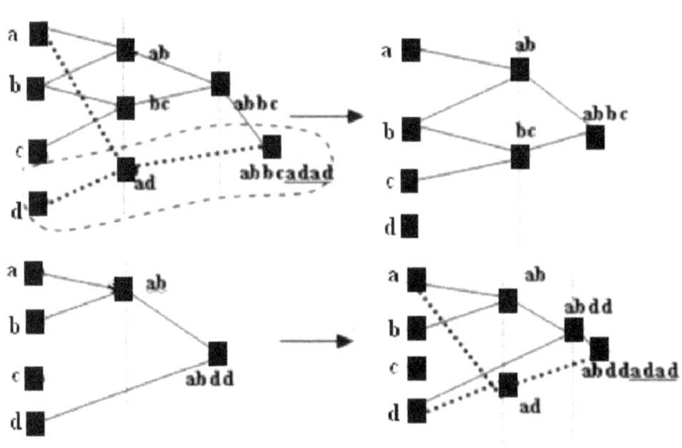

Fig. 6 Crossover operation on two GS-GMDH networks

$\{2^1, 2^2, ..., 2^{n_l+1}\}$ where n_l is the number of hidden layers of the chromosome with the smaller length. It is very evident from figures 5 and 6 that the crossover operation can certainly exchange the building blocks information of such GS-GMDH neural networks. In addition, such crossover operation can also produce different length of chromosomes which in turn leads to different size of GS-GMDH network structures. Similarly, the mutation operation can contribute effectively to the diversity of the population. This operation is simply accomplished by changing one or more symbolic digits as genes in a chromosome to another possible symbol, for example, *abbcadad* to *abbccdad*. It should be noted that such evolutionary operations are acceptable only if a valid chromosome is produced. Otherwise, these operations are simply repeated until a valid chromosome is constructed.

4 Single-Objective Hybrid Genetic Design of GMDH-Type Neural Networks Modelling and Prediction of Complex Processes

In this section, the described GMDH-type neural networks in previous sections have been used for modeling and prediction of the Caspiansea level change, and for modelling and prediction of an explosive cutting process.

4.1 Application to the Modelling and Prediction of Level Variations of the Caspian Sea

The data used in this work for modelling of the level fluctuations of the Caspian Sea relate to the recorded levels in years 1845 to 1993 [22] - [23]. However, in order to construct an input-output table be used by such evolutionary method for GMDH-type neural net-work model, 50 various inputs have been considered for possible contribution to represent the model of next year level of the Caspian sea. Such 50 inputs consist of 10 previous years of level, 10 increments of previous years of level and 20 moving average of previous years of level, and 10 moving average of previous years of incre-ment. Therefore, the first 10 columns of the input-output data table consist of the level of the Caspian Sea in the 1st, 2nd, . . ., 10th previous years denoted by *Level(i-1)*, *Level(i-2)*, . . . respectively. The next 10 columns of the input-output data table consist of increment values, denoted by *Inc-1(i)*, *Inc-2(i)*, . .. which is defined as

$$Inc\_j(i) = Level(i-j) - Level(i-j-l) \tag{14}$$

where *i* is the index of current year and *j* is the index of a particular increment. The next 20 columns of the input-output data table consist of moving average of previous years of level which is defined as

$$MA\_L\_j(i) = \sum_{k=1}^{j} \frac{Level(i-k)}{j} \tag{15}$$

where i is the index of current year and j is the index of a particu-lar moving average of level. The last 10 columns of the input-output data table consist of moving average of previous years of increment which is defined as

$$MA\_Inc\_j(i) = \sum_{k=1}^{j} \frac{Inc\_k(i)}{j} \tag{16}$$

where i is the index of current year and j is the index of a particular moving average of increment Therefore, such 50-input-1-output data table has been used to obtain an optimal GMDH-type neural network for the next year modelling of the Caspian Sea's level.

The GAs are used to design GMDH-type network systems for modeling of input-output data is discussed above. The structures of the GMDH-type neural network are shown in figures 7 and 8 corresponding to the modeling and modeling-prediction, respectively. It is clear that for modelling and for modelling-prediction 2 inputs and 4 inputs, respectively, out of 50 different inputs have been automatically selected to build polynomial equations for the next year level modelling of the Caspian Sea.

The corresponding polynomial representation of the model whose structure is given in figure 7 is

$$\begin{aligned} Y1 = &-10.8563 - 0.64009 Level(i-1) + 0.8493 MA\_L\_2(i) - \\ &2.6899(Level(i-1))^2 - 2.6061(MA\_L\_2(i))^2 + \\ &0.28179 Level(i-1) MA\_L\_2(i) \\ \\ \mathrm{Y2} = &-22.8484 - 0.6395 * \mathrm{MA\_L\_10(i)} + 9.09578 * \mathrm{MA\_Inc\_10(i)} - \\ &0.029308(\mathrm{MA\_L\_10(i)})^2 + 0.89254(\mathrm{MA\_Inc\_10(i)})^2 + \\ &0.14066 * \mathrm{MA\_L\_10(i)} * \mathrm{MA\_Inc\_10(i)} \\ \\ \mathrm{Level(i)} = &-7.3621 + 0.8547\mathrm{Y1} - 0.3967\mathrm{Y2} - 0.61913(\mathrm{Y1})^2 - \\ &0.62551 * (\mathrm{Y2})^2 + 1.2347(\mathrm{Y2})(\mathrm{Y1}) \end{aligned} \tag{17}$$

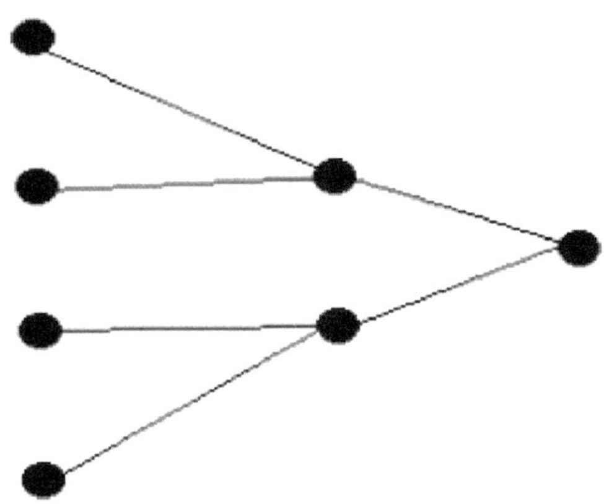

Fig. 7 Evolved structure of generalized GMDH neural network (modelling)

Fig. 8 Evolved structure of generalized GMDH neural network (modelling-prediction)

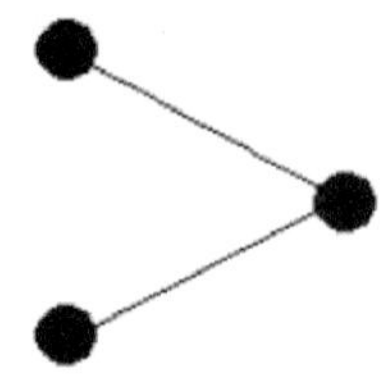

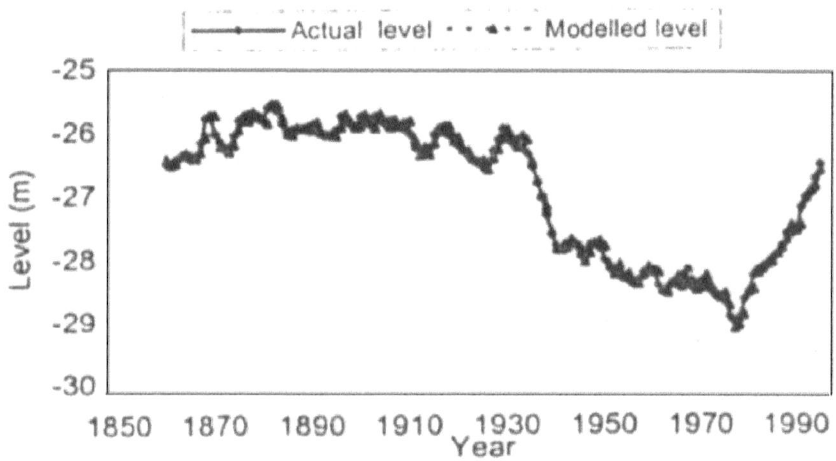

Fig. 9 Time-series comparison of actual level and the evolved GMDH neural model of the Caspian sea (modelling)

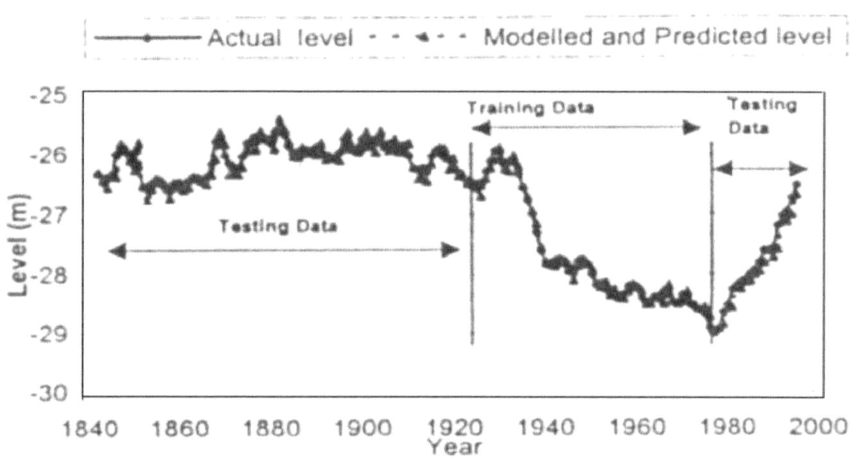

Fig. 10 Time-series comparison of actual level and 1-hidden evolved GMDH neural model of the Caspian sea (modelling-prediction)

The corresponding polynomial representation of the model whose structure is given in figure 8 is

$$\begin{aligned} Level(i) = {} & 22.0189 + 5.9986 Inc\_1(i) + 2.6287 MA\_L\_3(i) - \\ & 0.546285(Inc\_1(i))^2 + 0.03004(MA\_L\_3(i))^2 + \\ & 0.17256 * MA\_L\_3(i) * Inc\_1(i) \end{aligned} \quad (18)$$

The very good behaviour of such GMDH-type neural network model is also depicted in figures 9 and 10.

4.2 Application to the Modelling and Prediction of the Explosive Cutting Process

Explosive cutting of plates using shaped charges is one of the processes in mechanical engineering in which the physical interactions of various involved parameters are rather complex. During the last few decades the use of explosives as a source of energy has found many applications in engineering. The main difference be-tween explosives, magnetomotive forces, impact and any other source of energy is that a very large amount of energy is made avail-able to do work in a very short period of time. Explosives are now used in such diverse fields as welding, bulk cladding of plates, forming, sizing, powder compaction, hardening, and cutting. In some cases, there may be no other way of achieving the same results as in the explosive welding of dissimilar metals. In cutting metals using linear shaped charge, an explosive charge with a metallic liner is placed at a specific distance from the metal part. The cutting action is the consequence of the development of a very high-speed jet of molten metal produced by the collapse of the liner. A linear shaped charge consists of long metal liner backed with an explosive charge as shown in figure 11.The parameters of interest in this multi-input single-output system that affect both the

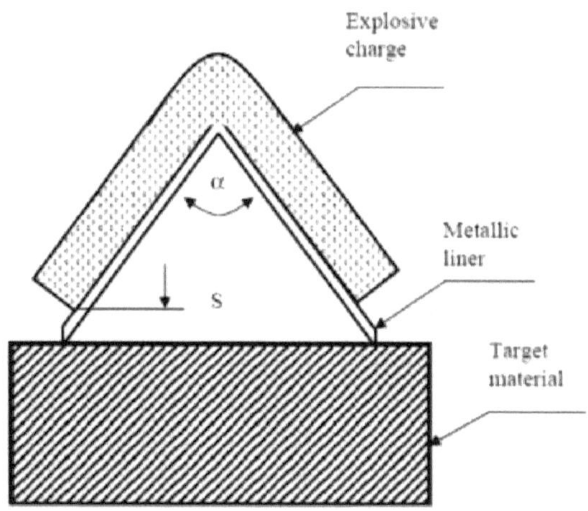

Fig. 11 A linear shape charge: S=Standoff distance; α=Apex angle

performance of the shaped charge and the depth of penetration are the apex angle, the liner thickness, the explosive weight and distribution, and the standoff distance. Accordingly, there has been a total number of 43 input-output experimental data considering 4 input parameters [14] which has been shown in Table 1.

A population size of n_popsize=20 was employed together with a crossover probability of p_cross=0.7 and a mutation probability of p_mutate=0.07 in a generation number of 200 after which no further improvement has been achieved for such population size. The structure of such evolved 4-hidden layer GMDH-type neural network is also shown in figure 12 corresponding to the genome representation of ***abbcadbdacbdbdcdacbcbcbdadcdbcbd***. The very good behaviour of such GMDH-type network model in conjunction with singular value decomposition approach for the coefficient of the quadratic polynomials is also depicted in figure 13.

However, in order to demonstrate the prediction ability of such evolved GMDH-type neural networks, the data has been divided into two different sets, namely, training and testing sets. The training set, which consists of 30 out of 43 inputs-output data pairs, is used for training the neural network models using the evolutionary method of this paper. The testing set, which consists of 13 unforeseen inputs-output data samples during the training process, is merely used for testing to show the prediction ability of such evolved GMDH-type neural network models during the training process. Again, the com-bination of evolutionary and SVD methods discussed above are used to design GMDH-type network systems for the training set of experimental input-output data. The results show that SVD approach for finding the quadratic polynomial coefficients is superior to direct solving of normal equations particularly in cases that the number of layers and/or neurons increases. A population size of n_popsize=20 was employed together with a crossover probability of p_cross=0.7 and a mutation probability of p_mutate=0.07 in a generation number of 200 after which no further improvement has been achieved for such population size.

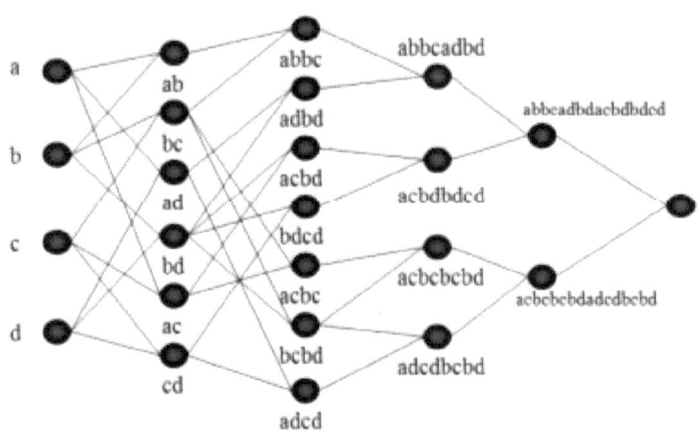

Fig. 12 The Evolved structure of GMDH-type network with 4 hidden layers

Table 1 Input-output data of explosive cutting process

	Inputs				Output
	Apex Angle	Standoff	Charge mass	Liner thickness	Depth of Penetration
1	45	0	50	0.9	2.4
2	60	0	50	0.9	3.1
3	75	0	50	0.9	4.7
4	90	0	50	0.9	6.1
5	95	0	50	0.9	8
6	100	0	50	0.9	8.2
7	105	0	50	0.9	7.1
8	120	0	50	0.9	5.4
9	135	0	50	0.9	4.7
10	100	-0.4	50	0.9	7
11	100	-0.2	50	0.9	8.2
12	100	0	50	0.9	9.05
13	100	0.2	50	0.9	8.25
14	100	0.4	50	0.9	8.25
15	100	1	50	0.9	7.8
16	100	0	50	0.9	9
17	100	0	50	1.38	8.3
18	100	0	50	1.26	8.3
19	100	0	50	1.13	9.3
20	100	0	50	1	9.3
21	100	0	50	0.74	6.9
22	100	0	50	0.61	7.1
23	100	0	50	0.48	5.9
24	100	0	50	0.35	6
25	100	0	150	2.4	16.3
26	100	0	150	1.95	17.1
27	100	0	150	1.5	16
28	100	0	150	1.05	13.1
29	100	0	150	0.83	11.1
30	100	0	150	0.6	10.3
31	100	0	250	3	21.9
32	100	0	250	2.81	22.2
33	100	0	250	2.52	21.1
34	100	0	250	2.23	21.9
35	100	0	250	1.9	22.4
36	100	0	250	1.65	22.4
37	100	0	250	1.36	21.5
38	100	0	12.25	0.4	2
39	100	0	100	1.2	12.1
40	100	0	150	1.5	16.1
41	100	0	200	1.7	19.6
42	100	0	250	1.9	22.4
43	100	0	300	2	25

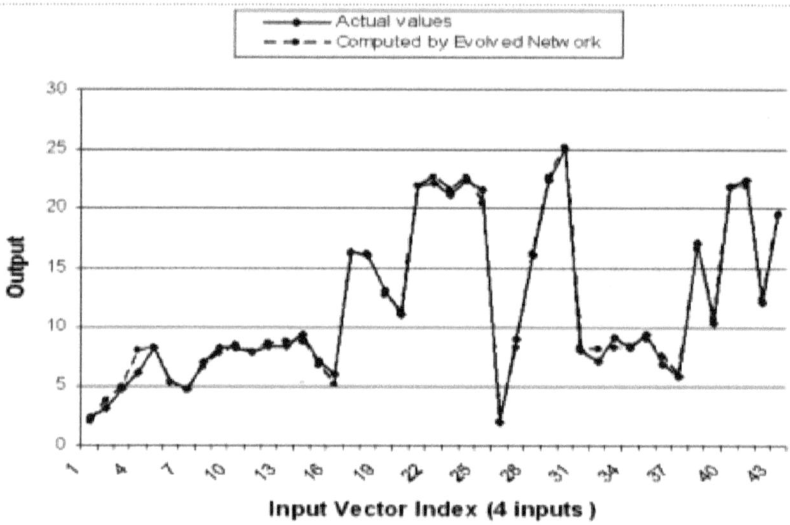

Fig. 13 Variation of Penetration with Input Data

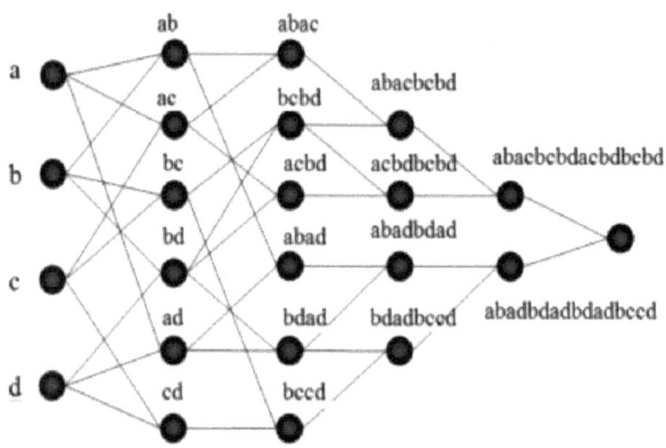

Fig. 14 Evolved structure of GMDH-type network with 4 hidden layers for modelling & prediction

The structure of such evolved 4-hidden layer GMDH-type neural network is also shown in figure 14 corresponding to the genome representation of ***abacbcbdacbdbcbdabadadbdadbdbccd***. The very good behaviour of such GMDH-type network in modelling and prediction in conjunction with singular value decomposition approach for the coefficient of the quadratic polynomials is also depicted in figure 15. It is clearly evident that the evolved GMDH-type neural network can successfully predict the output of testing data which has not been used during the training process.

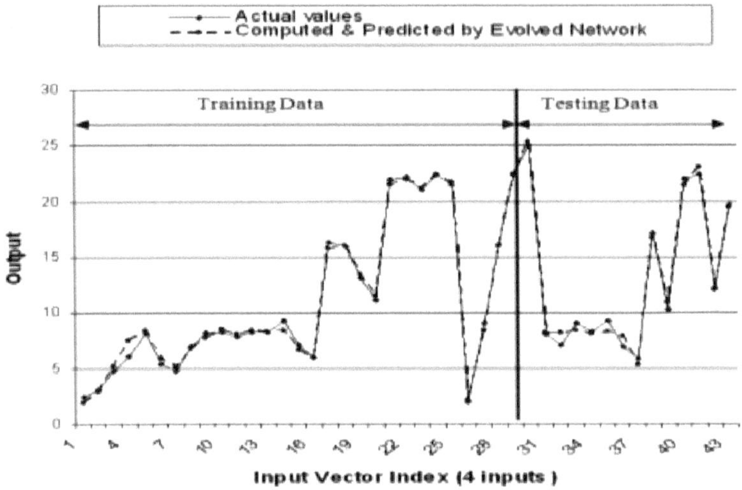

Fig. 15 Variation of Penetration with Input Data; modelling & prediction

5 Multi-objective Hybrid Genetic Design of GMDH-Type Neural Networks Modelling and Prediction of Complex Processes

Evolutionary algorithms have been widely used for multi-objective optimization because of their natural properties suited for these types of problems. This is mostly because of their parallel or population-based search approach. Therefore, most difficulties and deficiencies within the classical methods in solving multi-objective optimization problems are eliminated. For example, there is no need for either several runs to find the Pareto front or quantification of the importance of each objective using numerical weights. It is very important in evolutionary algorithms that the genetic diversity within the population be preserved sufficiently. This main issue in MOPs has been addressed by much related research work [24]. Consequently, the premature convergence of MOEAs is prevented and the solutions are directed and distributed along the true Pareto front if such genetic diversity is well provided. The Pareto-based approach of NSGA-II [25] has been recently used in a wide range of engineering MOPs because of its simple yet efficient non-dominance ranking procedure in yielding different levels of Pareto frontiers. However, the crowding approach in such a state-of-the-art MOEA [26] works efficiently for two-objective optimization problems as a diversity-preserving operator which is not the case for problems with more than two objective functions. The reason is that the sorting procedure of individuals based on each objective in this algorithm will cause different enclosing hyper-boxes. Thus, the overall crowding distance of an individual computed in this way may not exactly reflect the true measure of diversity or crowding property. In order to show this issue more clearly, some basics of NSGA-II are now represented. The entire population R_t is simply the current parent population P_t plus its offspring population Q_t which is created from the parent population P_t by using usual genetic operators. The selection is based on non-dominated sorting procedure which is used to classify the entire population R_t according to increasing order of dominance [26]. Thereafter, the best Pareto fronts from the top of the sorted list is transferred to create the new parent population P_{t+1} which is half the size of the entire population R_t. Therefore, it should be noted that all the individuals of a certain front cannot be accommodated in the new parent population because of space. In order to choose exact number of individuals of that particular front, a crowded comparison operator is used in NSGA-II to find the best solutions to fill the rest of the new parent population slots. The crowded comparison procedure is based on density estimation of solutions surrounding a particular solution in a population or front. In this way, the solutions of a Pareto front are first sorted in each objective direction in the ascending order of that objective value. The crowding distance is then assigned equal to the half of the perimeter of the enclosing hyper-box (a rectangular in bi-objective optimization problems). The sorting procedure is then repeated for other objectives and the overall crowding distance is calculated as the sum of the crowding distances from all objectives. The less crowded non-dominated individuals of that particular Pareto front are then selected to fill the new parent population. It must be noted that, in a *two-objective* Pareto optimization, if the solutions of a Pareto front are sorted in a *decreasing* order of importance to one objective, these solutions are then automatically ordered in an *increasing* order of importance to the second objective. Thus, the hyper-boxes surrounding an individual solution remain unchanged in the objective-wise sorting procedure of the crowding distance of NSGA-II in the two-objective Pareto optimization problem. However, in multi-objective Pareto optimization problems with more than two objectives, such sorting procedure of individuals based on each objective in this algorithm will cause different enclosing hyper-boxes. Thus, the overall crowding distance of an individual computed in this way may not exactly reflect the true measure of diversity or crowding property for the multi-objective Pareto optimization problems with more than two objectives. A new method is presented to modify NSGA-II so that it can be safely used for any number of objective functions (par-ticularly for more than two objectives) in MOPs.

5.1 Multi-objective Optimization

Multi-objective optimization which is also called multicriteria optimization or vector optimization has been defined as finding a vector of decision variables satisfying constraints to give optimal values to all objective functions [25]- [27]. In general, it can be mathematically defined as: find the vector $X^* = [x_1^*, x_2^*, \ldots, x_n^*]^T$ to optimize

$$F(X) = [f_1(X), f_2(X), \ldots, f_k(X)]^T \tag{19}$$

subject to m inequality constraints

$$g_i(X) \leq 0 \quad , \quad i = 1\ to\ m \tag{20}$$

and p equality constraints

$$h_j(X) = 0, j = 1top \qquad (21)$$

where $X^* \in \Re^n$ is the vector of decision or design variables, and $F(X) \in \Re^k$ is the vector of objective functions. Without loss of generality, it is assumed that all objective functions are to be minimized. Such multi-objective minimization based on the Pareto approach can be conducted using some definitions:

Definition of Pareto dominance

A vector $U = [u_1, u_2, ..., u_k] \in \Re^k$ dominates to vector $V = [v_1, v_2, ..., v_k] \in \Re^k$ (denoted by $U \prec V$) if and only if $\forall i \in \{1, 2, ..., k\}$, $u_i \leq v_i \wedge \exists j \in \{1, 2, ..., k\} : u_j < v_j$. It means that there is at least one u_j which is smaller than v_j whilst the rest u's are either smaller or equal to corresponding v's.

Definition of Pareto optimality

A point $X^* \in \Omega$ (Ω is a feasible region in $\Re^n$ satisfying equations 20 and 21) is said to be Pareto optimal (minimal) with respect to all $X \in \Omega$ if and only if $F(X^*) \prec F(X)$. Alternatively, it can be readily restated as $\forall i \in \{1, 2, ..., k\}, \forall X \in \Omega - \{X^*\}$ $f_i(X^*) \leq f_i(X) \wedge \exists j \in \{1, 2, ..., k\} : f_j(X^*) < f_j(X)$. It means that the solution X^* is said to be Pareto optimal (minimal) if no other solution can be found to dominate X^* using the definition of Pareto dominance.

Definition of Pareto front

For a given MOP, the Pareto front PF^* is a set of vectors of objective functions which are obtained using the vectors of decision variables in the Pareto set P^* , that is $PF^* = \{F(X) = (f_1(X), f_2(X), ..., f_k(X)) : X \in P^*\}$. Therefore, the Pareto front PF^* is a set of the vectors of objective functions mapped from P^*.

Definition of Pareto Set

For a given MOP, a Pareto set P^* is a set in the decision variable space consisting of all the Pareto optimal vectors, $P^* = \{X \in ? \mid \hbar X' \in ? : F(X') \prec F(X)\}$. In other words, there is no other X' in Ω that dominates any $X \in P^*$ in terms of their objective functions.

5.2 *Multi-objective Uniform-Diversity Genetic Algorithm (MUGA)*

The multi-objective uniform-diversity genetic algorithm (MUGA) uses non-dominated sorting mechanism together with a ε-elimination diversity preserving algorithm to get Pareto optimal solutions of MOPs more precisely and uniformly.

The non-dominated sorting method

The basic idea of sorting of non-dominated solutions originally proposed by Goldberg [21] used in different evolutionary multi-objective optimization algorithms such as in NSGA-II by Deb et al.,[25] has been adopted here. The algorithm simply compares each individual in the population with others to determine its non-dominancy. Once the first front has been found, all its non-dominated individuals are removed from the main population and the procedure is repeated for the subsequent fronts until the entire population is sorted and non-dominatedly divided into different fronts.

A sorting procedure to constitute a front could be simply accomplished by comparing all the individuals of the population and including the non-dominated individuals in the front. Such procedure can be simply represented as following steps:

$$1.\,Get\ the\ population\ (pop)$$

$$2.\,Include\ the\ first\ individual\ \{ind(1)\}\ in\ the\ front\ P^*\ as\ P^*(1)$$

$$let\ \ P^*\_size\ =\ 1$$

$$3.\,Compare\ other\ individuals\ \{ind(j),\ j=2,\ Pop\_size\}\ of\ the$$

$$pop\ with\ \{P^*(K), K=1, P^*\_size\}\ of\ the\ P^*$$

$$If\ ind(j) < P^*(K)\ replace\ the\ P^*(K)\ with\ the\ ind(j)$$

$$If\ P^*(K)\ <\ ind(K), j=j+1,$$

$$continue\ comparison:$$

$$Else\ include\ ind(j)\ in\ P^*, P^*\_size = P^*\_size+1, j=j+1,$$

$$continue\ comparison:$$

$$4.\,End\ of\ front\ P^*$$

It can be easily seen that the number of non-dominated solutions in P^* grows until no further one is found. At this stage, all the non-dominated individuals so far found in P^* are removed from the main population and the whole procedure of finding another front may be accomplished again. This procedure is repeated until the whole population is divided into different ranked fronts. It should be noted that the first rank front of the final generation constitute the final Pareto optimal solution of the multi-objective optimization problem.

The ε-elimination Diversity Preserving Approach

In the ε-elimination diversity approach that is used to replace the crowding distance assignment approach in NSGA-II [25], all the clones and ε-similar individuals are recognized and simply eliminated from the current population. Therefore, based on a value of ε as the elimination threshold, all the individuals in a front within this limit of a particular individual are eliminated. It should be noted that such ε-similarity must exist both in the space of objectives and in the space of the associated design variables. This will ensure that very different individuals in the space of design variables having ε-similarity in the space of objectives will not be eliminated from the population. The pseudo-code of the ε-elimination approach is depicted in figure 16. Evidently, the clones and ε-similar individuals are replaced from the population by the same number of new randomly generated individuals. Meanwhile, this will additionally help to explore the search space of the given MOP more effectively. It is clear that such replacement does not appear when a front rather than the entire population is truncated for ε-similar individual.

```
ε-elim= ε-elimination(pop)      // pop includes design variables and objective function
i=1; j=1;
get K (K=1 for the first front);
While i,j <pop_size
 e(i,j)= ||X(i,:),X(j,:) ||/||X(i,:) ||; X(i),X(j) □ P*k Ụ PF*k  //finding mean value of ε within pop.
end
ε=mean(e);
i=1;
until i+1<pop_size;
j=i+1
        until j<pop_size
        if e(i,j)<ε
        then {pop}={pop}/ {pop(j)}          //remove the ε-similar individual
        j=j+1
end
i=i+1
end
```

Fig. 16 The ε-elimination diversity preserving pseudo-code

The Main Algorithm of MUGA

It is now possible to present the main algorithm of MUGA which uses both non-dominated sorting procedure and ε-elimination diver-sity preserving approach which is given in figure 17. It first initiates a population randomly. Using genetic operators, another same size population is then created. Based on the ε-elimination algorithm, the whole population is then reduced by removing ε-similar individuals. At this stage, the population is re-filled by randomly generated individuals which helps to explore the search space more effectively. The whole population is then sorted using non-dominated sorting procedure. The obtained fronts are then used to constitute the main population. It must be noted that the front which must be truncated to match the size of the population is also evaluated by ε-elimination procedure to identify the ε-similar individuals. Such procedure is only performed to match the size of population within ± 10 present deviation to prevent excessive computational effort to population size adjustment. Finally, unless the number of

```
Get N                  //population size
t=1;                   //set generation number
Random_N(Pt);                   //generate the first population (P1) randomly
Qt=Recomb(Pt)                   //generate population Qt from Pt by genetic operators
Rt=Pt U Qt             //union of both parent and offspring population
Rt'=ε-elimination (Rt)// remove ε-similar individuals in Rt
Rt''= Rt'     U Random_(Rt size-R't size) (Pt')      //add random individuals to fill Rt to 2N

Do non-dominate sorting procedure (Rt'')   //Rt''=P*1U ...UP*k  where k is total number of fronts
   i=1
   Pt+1=Θ
   While not Pt+1_size>N              //includes fronts into new population
               Pt+1= Pt+1U P*i
               i=i+1
   end
   N'=N- Pt+1_size
 While not (0.9 N'< Pt+1_size<1.1 N
                       // remove the ε-similar individuals within the tolerance of ±10 percent
       F'=ε-elimination (P*i-1)
        If      F'_size< N'
       e=1.1*e
       else
       e=0.9 * e    // adjust the value of threshold e to get the right population size of the last front
       end
   end
   t=t+1                              //Start next generation
```

Fig. 17 The pseudo-code of the main algorithm of MUGA

individuals in the first rank front is changing in certain number of generations, randomly created individuals are inserted in the main population occasionally (e.g. every 20 generations of having non-varying first rank front).

5.3 *Multi-objective Genetic Design of GMDH-Type Neural Networks for a Variable Valve-Timing Spark-Ignition Engine*

The input-output data pairs used in such modelling involve two different data tables are given in [28]. The first table consist of two variables as inputs namely, intake valve-timing (V_1) and engine-speed (N) and one output which is fuel consumption (Fc) for the single-cylinder four-stroke spark-ignition engine. The second table consists of the same two variables as inputs and another output which is torque (T) of the engine. These tables consist of the total 77 pattern numbers, given in Table 2, have been obtained from the experiments to train GMDH-type neural networks [28]. However, in order to demonstrate the prediction ability of the evolved GMDH-type neural networks, the data have been divided into two different sets, namely, training and testing sets. The training set, which consists of 62 out of 77 inputs-output data pairs, is used for training the neural network models using the evolutionary method of this paper.

Table 2 Input-output Experimental data of the variable valve-timing spark-ignition engine

	Inputs		Outputs	
	Intake valve timing	Engine speed	Torque	Fuel consumption
1	30	1600	10.07	645.58
2	30	2000	10.88	843.6
3	30	2200	11.05	956.25
4	30	2400	10.75	1028.7
5	30	2600	9.75	1033.5
6	30	2800	9.12	1029.6
7	30	3000	8.35	1041.18
8	30	3200	8.15	1132.95
9	30	3600	6.15	1076.48
10	20	1600	10	646.8
11	20	1800	10.23	729.54
12	20	2000	10.58	821.4
13	20	2200	11.1	944.64
14	20	2600	10.13	1043.28
15	20	2800	9.88	1110.7
16	20	3000	9.22	1116.5
17	20	3400	8.25	1223.04
18	10	3600	7.47	1269
19	10	1600	9.83	641.85
20	10	1800	10.15	735.35
21	10	2200	10.97	956.34
22	10	2400	11.2	1049.93
23	10	2600	10.83	1109.2
24	10	2800	10.2	1157.13
25	10	3200	9.7	1283.75
26	10	3400	9.1	1321.92
27	10	3600	8.2	1285.44
28	0	1800	10.03	738.99
29	0	2000	10.47	847.53
30	0	2200	11.07	971.55
31	0	2400	11.1	1049.04
32	0	2600	11.25	1123.02
33	0	3000	10.97	1266.15
34	0	3200	10.45	1319.22
35	0	3400	10.22	1453.02
36	0	3600	9.43	1466.72
37	-10	1600	9.53	660.8
38	-10	1800	10	759.78
39	-10	2000	10.15	830.7

Table 2 (*continued*)

	Inputs		Outputs	
	Intake valve timing	Engine speed	Torque	Fuel consumption
40	-10	2200	10.52	938.96
41	-10	2600	10.72	1118.36
42	-10	2800	10.97	1194.62
43	-10	3000	11	1262.9
44	-10	3200	10.63	1302.96
45	-10	3400	10.5	1439.9
46	-20	1600	9.1	661.2
47	-20	1800	9.43	758.28
48	-20	2000	9.75	869.44
49	-20	2200	10.1	955.3
50	-20	2600	10.5	1109.68
51	-20	2800	10.8	1195.09
52	-20	3000	11.2	1291.84
53	-20	3200	11	1368.99
54	-20	3600	9.9	1443.51
55	-30	2000	8.92	847.88
56	-30	2200	9.22	937.04
57	-30	2400	9.4	1017.16
58	-30	2600	9.52	1067.08
59	-30	2800	9.88	1157.1
60	-30	3000	10.05	1235.56
61	-30	3200	10.1	1321.58
62	-30	3400	10.13	1397.07
63	30	1800	10.48	738.54
64	30	3400	7.25	1173.9
65	20	2400	10.85	1010.1
66	20	3200	9	1238.2
67	10	2000	10.27	821.3
68	10	3000	9.92	1204.32
69	0	1600	9.78	649.44
70	0	2800	11.43	1226.4
71	-10	2400	10.63	2182.95
72	-10	3600	9.85	1491.42
73	-20	2400	10.25	1050.06
74	-20	3400	10.63	1455.3
75	-30	1600	8.6	668.16
76	-30	1800	8.72	751.12
77	-30	3600	10.2	1528.45

The testing set, which consists of 15 unforeseen input-output data samples during the training process, is merely used for testing to show the prediction ability of such evolved GMDH-type neural network models during the training process. The GMDH-type neural networks are now used for such input-output data to find the polynomial model of fuel consumption and torque in respect to their effective input parameters. In order to genetically design such GMDH-type neural network described in previous section a population of 20 individuals with a crossover probability of 0.7 and mutation probability of 0.07 has been used in 250 generation that no further improvement has been achieved for such population size. The structure of the evolved 2-hidden layer GMDH-type neural networks are shown in figures 18 and 19 corresponding to the genome representations of ***aaababbb*** for engine-torque and ***aaaaabbb*** for fuel consumption in which ***a*** and ***b*** stand for engine-speed and valve-timing, respectively. The corresponding polynomial representation of such model for engine-torque is as follows

$$\begin{aligned} Y_1 &= 0.00009 + 0.138\, V_t + 0.0089N - 0.0012\, V_t^2 - 0.00005 N V_t \\ Y_2 &= 0.21 + 1.36\, Y_1 - 0.0013 N - 0.028\, Y_1^2 + 0.00005 N Y_1 \\ \mathrm{ET} &= 5.50 - 4.12 Y_1 + 3.84\, Y_2 + 6.98 Y_1^2 + 6.73 Y_2^2 - 13.63 Y_1 Y_2 \end{aligned} \tag{22}$$

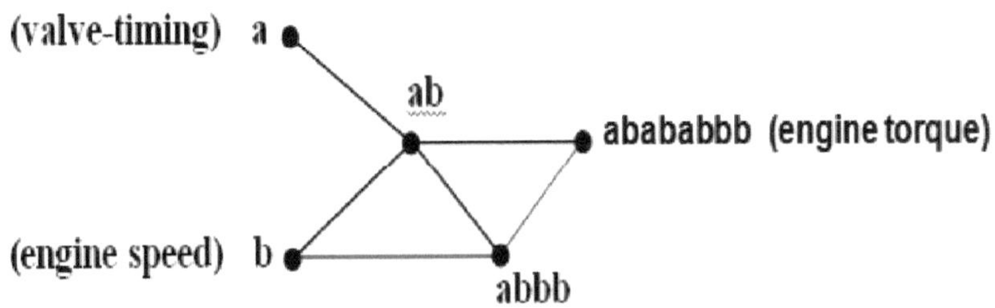

Fig. 18 Evolved structure of generalized GMDH neural network for engine torque

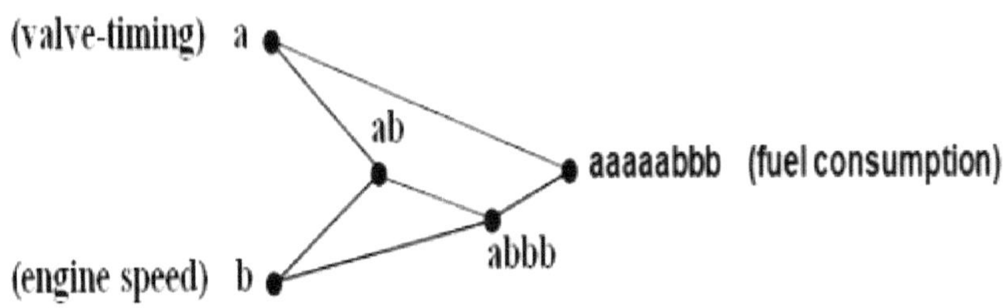

Fig. 19 Evolved structure of generalized GMDH neural network for fuel consumption

Similarly, the corresponding polynomial representation of the model for fuel consumption is in the form of

$$Y_3 = 0.0044 + 6.472\,V_t + 0.4994\,N - 0.07359\,V_t^2 - 00002\,N^2 - 0.00322\,V_t N$$

$$Y_4 = -0.00659 - 1.378\,Y_3 + N - 0.00124 Y_3^2 - 0.0005\,N^2 + 0.00172\,Y_3 N \tag{23}$$

$$\mathrm{FC} = -131.44 - 0.32 Y_4 + 1.28\,V_t - 0.025 Y_4^2 - 0.00013 V_t^2 - 0.00031 Y_4 V_t$$

The very good behaviour of the GMDH-type neural network models are also depicted in figures 20 and 21 for testing data of both fuel consumption and

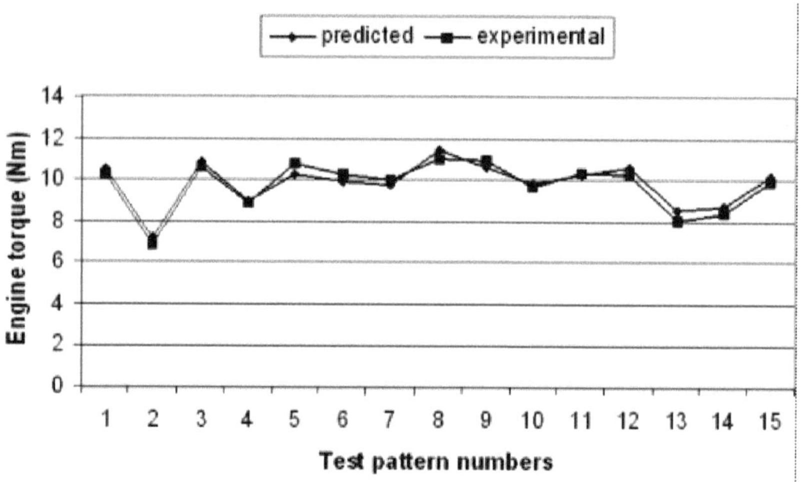

Fig. 20 Comparison of experimental values of engine torque with the predicted values using evolved GMDH neural network for testing data

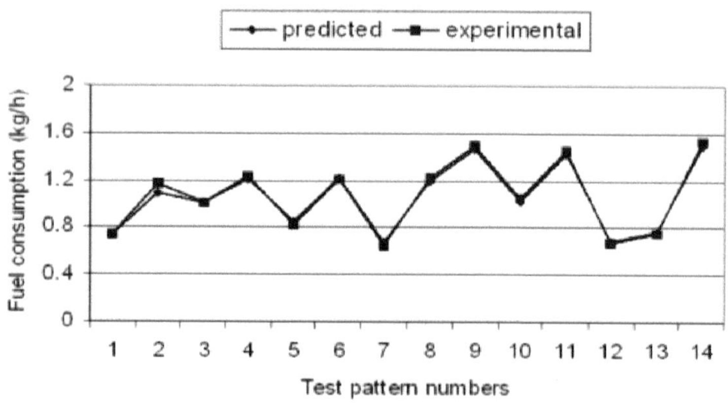

Fig. 21 Comparison of experimental values of fuel consumption with the predicted values using evolved GMDH neural network for testing data

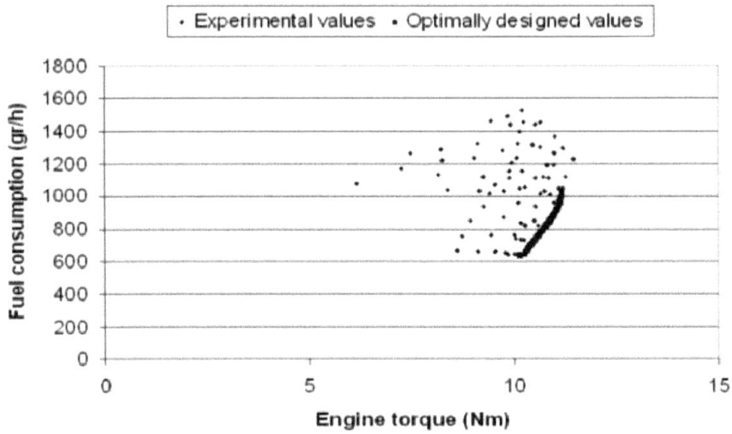

Fig. 22 Overlay graph of the obtained optimal Pareto front with the Exp. data

engine-torque, respectively. It is clearly evident that the evolved GMDH-type neural network in terms of simple polynomial equations could successfully model and predict the output of testing data that has not been used during the training process.

The Pareto front obtained from the GMDH-type neural network model has been superimposed with the corresponding experimental results in figure 22. It can be clearly seen that such obtained Pareto front lies on the best possible combination of the objective values of experimental data (except for two data samples) which demonstrates the effectiveness of the multi-objective approach both in deriving the model and in obtaining the Pareto front.

5.4 Multi-objective Genetic Design of GMDH-Type Neural Networks for a Nonlinear System

The input-output data used in such modelling involve 100 data pairs randomly generated from a nonlinear system [29] with three inputs x1, x2, x3, and a single output y given by

$$y = (1 + x_1^{0.5} + x_2^{-1.} + x_3^{-1.5}) 1 \leq x_1, x_2, x_3 \leq 5 \tag{24}$$

Which have been given in Table 3. There are 50 pattern num-bers which have been randomly selected from those data pairs to train such GMDH-type neural networks. However, a testing set which consists of 50 unforeseen input-output data samples during the training process is merely used for testing to show the prediction ability of such evolved GMDH-type neural network models during the training process.

The GMDH-type neural networks are now used for such input-output data to find the polynomial model of y in such nonlinear sys-tem process with respect to their input parameters. In order to design GMDH-type neural network described in previous section from a multi-objective optimum point of view, a population of 60 individuals with a crossover probability of 0.95 and mutation probability of 0.1 has

Table 3 Input-output data of the nonlinear process

	Inputs		Outputs	
	x_1	x_2	x_3	y
1	4.96	3	3.6667	3.7029
2	4.92	4	2.3333	3.7487
3	4.88	2	4.5556	3.8119
4	4.84	4.5	3.2222	3.5951
5	4.8	2.5	1.8889	3.9761
6	4.76	3.5	4.1111	3.5874
7	4.72	1.5	2.7778	4.0552
8	4.68	4.75	1.4444	3.9499
9	4.64	2.75	4.8519	3.6113
10	4.6	3.75	3.5185	3.5629
11	4.56	1.75	2.1852	4.0164
12	4.52	4.25	4.4074	3.4694
13	4.48	2.25	3.0741	3.7466
14	4.44	3.25	1.7407	3.8502
15	4.4	1.25	3.963	4.0244
16	4.36	4.875	2.6296	3.5277
17	4.32	2.875	1.2963	4.1038
18	4.28	3.875	4.7037	3.4249
19	4.24	1.875	3.3704	3.7541
20	4.2	4.375	2.037	3.6219
21	4.16	2.375	4.2593	3.5744
22	4.12	3.375	2.9259	3.5259
23	4.08	1.375	1.5926	4.2447
24	4.04	4.625	3.8148	3.3604
25	4	2.625	2.4815	3.6368
26	3.96	3.625	1.1481	4.0787
27	3.92	1.625	4.9506	3.6861
28	3.88	4.125	3.6173	3.3575
29	3.84	2.125	2.284	3.7199
30	3.8	3.125	4.5062	3.3739
31	3.76	1.125	3.1728	4.0049
32	3.72	4.9375	1.8395	3.5321
33	3.68	2.9375	4.0617	3.3809
34	3.64	3.9375	2.7284	3.3837
35	3.6	1.9375	1.3951	4.0204
36	3.56	4.4375	4.8025	3.2072
37	3.52	2.4375	3.4691	3.4412
38	3.48	3.4375	2.1358	3.4768
39	3.44	1.4375	4.358	3.6603
40	3.4	4.6875	3.0247	3.2473

Table 3 (*continued*)

	Inputs		Outputs	
	x_1	x_2	x_3	y
41	3.36	2.6875	1.6914	3.6597
42	3.32	3.6875	3.9136	3.2224
43	3.28	1.6875	2.5802	3.6449
44	3.24	4.1875	1.2469	3.757
45	3.2	2.1875	4.6543	3.3456
46	3.16	3.1875	3.321	3.2566
47	3.12	1.1875	1.9877	3.9653
48	3.08	4.8125	4.2099	3.0786
49	3.04	2.8125	2.8765	3.3041
50	3	3.8125	1.5432	3.516
51	2.96	1.8125	3.7654	3.409
52	2.92	4.3125	2.4321	3.2043
53	2.88	2.3125	1.0988	3.9977
54	2.84	3.3125	4.9012	3.0793
55	2.8	1.3125	3.5679	3.5836
56	2.76	4.5625	2.2346	3.1799
57	2.72	2.5625	4.4568	3.1458
58	2.68	3.5625	3.1235	3.0989
59	2.64	1.5625	1.7901	3.6823
60	2.6	4.0625	4.0123	2.983
61	2.56	2.0625	2.679	3.3129
62	2.52	3.0625	1.3457	3.5546
63	2.48	1.0625	4.7531	3.6125
64	2.44	4.9688	3.4198	2.9214
65	2.4	2.9688	2.0864	3.2179
66	2.36	3.9688	4.3086	2.9
67	2.32	1.9688	2.9753	3.2259
68	2.28	4.4688	1.642	3.209
69	2.24	2.4688	3.8642	3.0334
70	2.2	3.4688	2.5309	3.0199
71	2.16	1.4688	1.1975	3.9136
72	2.12	4.7188	4.6049	2.7691
73	2.08	2.7188	3.2716	2.979
74	2.04	3.7188	1.9383	3.0678
75	2	1.7188	4.1605	3.1139
76	1.96	4.2188	2.8272	2.8474
77	1.92	2.2188	1.4938	3.3841
78	1.88	3.2188	3.716	2.8214
79	1.84	1.2188	2.3827	3.4489
80	1.8	4.8438	1.0494	3.4783

Table 3 (*continued*)

	Inputs			Outputs
	x_1	x_2	x_3	y
81	1.76	2.8438	4.9835	2.7682
82	1.72	3.8438	3.6502	2.715
83	1.68	1.8438	2.3169	3.1221
84	1.64	4.3438	4.5391	2.6142
85	1.6	2.3438	3.2058	2.8658
86	1.56	3.3438	1.8724	2.9384
87	1.52	1.3438	4.0947	3.0978
88	1.48	4.5938	2.7613	2.6522
89	1.44	2.5938	1.428	3.1716
90	1.4	3.5938	4.8354	2.5555
91	1.36	1.5938	3.5021	2.9462
92	1.32	4.0938	2.1687	2.7063
93	1.28	2.0938	4.3909	2.7177
94	1.24	3.0938	3.0576	2.6238
95	1.2	1.0938	1.7243	3.4514
96	1.16	4.9063	3.9465	2.4084
97	1.12	2.9063	2.6132	2.6391
98	1.08	3.9063	1.2798	2.9859
99	1.04	1.9063	4.6872	2.6429
100	1	4.4063	3.3539	2.3898

been used in 250 generation that no further improvement has been achieved for such population size. A multi-objective optimization of GMDH-type neural networks including all three objectives can offer more choices for a designer. Figure 23 depicts the non-dominated points of 3-objective optimization process in the plane of (TE-PE). It should be noted that there is a single set of non-dominated points as a result of 3-objective Pareto optimization of TE, PE and N that are shown in that plane. Therefore, there are some points in the plane that may dominate others in the case of 3-objective optimization. However, these points are all non-dominated when considering all "three" objectives simultaneously. By careful investigation of the results of 3-objective optimization in that plane, the Pareto front of the corresponding 2-objective optimization (TE-PE) can now be observed. In this figure, points A and B stand for the best (TE) and the best (PE), respectively.

The corresponding values of errors, number of neurons, and the structure of these extreme optimum design points are given in Table 4. Clearly, there is an important optimal design fact between these two objective functions which has been discovered by the Pareto optimum design of GMDH-type neural networks. Such important design fact could not have been found without the multi-objective Pa-reto optimization of those GMDH-type neural networks. From figure 23 points C is the

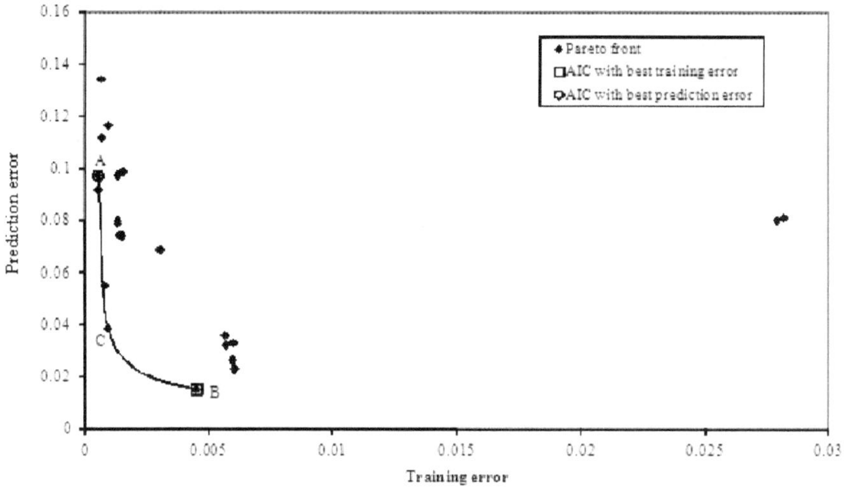

Fig. 23 Prediction error variation with training error in 3-objective optimization

Table 4 Objective functions and structure of networks of different points shown on figure 23

	Network's chromosome	No. of Neurons	Training Error	Prediction Error
Point A	bbbbbbbbaabcabab	5	0.000545619	0.097273886
Point B	Bbbbabacbbabaaaa	7	0.004518445	0.015062286
Point C	bbbbbbbbaabcabac	7	0.000938418	0.038422696

point which demonstrates such important optimal design fact. Point C in the Pareto front of optimum design of TE and PE, exhibits small increase in the value of TE in comparison with that of point A whilst its PE shows significant improvement (about 150 times better prediction error). Therefore, point C could be a trade-off optimum choice when considering the minimum values of both PE and TE simultaneously. The structure and network configuration corresponding to point C is shown in figure 24.

In order to compare these results, AIC [30] has been used both for training and testing data in two different single objective optimiza-tion processes. AIC is defined by

$$AIC = n\log_e(E) + 2(N+1) + C \tag{25}$$

where E, the mean square of error, is computed using equation 6 , N is the number of neurons, n is number of training/testing error, and C is a constant.

Therefore, two optimum points have been found using AIC and are shown in figure 23. Clearly, these two points coincide with the points A and B correspondingly. It is then evident that the Pareto optimum design of GMDH-type neural networks presented in this paper are inclusive of those obtained by AIC and also

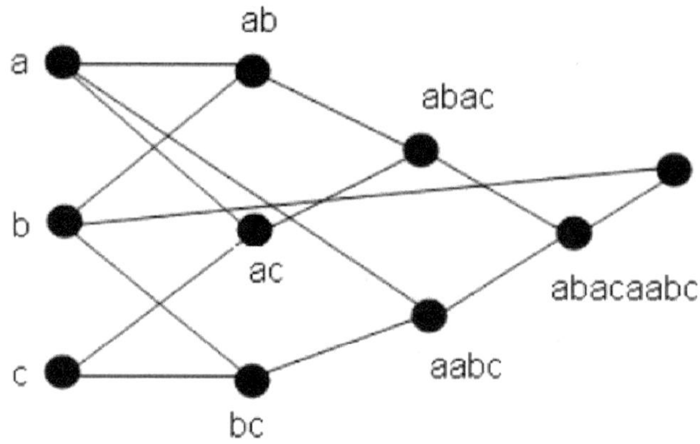

Fig. 24 The network's structure of point C in which a, b, c and d stand for x_1, x_2, x_3 respectively

presents more effective way of choosing trade-off optimum models with respect to conflicting objective functions. It should be noted that point C could be achieved by a proper weighting coefficients (which is not know *a priori*) of prediction and training errors using AIC in only convex programming problems.

5.5 *Multi-objective Genetic Design of GMDH-type Neural Networks for Modelling and Prediction of Explosive Cutting Process*

The input-output data that have been used in this section have been described in section 4.2. In order to design GMDH-type neural network described in previous section from a multi-objective optimum point of view, a population of 60 individuals with a crossover probability of 0.95 and mutation probability of 0.1 has been used in 250 generation that no further improvement has been achieved for such population size. In the multi-objective optimization design of such GMDH-type neural networks [31], different pairs of conflicting objectives (*TE, PE*), (*TE, N*) and (*PE, N*) are selected for 2-objective optimization design of neural networks. The obtained Pareto front for each pair of 2-objective optimization have been shown through figures 25, 26 and 27 for (*TE, PE*), (*TE, N*) and (*PE, N*), respectively It is clear from these figures that all design points representing different GMDH-type neural networks are non-dominated with respect to each other corresponding to that pair of conflicting objectives. Figure 25 depicts the Pareto front of 2-objective optimization of training error (*TE*) and prediction error (*PE*) representing different non-dominated optimum points. In this figure, points A and B stand for the best (*PE*) and the best (*TE*), respectively. It must be noted that the number of neurons (*N*) is not an objective function in this case and only (*TE*) and (*PE*) have been accounted in such 2-objective optimum design of GMDH-type neural networks. Similarly,

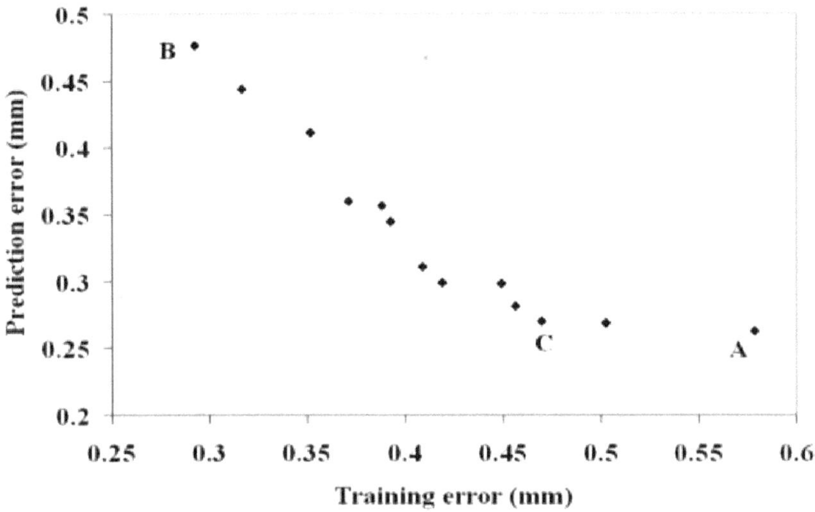

Fig. 25 Pareto front of prediction error and training error in 2-objective optimization

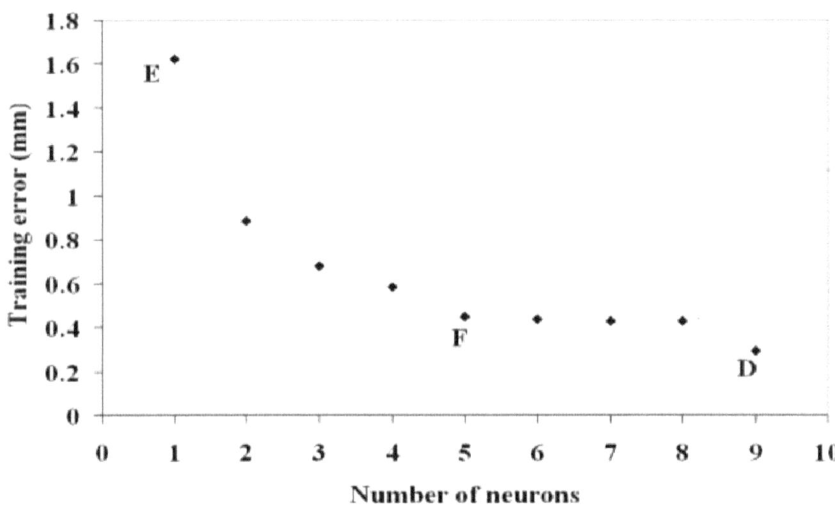

Fig. 26 Pareto front of training error and number of neurons in 2-objective optimization

figures 26 and 27 depict the Pareto front of 2-objective optimization of training error and number of neurons (*TE, N*) and prediction error and number of neurons (*PE, N*), respectively. In this figures, points D and G stand for the best optimum values obtained for *TE* and *PE* in their corresponding 2-objective optimization proc-ess with respect to the number of neurons (*N*). On the other hand, points *E* and *H* stands for the simplest structure of GMDH-type neural networks (*N*=1) with their corresponding values of (*TE*) and (*PE*). It is clear from these figures that all the optimum design

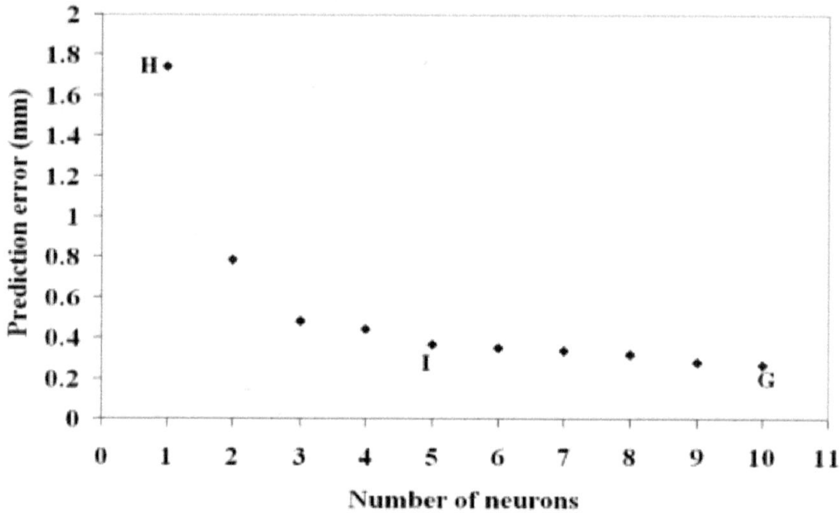

Fig. 27 Pareto front of prediction error and number of neurons in 2-objective optimization

points (GMDH-type neural networks) in a Pareto front are non-dominated and could be chosen by a designer for modelling and prediction of explosive cutting process. It is clear from the these figures that choosing a better value for any objective function in a Pareto front would cause a worse value for another objective. However, if the set of decision variables (genome structure of GMDH-type neural networks and the associated coefficients) is selected based on each of the corresponding sets, it will lead to the best possible combination of those two objectives as shown in figures 25, 26 and 27. In other words, if any other set of decision variables is chosen, the cor-responding values of the pair of objectives will locate a point inferior to the corresponding Pareto front. Such inferior area in the space of the two objectives is in fact top/right side of figures25, 26 and 27. Clearly, there are some important optimal design facts between the two objective functions which have been discovered by the Pareto optimum design of GMDH-type neural networks. Such important design facts could not have been found without the multi-objective Pareto optimization of those GMDH-type neural networks. From figures 25, 26 and 27 points C, F, and I are the points which demonstrate these important optimal design facts. Point C in the 2-objective Pareto optimum design of TE and PE, exhibit a very small increase in the value of *PE* (about 3%) in comparison with that point of A except that its training error is about 24% better than that of point A. Therefore, point C could be a trade-off optimum choice when considering the minimum values of both *PE* and *TE*. The structure and network configuration corresponding to point C is shown in figure 28 whose good behaviour of such GMDH-type neural networks model in training and prediction data are shown in figure 29.

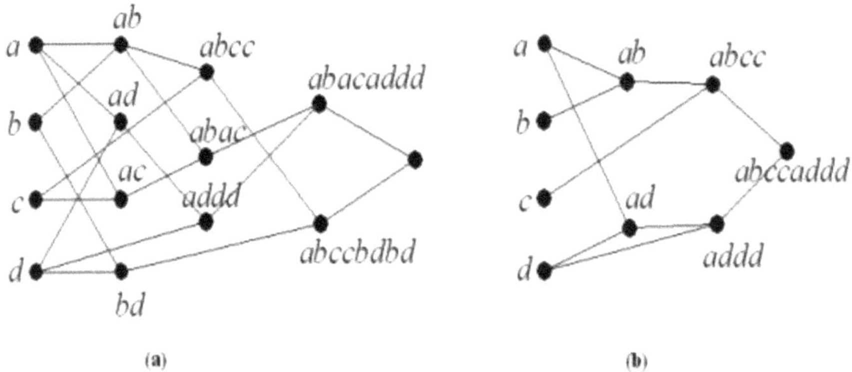

Fig. 28 The structure of network corresponding to (a) point C on figure 25, (b) point I on figure 27, in which a, b, c and d stand for apex angle, standoff, charge mass, and liner thickness, respectively

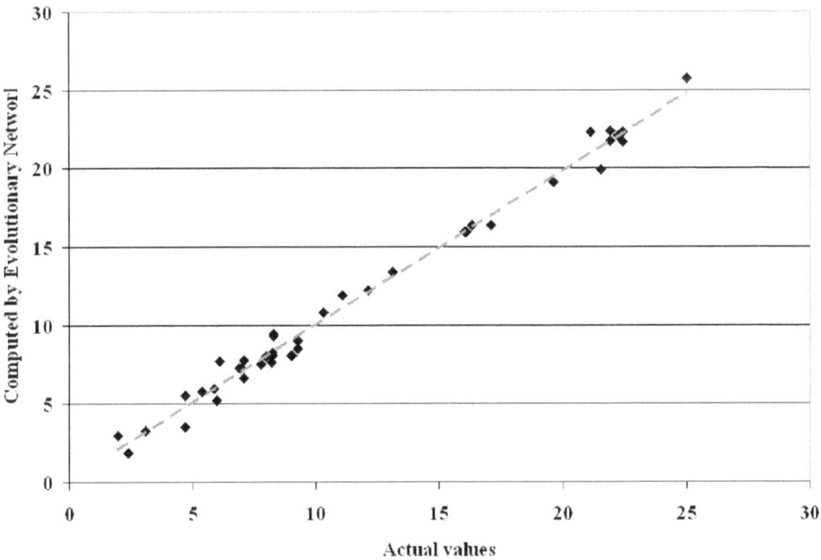

Fig. 29 Comparison of actual values with the evolved GMDH model corresponding to optimum point C

Similarly, points F and I of figures 26 and 27 demonstrate the trade-offs between the complexity of networks (number of neurons) and training error and prediction error, respectively. For example, point I exhibits a very small increase in *PE* in comparison with that of point G whilst its number neurons is 50% less than that of G which corresponds to a much simpler structure of neural network. The

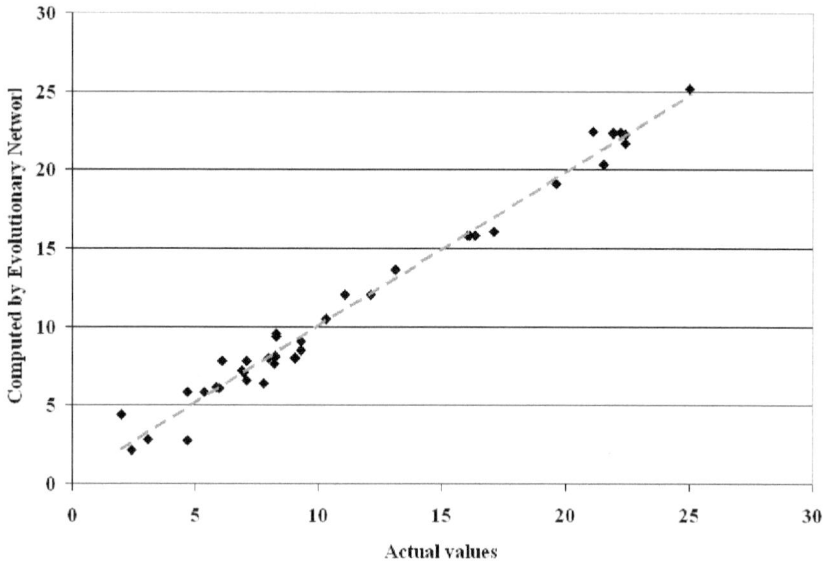

Fig. 30 Comparison of actual values the with evolved GMDH model corresponding to optimum point I

corresponding structure of point I is shown in figure 28 whose good behaviour of such GMDH-type neural network model both in training and prediction data are shown in figure 30.

A multi-objective optimization of GMDH-type neural networks including all three objectives can offer more choices for a designer. Moreover, such 3-objective optimization result can subsume all the 2-objective optimization results presented in previous section. In this case, the computation time needed for the 3-objective optimization process was approximately 730 seconds using a computer with Intel Pentium 4 CPU processor. Such non-dominated individuals of 3-objective optimization process have been shown in the planes of (*N -TE*) and (*N-PE*) in figures 31 and 32, respectively. It should be noted that there is a single set of points as the result of 3-objective Pareto optimization of *TE*, *PE* and *N* that are shown in different planes together with their corresponding 2-objective optimization results. Therefore, there are some points in each plane that may dominate others in the same plane in the case of 3-objective optimization. However, these points are all non-dominated when considering all three objectives simultaneously. By careful investigation of the results of 3-objective optimization in each plane, the Pareto fronts of the corresponding 2-objective optimization obtained previously can now be observed in these figures. It can be readily seen that the results of such 3-objective optimization include the Pareto fronts of each 2-objective optimization and thus provide more optimal choices for designer. Consequently, the Pareto optimization of GMDH-type

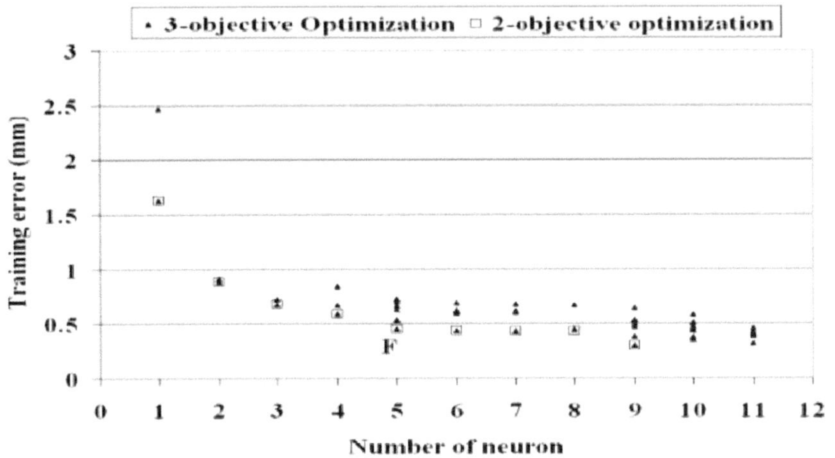

Fig. 31 Number of neurons variation with training error in both 3-objective and 2-objective optimization

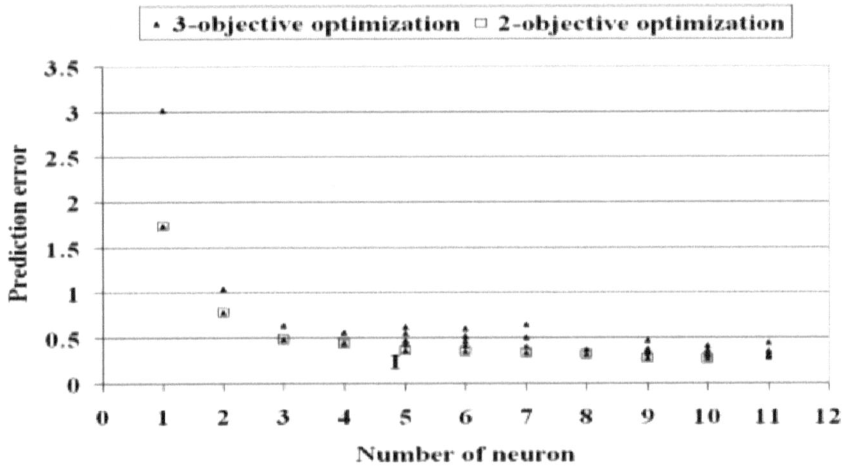

Fig. 32 Number of neuron variation with prediction error in both 3-objective and 2-objective optimization

neural networks reveals that the models corresponding to the C or F or I could be compromisingly chosen via a trade-off point of view regarding *TE*, *PE* and *N*.

In order to compare the effectiveness of the ε-elimination diversity algorithm presented here, the 3-objective optimum design problem (*TE*, *PE* and *N*) has also been solved using the crowding dis-tance sorting procedure of NSGA-II [25]. Figure 33 shows all the Pareto points obtained using both procedures of the modified NSGA-II of this work and that of NSGA-II reported in [25] in the plane of training

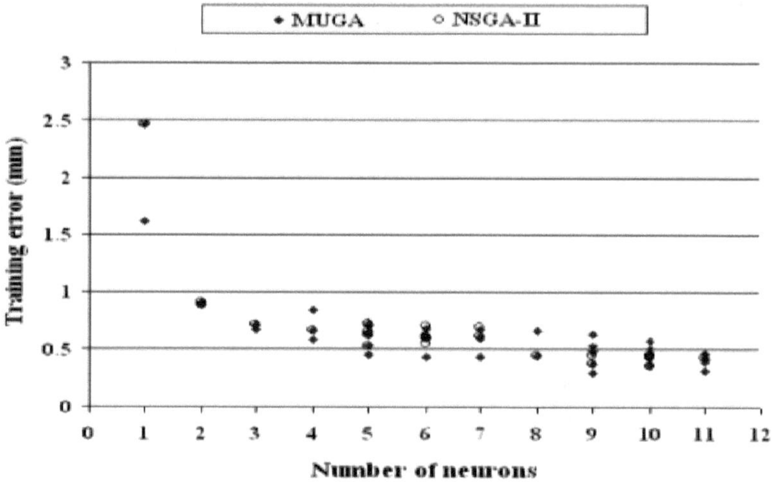

Fig. 33 Comparison of Pareto points using two methods for the 3-objective optimization problem shown in the plane of *TE* & *N*

error (*TE*) and number of neurons (*N*). It is evident from this figure that the Pareto front obtained by the method of this work is better than that of NSGA-II both in the values of training errors and diversity.

6 Conclusion

Hybrid Genetic Algorithms and SVD have been successfully used to design the coefficients as well as the connectivity configuration of GMDH-type neural networks used for modelling and prediction of various complex models in both single and multi-objective Pareto based optimization processes. In this way, a specific encoding scheme has been presented to genetically design GMDH-type neural networks. Such generalization of network's topology provides near optimal networks in terms of hidden layers and/or number of neurons and their connectivity configuration, so that a polynomial expression for dependent variable of the process can be achieved consequently. The multi-objective optimization led to the discovering of useful optimal design principles in the space of objective functions.

The important conflicting objective functions of GMDH-type neural networks have been selected as Training Error (TE), Prediction Error (PE) and Number of Neurons (N) of such neural networks. Therefore, optimal Pareto fronts of such models have been obtained in each case which exhibit the trade-offs between the corresponding pair of conflicting objectives and, thus, provide different non-dominated optimal choices of GMDH-type neural networks models. In addition to discovering the trade-off optimum points, it has been shown that the Pareto front obtained by the approach of this chapter involves those that can be found by Akaike's Information

Criterion which thus exhibits the effectiveness of the Pareto optimum design of GMDH-type neural networks presented in this chapter.

References

1. Astrom, K.J., Eykhoff, P.: System identification a survey. Automatica 7, 123–162 (1971)
2. Sanchez, E., Shibata, T., Zadeh, L.A.: Genetic algorithms and fuzzy logic systems. World Scientific, Riveredge (1997)
3. Kristinson, K., Dumont, G.: System identification and control using genetic algorithms. IEEE Trans. On Sys., Man, and Cybern. 22, 1033–1046 (1992)
4. Koza, J.: Genetic programming, on the programming of computers by means of natural selection. MIT Press, Cambridge (1992)
5. Iba, H., Kuita, T., deGaris, H.: System identification using structured genetic algorithms. In: Proc. of 5th Int. Conf. on Genetic Algorithms, ICGA 1993, USA (1993)
6. Rodriguez-Vazquez, K.: Multi-objective evolutionary algorithms in non-linear system identification. PhD thesis, Department of Automatic Control and Systems Engineering, The University of Sheffield, Sheffield, UK (1999)
7. Fonseca, C.M., Fleming, P.J.: Nonlinear system identification with multi-objective genetic algorithms. In: Proceedings of the 13th World Congress of the International Federation of Automatic Control, pp. 187–192. Pergamon Press, San Francisco (1996)
8. Liu, G.P., Kadirkamanathan, V.: Multi-objective criteria for neural network structure selection and identification of nonlinear systems using genetic algorithms. IEE Proceedings on Control Theory and Applications 146, 373–382 (1999)
9. Ivakhnenko, A.G.: Polynomial theory of complex systems. IEEE Trans. Syst. Man & Cybern. SMC-1, 364–378 (1971)
10. Farlow, S.J.: Self-organizing method in modeling: GMDH type algorithm. Marcel Dekker Inc., New York (1984)
11. Mueller, J.A., Lemke, F.: Self-organizing data mining: An intelligent approach to extract knowledge from data. Pub. Libri, Hamburg (2000)
12. Iba, H., deGaris, H., Sato, T.: A numerical approach to genetic programming for system identification. Evolutionary computation 3(4), 417–452 (1996)
13. Nariman-Zadeh, N., Darvizeh, A., Felezi, M.E., Gharababei, H.: Polynomial modelling of explosive compaction process of metallic powders using GMDH-type neural networks and singular value decomposition. Modelling and Simulation in Materials Science and Engineering 10, 727–744 (2002)
14. Nariman-Zadeh, N., Darvizeh, A., Ahmad-Zadeh, G.R.: Hybrid genetic design of GMDH-type neural networks using singular value decomposition for modelling and prediction of the explosive cutting process. Proceedings of the I MECH E Part B Journal of Engineering Manufacture 217, 779–790 (2003)
15. Nariman-Zadeh, N., Darvizeh, A., Darvizeh, M., Gharababei, H.: Modelling of explosive cutting process of plates using GMDH-type neural network and singular value decomposition. Journal of Materials Processing Technology 128, 80–87 (2002)
16. Porto, V.W.: Evolutionary computation approaches to solving problems in neural computation. In: Back, T., Fogel, D.B., Michalewicz, Z. (eds.) Handbook of Evolutionary Computation Back, pp. D1.2:1–D1.2:6. Institute of Physics Publishing / Oxford University Press, New York (1997)
17. Yao, X.: Evolving artificial neural networks. Proceedings of IEEE 87(9), 1423–1447 (1999)

18. Vasechkina, E.F., Yarin, V.D.: Evolving polynomial neural network by means of genetic algorithm: some application examples. Complexity International 9 (2001)
19. Nariman-Zadeh, N., Darvizeh, A., Jamali, A., Moeini, A.: Evolutionary design of generalized polynomial neural networks for modelling and prediction of explosive forming process. Journal of Material Processing and Technology 164-165, 1561–1571 (2005)
20. Press, W.H., Teukolsky, S.A., Vetterling, W.T., Flannery, B.P.: Numerical Recipes in FORTRAN: The art of scientific computing, 2nd edn. Cambridge University Press, Cambridge (1992)
21. Goldberg, D.E.: Genetic algorithms in search, optimization, and machine learning. Addison-Wesley, Reading (1989)
22. Felezi, M.E., Nariman-Zadeh, N., Darvizeh, A., Jamali, A., Teymoorzadeh, A.: A Polynomial model for the level variations of the Caspian sea using evolutionary design of general-ized GMDH-type neural networks. WSEAS Transactions on CIRCUIT and SYSTEMS 3(2) (2004)
23. `http://www.inco.ac.ir`
24. Toffolo, A., Benini, E.: Genetic diversity as an objective in multi-objective evolutionary algorithms. Evolutionary Computation 11(2), 151–167 (2003)
25. Deb, K., Agrawal, S., Pratap, A., Meyarivan, T.: A fast and elitist multi-objective genetic algorithm: NSGA-II. IEEE Trans. On Evolutionary Computation 6(2), 182–197 (2002)
26. Coello Coello, C.A., Becerra, R.L.: Evolutionary multi-objective optimization using a cultural algorithm. In: IEEE Swarm Intelligence Symp., pp. 6–13 (2003)
27. Nariman-Zadeh, N., Atashkari, K., Jamali, A., Pilechi, A., Yao, X.: Inverse modelling of multi-objective thermodynamically optimized turbo engines using GMDH-type neural networks and evolutionary algorithms. Engineering Optimization 37, 437–462 (2005)
28. Atashkari, K., Nariman-Zadeh, N., Golcu, M., Khalkhali, A., Jamali, A.: Modelling and multi-objective optimization of a variable valve-timing spark-ignition engine using polynomial neural networks and evolutionary algorithms. Energy Conversion and Management 48(3), 1029–1041 (2007)
29. Nariman-Zadeh, N., Jamali, A.: Pareto genetic design of GMDH-type neural networks for nonlinear systems. In: IWIM 2007, Prague, Czech Rep (2007)
30. Akaike, H.: A new look at the statistical model identification. IEEE Trans. Automatic Control AC-19(6), 716–723 (1974)
31. Nariman-Zadeh, N., Jamali, A., Darvizeh, A., Masoumi, A., Hamrang, S.: Multi-objective evolutionary optimization of polynomial neural networks for modelling and prediction of explosive cutting process. In: Engineering Applications of Artificial Intelligence (accepted) (2008)

Hybrid Differential Evolution and GMDH Systems

Godfrey Onwubolu

Abstract. This chapter describes the newly proposed design methodology of the hybrid differential evolution, DE and GMDH. The architecture of model is not predefined, but can be self-organized automatically during the design process. The hybrid DE and SVD is used for simultaneous parametric and structural design of GMDH networks used for modelling and prediction of various complex models. The *DE-GMDH* approach has been applied to the problem of developing predictive model for tool-wear in turning operations; the exchange rate problem; the Box-Jenkins gas furnace data, with experimental results clearly demonstrating that the proposed DE-GMDH-type network outperforms the existing models both in terms of better approximations capabilities as well as generalization abilities.

1 Introduction

The GMDH is a heuristic self-organizing modeling method which Ivakhnenko [1] - [3] introduced as a rival to the method of stochastic approximations. The method is particularly useful in solving the problem of modeling multi-input to single-output data. In GMDH-type self-organizing modeling algorithm, models are generated adaptively from data in form of networks of active neurons in a repetitive generation of populations of competing models of growing complexity and corresponding validation and selection model until an optimal complex model which is not too simple and not too complex have been realized. The modeling approach grows a tree-like network out of data of input and output variables (seed information) in a pair-wise

Godfrey Onwubolu
Knowledge Management & Mining, Richmond Hill, Canada
e-mail: onwubolu_g@dsgm.ca

G.C. Onwubolu (Ed.): Hybrid Self-Organizing Modeling Systems, SCI 211, pp. 139–191.
springerlink.com

combination and competitive selection from a simple single individual (neuron) to a desired final solution, which does not have an overspecialized behavior (model). In this approach, neither the number of neurons and the number of layers in the network, nor the actual behavior of each created neuron is predefined. The modeling is self-organizing because the number of neurons, the number of layers and the actual behavior of each created neuron are adjusting during the process of self-organization [4] - [6].

2 Inductive Modeling: Group Method of Data Handling (GMDH)

The causality relationship between the inputs and the output of a multiple inputs single output self-organizing network can be represented by an infinite Volterra-Kolmogorov-Gabor (VKG) polynomial of the form 1:

$$y_n = a_0 + \sum_{i=1}^{M} a_i x_i + \sum_{i=1}^{M}\sum_{j=1}^{M} a_{ij} x_i x_j + \sum_{i=1}^{M}\sum_{j=1}^{M}\sum_{k=1}^{M} a_{ijk} x_i x_j x_k \ldots \tag{1}$$

where $X = (x_1, x_2, ..., x_M)$ is the vector of input variables and $A = (a_0, a_i, a_{ij}\ a_{ijk} ...)$ is the vector of coefficients or weights.

This is the discrete-time analogue of a continuous time Volterra series and can be used to approximate any stationary random sequence of physical measurements. Ivakhnenko showed that the VKG series can be expressed as a cascade of second order polynomials using only pairs of variables [2] - [5] as shown in Figure 1. The corresponding network can be constructed from simple polynomial and delay elements. As the learning procedure evolves, branches that do not contribute significantly to the specific output can be pruned, thereby allowing only the dominant causal relationship to evolve.

2.1 *GMDH Layers*

When constructing a GMDH network, all combinations of the inputs are generated and sent into the first layer of the network. The outputs from this layer are then classified and selected for input into the next layer with all combinations of the selected outputs being sent into layer 2. This process is continued as long as each subsequent $layer_{(n+1)}$ produces a better result than $layer_{(n)}$. When $layer_{(n+1)}$ is found to not be as good as the $layer_{(n)}$, the process is stopped.

2.2 *GMDH Nodes*

Self-organizing networks are constructed from elemental polynomial neurons each of which possesses only a pair of dissimilar inputs $(x_i,\ x_j)$. Each layer consists of nodes generated to take a specific pair of the combination of inputs as its source. Each node produces a set of coefficients ai where i ε 0, 1, 2,.., m such that equation 2 is estimated using the set of training data. This equation is tested for fit by determining the mean square error of the predicted $\hat{y}$ and actual y values as shown in equation 3 using the set of testing data.

$$\hat{y}_n = a_0 + a_1 x_{i_n} + +a_2 x_{j_n} + a_3 x_{i_n} x_{j_n} + a_4 x_{i_n}^2 + a_5 x_{jn}^2 \tag{2}$$

$$e = \sum_{n=1}^{N} (\hat{y}_n - y_n)^2 \tag{3}$$

In determining the values of a that would produce the "best fit", the partial derivatives of equation 3 are taken with respect to each constant value a_i and set equal to zero.

$$\frac{\partial e}{\partial a_i} = 0 \tag{4}$$

Expanding equation 4 results in the following system of equations that are solved using the training data set.

$$\sum_{n=1}^{N} y = \sum_{n=1}^{N} a_0 + a_1 x_i + +a_2 x_j + a_3 x_i x_j + a_4 x_i^2 + a_5 x_j^2 \tag{5}$$

$$\sum_{n=1}^{N} y x_i = \sum_{n=1}^{N} a_0 x_i + a_1 x_i^2 + +a_2 x_i x_j + a_3 x_i^2 x_j + a_4 x_i^3 + a_5 x_i x_j^2 \tag{6}$$

$$\sum_{n=1}^{N} y x_j = \sum_{n=1}^{N} a_0 x_j + a_1 x_i x_j + +a_2 x_j^2 + a_3 x_i x_j^2 + a_4 x_i^2 x_j + a_5 x_j^3 \tag{7}$$

$$\sum_{n=1}^{N} y x_i x_j = \sum_{n=1}^{N} a_0 x_i x_j + a_1 x_i^2 x_j + +a_2 x_i x_j^2 + a_3 x_i^2 x_j^2 + a_4 x_i^3 x_j + a_5 x_i x_j^3 \tag{8}$$

$$\sum_{n=1}^{N} y x_i^2 = \sum_{n=1}^{N} a_0 x_i^2 + a_1 x_i^3 + +a_2 x_i^2 x_j + a_3 x_i^3 x_j + a_4 x_i^4 + a_5 x_i^2 x_j^2 \tag{9}$$

$$\sum_{n=1}^{N} y x_j^2 = \sum_{n=1}^{N} a_0 x_j^2 + a_1 x_i x_j^2 + +a_2 x_j^3 + a_3 x_i x_j^3 + a_4 x_i^2 x_j^2 + a_5 x_j^4 \tag{10}$$

The equations can be simplified using matrix mathematics as follows

$$W = \begin{pmatrix} 1 & x_i & x_j & x_i x_j & x_i^2 & x_j^2 \end{pmatrix} \tag{11}$$

$$X = W^T W \tag{12}$$

$$X = \begin{pmatrix} 1 & x_i & x_j & x_ix_j & x_i^2 & x_j^2 \\ x_i & x_i^2 & x_ix_j & x_i^2x_j & x_i^3 & x_ix_j^2 \\ x_j & x_ix_j & x_j^2 & x_ix_j^2 & x_i^2x_j & x_j^3 \\ x_ix_j & x_i^2x_j & x_ix_j^2 & x_i^2x_j^2 & x_i^3x_j & x_ix_j^3 \\ x_i^2 & x_i^3 & x_i^2x_j & x_i^3x_j & x_i^4 & x_i^2x_j^2 \\ x_j^2 & x_ix_j^2 & x_j^3 & x_ix_j^3 & x_i^2x_j^2 & x_j^4 \end{pmatrix} \tag{13}$$

$$\mathrm{a} = \begin{pmatrix} a_0 \; a_1 \; a_2 \; a_3 \; a_4 \; a_5 \end{pmatrix} \tag{14}$$

$$b = (yW)^T \tag{15}$$

This system of equations then can be written as:

$$\sum_{n=1}^{N} aX = \sum_{n=1}^{N} b \tag{16}$$

The node is now responsible for evaluating all inputs of x_{i_n}, x_{j_n}, y_n data values in **A** and **b** for the *training* set of data. Solving the system of equations results in x being the node's computed set of coefficients. Using these coefficients in equation 2, the node then computes its error by processing the set of testing data in equations 2 and 3. The error is the measure of fit that this node achieved.

2.3 GMDH Connections

A GMDH layer sorts its nodes based on the error produced, saving the best N nodes. The generated y_n values of each node become one set of inputs to be used by the next layer when it combines all outputs from the previous layer's nodes assigning them to the new layer's nodes (See Figure 1). The layer must remember the nodes that were saved so that other data submitted to the network will follow the same generated path to the output.

2.4 GMDH Network

When the GMDH network is completed, there is a set of original inputs that filtered through the layers to the optimal output node. This is the computational network that is to be used in computing predictions.

The nodes in the input layer that are "winners" (starred nodes in Figure 1) at modeling the system output are retained and form the input to the next layer. The inputs for layer 1 are formed by taking all combinations of the surviving output approximations from the input layer nodes. It is seen that at each layer the order of the polynomial approximation is increased by two. The layer 2 nodes that are "winners" at approximating the system output are retained and form the layer 3 inputs.

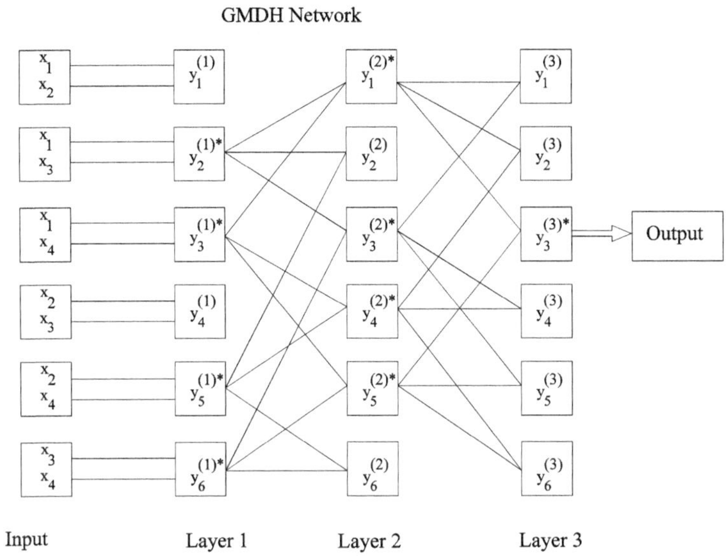

Fig. 1 GMDH forward feed functional network

This process is repeated until the current layer's best approximation is inferior to the previous layer's best approximation [7].

2.5 Advantages of GMDH Technique

The advantage of using pairs of input is that only six weights (coefficients) have to be computed for each neuron. The number of neurons in each layer increases approximately as the square of the number of inputs. During each training cycle, the synaptic weights of each neuron that minimize the error norm between predicted and measured values are computed and those branches that contribute least to the output of the neuron are discarded, the remaining branches being retained and their synaptic weights kept unchanged thereafter. A new layer is subsequently added and the procedure is repeated until the specified termination conditions are met.

2.6 Limitations of GMDH Technique

Although GMDH provides for a systematic procedure of system modeling and prediction, it has also a number of shortcomings. Among the most problematic can be stated:

- A tendency to generate quite complex polynomial (since the complexity of the network increases with each training and selection cycle through addition of new layers) for relatively simple systems (data input);.

- An inclination to producing overly complex network (model) when dealing with highly nonlinear systems owing to its limited generic structure (quadratic two-variable polynomial).

In order to alleviate these problems, a number of researchers have attempted to hybridize GMDH with some evolutionary optimization techniques. Amongst them, Iba et al. [8] presented the GP-GMDH (Genetic Programming-GMDH) algorithm and showed that it performs better than the conventional GMDH algorithm. Nariman-Zadeh et al. [9] proposed a hybrid of genetic algorithm (GA) and GMDH which outperforms conventional GMDH approach. Other related research work include that of Hiassat et al. [21] who used genetic programming to find the best function that maps the input to the output in each layer of the GMDH algorithm, and showed that it performs better than the conventional GMDH algorithm in time series prediction using financial and weather data; and Oh et al. [22] also realized the genetically optimized polynomial neural network (g-PNN). Onwubolu [10][11] proposed a hybrid of differential evolution (DE) and GMDH and clearly showed that this framework outperforms conventional GMDH approach. The work reported in this chapter adopted the DE-GMDH reported in Onwubolu [10][11] with some enhancement to support high dimensionality problems which are common in bioinformatics applications.

Consequently, it could be inferred that so far only genetic population-based optimization techniques have been mainly hybridized with GMDH to improve the performance of the standard GMDH approach. The main focus of this chapter therefore is to extend the hybridization spectrum to include DE which is one of the population-based optimization methods. In this paper, we introduce a hybrid modeling paradigm based on DE and GMDH for modeling and prediction of complex systems. The remaining sections are organized as follows. Section 3 presents the classical DE approach. Section 3.3 presents the discrete DE approach. Section 4 presents the proposed hybrid system. Section 5 presents the methodology for modeling linear and nonlinear functions. Section 6 presents the results of experiments carried out. The conclusions from this study are given in Section 7.

3 Classical Differential Evolution Algorithm

The Differential Evolution (Exploration) [DE] algorithm introduced by Storn and Price [12] is a novel parallel direct search method, which utilizes *NP* parameter vectors as a population for each generation *G*. DE can be categorized into a class of floating-point encoded, evolutionary optimization algorithms. Detailed descriptions of DE are provided. DE algorithm was originally designed to work with continuous variables, and the inventors, state that [13]:

"*Even if a variable is discrete or integral, it should be initialized with a real value since DE internally treats all variables as floating-point values regardless of their type*".

To solve discrete or combinatorial problems in general, Onwubolu [14] introduced the forward/backward transformation techniques, which facilitates solving any discrete or combinatorial problem. Price et al.[15] further state in their recent book that:

"*..its (DE) suitability as a combinatorial optimizer is still a topic of considerable debate and a definite judgment cannot be given at this time .. Certainly in the case of the TSP, the most successful strategies for the TSP continue to be those that rely on special heuristics.*" [14].

Successful applications of the DE to a number of combinatorial problems are found in [16][17][18]. Generally, the function to be optimized, $\Im$, is of the form $\Im(X) : R^D \to R$. The optimization target is to minimize the value of this objective function $\Im(X)$, $\min(\Im(X))$ by optimizing the values of its parameters $X = \{x_1, x_2, ..., x_D\}$, $X \in R^D$, where X denotes a vector composed of D objective function parameters. Usually, the parameters of the objective function are also subject to lower and upper boundary constraints, $x^{(L)}$ and $x^{(U)}$, respectively: $x_j^{(L)} \leq x_j \leq x_j^{(U)}\ \forall j \in [1, D]$.

3.1 The Steps Involved in Classical Differential Evolution

The steps involve in classical DE are outlined as follows [12]:

Step 1: Initialization

As with all evolutionary optimization algorithms, DE works with a population of solutions, not with a single solution for the optimization problem. Population P of generation G contains NP solution vectors called individuals of the population and each vector represents potential solution for the optimization problem:

$$P^{(G)} = X_i^{(G)} \quad i = 1, ..., NP;\ G = 1, ..., G_{\max} \tag{17}$$

In order to establish a starting point for optimum seeking, the population must be initialized. Often there is no more knowledge available about the location of a global optimum than the boundaries of the problem variables. In this case, a natural way to initialize the population $P^{(0)}$ (initial population) is to seed it with random values within the given boundary constraints:

$$P^{(0)} = x_{j,i}^{(0)} = x_j^{(L)} + rand_j[0,1] \bullet \left(x_j^{(U)} - x_j^{(L)}\right) \quad \forall i \in [1, NP];\ \forall j \in [1, D] \tag{18}$$

where $rand_j[0, 1]$ represents a uniformly distributed random value that ranges from zero to one.

Step 2: Mutation

The self-referential population recombination scheme of DE is different from the other evolutionary algorithms. From the first generation onward, the population of the subsequent generation is obtained on the basis of the current population. First a temporary or trial population of *candidate vectors* for the subsequent generation, $V^{(G)} = v_{j,i}^{(G)}$, is generated as follows:

$$v_{j,i}^{(G)} = x_{j,r3}^{(G)} + F \bullet \left(x_{j,r1}^{(G)} - x_{j,r2}^{(G)} \right) \tag{19}$$

where $i = [1, NP]$; $j = [1, D]$, $r1$, $r2$, $r3 \in [1, NP]$, randomly selected, except: $r1 \neq r2 \neq r3 \neq i$, and $CR \in [0, 1]$, $F \in (0, 1]$.

Three randomly chosen indexes, *r1*, *r2*, and *r3* refer to three randomly chosen vectors of population. They are mutually different from each other and also different from the running index *i*. New random values for *r1*, *r2*, and *r3* are assigned for each value of index *i* (for each vector). A new value for the random number *rand*[0, 1] is assigned for each value of index *j* (for each vector parameter).

Step 3: Crossover

To compliment the differential mutation strategy, DE also employs uniform crossover. Sometimes referred to as discrete recombination, (dual) crossover builds trial vectors out of parameter values that have been copied from two different vectors. In particular, DE crosses each vector with a mutant vector:

$$u_{j,i}^{(G)} = \begin{cases} v_{j,i,g} \; if \; rand_j[0,1] < CR \vee j = j_{rand} \\ \qquad x_{i,j}^{(G)} \; if \; otherwise \end{cases} \tag{20}$$

where $j_{rand} \in [1, D]$. *F* and *CR* are DE control parameters. Both values remain constant during the search process. Both values as well as the third control parameter, *NP* (population size), remain constant during the search process. *F* is a real-valued factor in range [0.0, 1.0] that controls the amplification of differential variations. *CR* is a real-valued crossover factor in the range [0.0, 1.0] that controls the probability that a trial vector will be selected form the randomly chosen, mutated vector, $u_{j,i}^{(G)}$ instead of from the current vector, $x_{j,i}^{(G)}$. Practical advice on how to select control parameters *NP*, *F* and *CR* can be found in Storn and Price [12] [19].

Step 4: Selection

The selection scheme of DE also differs from the other evolutionary algorithms. If the trial vector, $u_{i,g}$, has an equal or lower objective function value than that of its target vector, $x_{i,g}$, it replaces the target vector in the next generation; otherwise, the target vector remains in place in the population for the least one more generation. These conditions are written as follows:

$$x_i^{(G+1)} = \begin{cases} u_i^{(G)} \; if \; \Im\left(u_i^{(G)}\right) \leq \Im\left(x_i^{(G)}\right) \\ x_i^{(G)} \; if \; otherwise \end{cases} \tag{21}$$

Step 5: Stopping criteria

Once the new population is installed, the process of mutation, recombination and selection is repeated until the optimum is located, or a pre-selected termination criterion such as the number of generations reaching a preset maximum, $g_{\max}$, is satisfied.

As already discussed, the improvement strategies of forward/backward transformation were proposed to facilitate solving discrete or combinatorial optimization problems using DE (for details, see [14] [17]).

3.2 Ten different Working Strategies in Differential Evolution

Price and Storn [13] have suggested ten different working strategies. It mainly depends on the problem on hand for which strategy to choose. The strategies vary on the solutions to be perturbed, number of difference solutions considered for perturbation, and finally the type of crossover used. The following are the different strategies being applied.

Strategy	Formulation
Strategy 1: DE/best/1/exp:	$\mathbf{u}_{i,G+1} = \mathbf{x}_{best,G} + F \bullet \left(\mathbf{x}_{r_1,G} - \mathbf{x}_{r_2,G}\right)$
Strategy 2: DE/rand/1/exp:	$\mathbf{u}_{i,G+1} = \mathbf{x}_{r_1,G} + F \bullet \left(\mathbf{x}_{r_2,G} - \mathbf{x}_{r_3,G}\right)$
Strategy 3: DE/rand-to-best/1/exp	$\mathbf{u}_{i,G+1} = \mathbf{x}_{i,G} + \lambda \bullet \left(\mathbf{x}_{best,G} - \mathbf{x}_{i,G}\right) + F \bullet \left(\mathbf{x}_{r_1,G} - \mathbf{x}_{r_2,G}\right)$
Strategy 4: DE/best/2/exp:	$\mathbf{u}_{i,G+1} = \mathbf{x}_{best,G} + F \bullet \left(\mathbf{x}_{r_1,G} - \mathbf{x}_{r_2,G} - \mathbf{x}_{r_3,G} - \mathbf{x}_{r_4,G}\right)$
Strategy 5: DE/rand/2/exp:	$\mathbf{u}_{i,G+1} = \mathbf{x}_{r_5,G} + F \bullet \left(\mathbf{x}_{r_1,G} - \mathbf{x}_{r_2,G} - \mathbf{x}_{r_3,G} - \mathbf{x}_{r_4,G}\right)$
Strategy 6: DE/best/1/bin:	$\mathbf{u}_{i,G+1} = \mathbf{x}_{best,G} + F \bullet \left(\mathbf{x}_{r_1,G} - \mathbf{x}_{r_2,G}\right)$
Strategy 7: DE/rand/1/bin:	$\mathbf{u}_{i,G+1} = \mathbf{x}_{r_1,G} + F \bullet \left(\mathbf{x}_{r_2,G} - \mathbf{x}_{r_3,G}\right)$
Strategy 8: DE/rand-to-best/1/bin:	$\mathbf{u}_{i,G+1} = \mathbf{x}_{i,G} + \lambda \bullet \left(\mathbf{x}_{best,G} - \mathbf{x}_{i,G}\right) + F \bullet \left(\mathbf{x}_{r_1,G} - \mathbf{x}_{r_2,G}\right)$
Strategy 9: DE/best/2/bin	$\mathbf{u}_{i,G+1} = \mathbf{x}_{best,G} + F \bullet \left(\mathbf{x}_{r_1,G} - \mathbf{x}_{r_2,G} - \mathbf{x}_{r_3,G} - \mathbf{x}_{r_4,G}\right)$
Strategy 10: DE/rand/2/bin:	$\mathbf{u}_{i,G+1} = \mathbf{x}_{r_5,G} + F \bullet \left(\mathbf{x}_{r_1,G} - \mathbf{x}_{r_2,G} - \mathbf{x}_{r_3,G} - \mathbf{x}_{r_4,G}\right)$

The convention shown is DE/x/y/z. DE stands for Differential Evolution, x represents a string denoting the solution to be perturbed, y is the number of difference solutions considered for perturbation of x, and z is the type of crossover being used (exp: exponential; bin: binomial)

DE has two main phases of crossover: binomial and exponential. Generally a child solution $\mathbf{u}_{i,G+1}$ is either taken from the parent solution $\mathbf{x}_{i,G}$ or from a mutated donor solution $\mathbf{v}_{i,G+1}$ as shown: $\mathbf{u}_{j,i,G+1} = v_{j,i,G+1} = x_{j,r_3,G} + F \bullet \left(x_{j,r1,G} - x_{j,r_2,G}\right)$.

The frequency with which the donor solution $\mathbf{v}_{i,G+1}$ is chosen over the parent solution $\mathbf{x}_{i,G}$ as the source of the child solution is controlled by both phases of crossover. This is achieved through a user defined constant, crossover *CR* which is held constant throughout the execution of the heuristic.

The *binomial* scheme takes parameters from the donor solution every time that the generated random number is less than the CR as given by $rand_j[0,1) < CR$, else all parameters come from the parent solution $\mathbf{x}_{i,G}$.

The *exponential* scheme takes the child solutions from $\mathbf{x}_{i,G}$ until the first time that the random number is greater than CR, as given by $rand_j[0,1) < CR$, otherwise the parameters comes from the parent solution $\mathbf{x}_{i,G}$.

To ensure that each child solution differs from the parent solution, both the exponential and binomial schemes take at least one value from the mutated donor solution $\mathbf{v}_{i,G+1}$.

3.3 Discrete Differential Evolution

The canonical DE cannot be applied to discrete or permutative problems without modification. The internal crossover and mutation mechanism invariably change any applied value to a real number. This in itself will lead to infeasible solutions. The objective then becomes one of transformation, either that of the population or that of the internal crossover/mutation mechanism of DE. A number of researchers have decided not to modify in any way the operation of DE strategies, but to manipulate the population in such a way as to enable DE to operate unhindered. Since the solution for a population is permutative, suitable conversion routines are required in order to change the solution from integer to real and then back to integer after crossover.

Application areas where DE for permutative-based combinatorial optimization problems can be applied include but not limited to the following:

1. Scheduling: Flow Shop, Job Shop, etc.
2. Knapsack
3. Linear Assignment Problem (LAP)
4. Quadratic Assignment Problem (QAP)
5. Traveling Salesman Problem (TSP)
6. Vehicle Routine Problem (VRP)
7. Dynamic pick-and-place model of placement sequence and magazine assignment in robots

Since the solution for the population is permutative, a suitable conversion routine was required in order to change the solution from integer to real and then back to integer after crossover. The population was generated as permutative string. Two conversions routines were devised, one was *Forward transformation* and the other *Backward transformation* for the conversion between integer and real values. This new heuristic was termed Discrete Differential Evolution (DDE) [17].

3.4 Permutative Population

The first part of the heuristic generates the permutative population. A permutative solution is one, where each value within the solution is unique and systematic. A basic description is given in 22. Examples of permutative-type problems include TSP, flow shop scheduling, clustering, etc.

$$\begin{aligned} P_G &= \{x_{1,G}, x_{2,G}, ..., x_{NP,G}\},\ x_{i,G} = x_{j,i,G} \\ x_{j,i,G=0} &= (\text{int})\left(rand_j[0,1] \bullet \left(x_j^{(hi)} + 1 - x_j^{(lo)}\right) + \left(x_j^{(lo)}\right)\right) \\ &\quad if\ x_{j,i} \notin \{x_{0,i}, x_{1,i}, ..., x_{j-1,i}\} \\ i &= \{1,2,3,...,NP\}, j = \{1,2,3,..,D\} \end{aligned} \tag{22}$$

where P_G represents the population, $x_{j,i,G=0}$ represents each solution within the population and $x_j^{(lo)}$ and $x_j^{(hi)}$ represents the bounds. The index i references the solution from 1 to NP, and j which references the values in the solution.

3.5 Forward Transformation

The transformation schema represents the most integral part of the code. [14] developed an effective routine for the conversion.

Let a set of integer numbers be represented as $x_i \in x_{i,G}$ which belong to solution $x_{j,i,G=0}$. The equivalent continuous value for x_i is given as $1 \bullet 10^2 < 5 \bullet 10^2 \leq 10^2$.

The domain of the variable x_i has length = 5 as shown in $5 \bullet 10^2$. The precision of the value to be generated is set to two decimal places (2 d.p.) as given by the superscript two (2) in 10^2 . The range of the variable x_i is between 1 and 10^3. The lower bound is 1 whereas the upper bound of 10^3 was obtained after extensive experimentation. The upper bound 10^3 provides optimal filtering of values which are generated close together [20].

The formulation of the forward transformation is given as:

$$x'_i = -1 + \frac{x_i \bullet f \bullet 5}{10^3 - 1} \tag{23}$$

Equation 23 when broken down, shows the value x_i multiplied by the length 5 and a scaling factor f. This is then divided by the upper bound minus one (1). The value computed is then decrement by one (1). The value for the scaling factor f was established after extensive experimentation. It was found that when f was set to 100, there was a "tight" grouping of the value, with the retention of optimal filtration's of values. The subsequent formulation is given as:

$$x'_i = -1 + \frac{x_i \bullet f \bullet 5}{10^3 - 1} = -1 + \frac{x_i \bullet f \bullet 5}{10^3 - 1} \tag{24}$$

Illustration:

Take a integer value 15 for example. Applying Equation 23, we get:

$$x_i' = -1 + \frac{15 \bullet 500}{999} = 6.50751$$

This value is used in the DE internal representation of the population solution parameters so that mutation and crossover can take place.

3.6 Backward Transformation

The reverse operation to forward transformation, backward transformation converts the real value back into integer as given in Equation 25 assuming x_i to be the real value obtained from Equation 24.

$$\text{int}[x_i] = \frac{(1+x_i) \bullet (10^3 - 1)}{5 \bullet f} = \frac{(1+x_i) \bullet (10^3 - 1)}{500} \tag{25}$$

The value x_i is rounded to the nearest integer.

Illustration:

Take a continuous value -0.17. Applying equation Equation 25:

$$\text{int}[x_i] = \frac{(1+-0.17) \bullet (10^3 - 1)}{500} = |3.3367| = 3$$

The obtained value is 3, which is the rounded value after transformation.
These two procedures effectively allow DE to optimise permutative solutions.

3.7 Recursive Mutation

Once the solution is obtained after transformation, it is checked for feasibility. Feasibility refers to whether the solutions are within the bounds and unique in the solution.

$$x_{i,G+1} = \begin{cases} u_{i,G+1} \text{ if } \begin{cases} u_{j,i,G+1} \neq \{u_{1,i,G+1}, \ldots, u_{j-1,i,G+1}\} \\ x^{(lo)} \leq u_{j,i,G+1} \leq x^{(lo)} \end{cases} \\ x_{i,G} \end{cases} \tag{26}$$

Recursive mutation refers to the fact that if a solution is deemed in-feasible, it is discarded and the parent solution is retained in the population as given in Equation 26.

3.8 *Discrete Differential Evolution (DDE)*

The basic outline DDE is given in Fig. 2.

1. **Initial Phase**
   a. *Population Generation*: An initial number of discrete trial solutions are generated for the initial population.
2. **Conversion**
   a. *Discrete to Floating Conversion*: This conversion schema transforms the parent solution into the required continuous solution.
   b. *DE Strategy*: The DE strategy transforms the parent solution into the child solution using its inbuilt crossover and mutation schemas.
   c. *Floating to Discrete Conversion*: This conversion schema transforms the continuous child solution into a discrete solution.
3. **Selection**
   a. *Validation*: If the child solution is feasible, then it is evaluated and accepted in the next population, if it improves on the parent solution.

Fig. 2 DDE outline

The general schematic of the discrete DE (DDE) is given in Figure 3.

$$\text{Input}: D, G_{\max}, NP \geq 4, F \in (0, 1+), CR \in [0,1], \text{bounds}: x^{(lo)}, x^{(hi)}.$$

$$\text{Initialize}: \begin{cases} \forall i \leq NP \wedge \forall j \leq D \begin{cases} x_{i,j,G=0} = x_j^{(lo)} + rand_j[0,1] \bullet \left(x_j^{(hi)} - x_j^{(lo)}\right) \\ if x_{j,i} \notin \{x_{0,i}, x_{1,i}, \ldots, x_{j-1,i}\} \end{cases} \\ i = \{1,2,\ldots,NP\}, j = \{1,2,\ldots,D\}, G = 0, rand_j[0,1] \in [0,1] \end{cases}$$

$$\text{Cost}: \forall i \leq NP : f(x_{i,G=0})$$

$$\begin{cases} \text{While} \quad G < G_{\max} \\ \forall i \leq NP \begin{cases} \text{Mutate and recombine}: \\ \quad r_1, r_2, r_3 \in \{1,2,\ldots,NP\}, \text{randomly selected, except}: r_1 \neq r_2 \neq r_3 \neq i \\ \quad j_{rand} \in \{1,2,\ldots,D\}, \text{randomly selected once each } i \\ \quad \forall j \leq D, u_{j,i,G+1} = \begin{cases} (\gamma_{j,r_3,G}) \leftarrow (x_{j,r_3,G}) : (\gamma_{j,r_1,G}) \leftarrow (x_{j,r_1,G}) : (\gamma_{j,r_2,G}) \leftarrow (x_{j,r_2,G}) \\ \quad \text{Forward Transformation} \\ \gamma_{j,r_3,G} + F \cdot (\gamma_{j,r_1,G} - \gamma_{j,r_2,G}) \\ \quad \text{if } (rand_j[0,1] < CR \vee j = j_{rand}) \\ (\gamma_{j,i,G}) \leftarrow (x_{j,i,G}) \text{ otherwise} \end{cases} \\ \left(u'_{i,G+1}\right) = (\rho_{j,i,G+1}) \leftarrow (\varphi_{j,i,G+1}) \quad \text{Backward Transformation} \\ \text{Recursive Mutation}: \\ u_{i,G+1} = \begin{cases} u_{i,G+1} \text{if} \begin{cases} u_{j,i,G+1} \neq \{u_{1,i,G+1}, \ldots, u_{j-1,i,G+1}\} \\ x^{(lo)} \leq u_{j,i,G+1} \leq x^{(hi)} \end{cases} \\ x_{i,G} \quad \text{otherwise} \end{cases} \\ \text{Select}: \\ x_{i,G+1} = \begin{cases} u_{i,G+1} \text{ if } f(u_{i,G+1}) \leq f(x_{i,G}) \\ x_{i,G} \quad \text{otherwise} \end{cases} \end{cases} \\ G = G + 1 \end{cases}$$

Fig. 3 DDE schematic

3.9 Enhanced Differential Evolution (EDE)

The advanced form of the basic discrete differential evolution is the enhanced differential evolution (EDE) covered in [16], [18]. A number of strategies were included to speed up computation as well as enhance memory management; otherwise the architecture of EDE is basically the same as that of the basic DDE. The EDE shows much promise than DDE, hence, the next logical step was to devise a method which would repair the infeasible solutions and hence add viability to the heuristic. To this effect, three different repairment strategies were developed, each of which used a different *index* to repair the solution. After repairment, three different enhancement features were added. This was done to add more depth to the code in order to solve permutative problems. The enhancement routines were *standard mutation*, *insertion* and *local search*. The basic outline of the EDE is given in Fig. 4.

1. **Initial Phase**
   a. *Population Generation*: An initial number of discrete trial solutions are generated for the initial population.
2. **Conversion**
   a. *Discrete to Floating Conversion*: This conversion schema transforms the parent solution into the required continuous solution.
   b. *DE Strategy*: The DE strategy transforms the parent solution into the child solution using its inbuilt crossover and mutation schemas.
   c. *Floating to Discrete Conversion*: This conversion schema transforms the continuous child solution into a discrete solution.
3. **Mutation**
   a. *Relative Mutation Schema*: Formulates the child solution into the discrete solution of unique values.
4. **Improvement Strategy**
   a. *Mutation*: Standard mutation is applied to obtain a better solution.
   b. *Insertion*: Uses a two-point cascade to obtain a better solution.
5. **Local Search**
   a. *Local Search*: 2 Opt local search is used to explore the neighborhood of the solution.

Fig. 4 EDE outline

4 The Hybrid Differential Evolution And GMDH System

It is evident from the previous two sections that both modeling methods have many common features, but, unlike the GMDH, DE does not follow a pre-determined path for input data generation. The same input data elements can be included or excluded at any stage in the evolutionary process by virtue of the stochastic nature of the selection process. A DE algorithm can thus be seen as implicitly having the capacity to learn and adapt in the search space and thus allows previously bad elements to be included if they become beneficial in the latter stages of the search process. The standard GMDH algorithm is more deterministic and would thus discard any underperforming elements as soon as they are realized.

Using DE in the selection process of the GMDH algorithm, the model building process is free to explore a more complex universe of data permutations. This selection procedure has three main advantages over the standard selection method.

- Firstly, it allows unfit individuals from early layers to be incorporated at an advanced layer where they generate fitter solutions;
- Secondly, it also allows those unfit individuals to survive the selection process if their combinations with one or more of the other individuals produce new fit individuals, and;
- Thirdly, it allows more implicit non-linearity by allowing multi-layer variable interaction.

The new DE-GMDH algorithm recently proposed [10] [11] for both *prediction* and *modeling* is constructed in exactly the same manner as the standard GMDH algorithm except for the selection process. We give more design details in this paper and we do extensive experimentation by applying the *DE-GMDH* algorithm to modeling tool wear, predicting the exchange rates of three international currencies, and also predicting the Box-Jenkins gas furnace time series data. In order to select the individuals that are allowed to pass to the next layer, all the outputs of the GMDH algorithm at the current layer are entered as inputs in the DE algorithm where they are allowed to propagate, mutate, crossover and combine with other individuals in order to prove their fitness as shown in Figure 5.

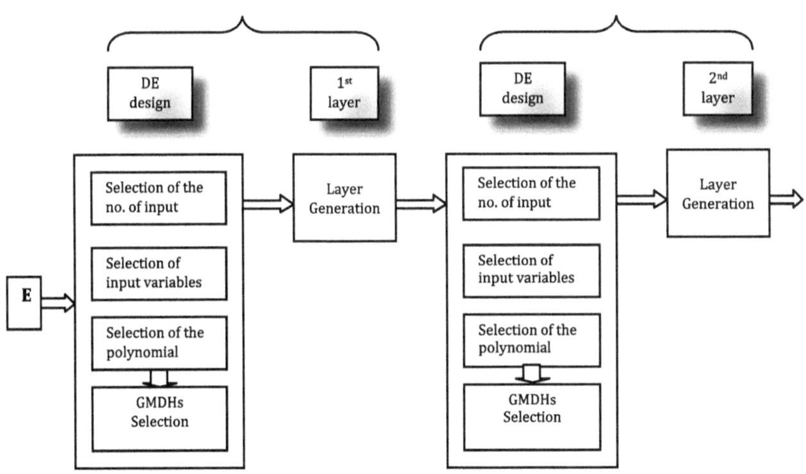

Fig. 5 Overall architecture of the DE-GMDH

The selected fit individuals are then entered in the GMDH algorithm as inputs at the next layer. The whole procedure is repeated until the criterion for terminating the GMDH run has been reached. The approach leads to some generalization because it becomes possible to predict not only the test data obtained during experimentation, but other test data outside the experimental results can also be used.

4.1 Structural Optimization: Representation of Encoding Strategy of Each Partial Descriptor (PD)

In this section, the structural optimization for the DE-GMDH for modeling is described in details. In the standard GMDH, the issues to address are: (i) how to determine the optimal *number of input variables*; (ii) how to select the *order of polynomial* forming a partial descriptor (PD) in each node; and (ii) how to determine which *input variables* are chosen. In this chapter these problems are resolved by some aspects of the DE-GMDH architecture described.

4.1.1 Structural Optimization: Representation of Encoding Strategy of Each Partial Descriptor (PD)

Determining the initial information for constructing the DE-GMDH structure is an important decision that should be initially made. Generally, there are two types of input-output relations. One of them is the multiple-input/single-output relationship. The other is the multiple-input/multiple-output relationship. The information describing the multiple-input scenario has to be carefully synthesised (see Figure 6). For example, when the problem size is small then all the inputs that describe a problem are used as inputs to the network. However, when the dimension of the inputs describe a problem is very high, pre-processing is essential.

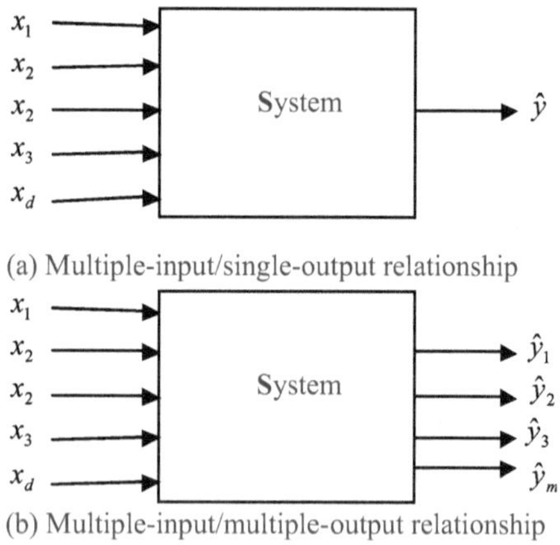

Fig. 6 Input-output relationships

Pre-processing for high dimensionality is an essential aspect of modelling because GMDH will over-learn and have oscillating, highly nonlinear solutions if the system being analyzed is complex. When the inputs are reduced so that $d \gg h$, then it would be easier for GMDH to be used for modelling the complex system (see Figure 7). For multiple outputs, it is necessary to consider multi-objective optimization. In this case the number of outputs is not reduced but some heuristics are used to solve the multi-objective optimization problem. The most popular method used is the Pareto-front approach. However, there are a number of other methods that are effective and efficient, yet simple to implement.

Fig. 7 Input-output relationships

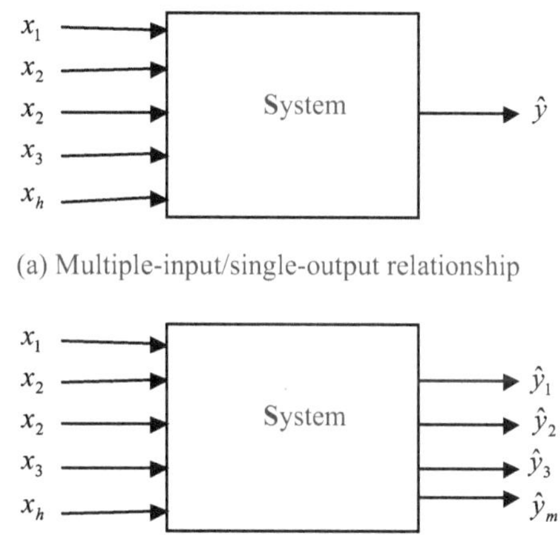

4.1.2 Form Training and Testing Data

There are different ways of forming the training and testing dataset. In some cases, the entire dataset is split in a ratio of 50:50 consecutively for training and testing respectively. In other cases, instead of consecutive splitting, 50% of the original dataset are random generated for training while the rest are used for testing purpose. The percentage could be different from the 50:50-ratio and it all depends on the application of interest. In some other cases, specific samples are designated for training while others are designated for testing purpose.

4.1.3 Determine Initial Information for Constructing the DE-GMDH Structure

The initial pieces of information for the DE-GMDH structural design are those in which the design of optimal parameters available within the polynomial neuron (viz. the number of input variables, the order of the polynomials, and input variables) at last leads to a structurally and parametrically optimized network, which is more flexible as well as simpler in architecture than the conventional GMDH. The polynomials differ according to the number of input variables and the polynomial order. Several types of polynomials are shown in Table 1.

Table 1 Different types of the polynomial in PDs

No. Of Inputs	1	2	3
Order of the polynomial			
1 (Type 1)	Linear	Bilinear	Trilinear
2 (Type 2)	Quadratic	Biquadratic	Triquadratic
3 (Type 3)	Modified quadratic	Modified biquadratic	Modified triquadratic

For an example, the specific forms of a PD in the case of two inputs are given as

- Bilinear = $c_0 + c_1x_1 + c_2x_2$
- Biquadratic = $c_0 + c_1x_1 + c_2x_2 + c_3x_1^2 + +c_4x_2^2 + c_5x_1x_2$
- Modified biquadratic = $c_0 + c_1x_1 + c_2x_2 + c_5x_1x_2$
- Trilinear = $c_0 + c_1x_1 + c_2x_2 + c_3x_3$
- Triquadratic = $c_0 + c_1x_1 + c_2x_2 + c_3x_3 + c_4x_1^2 + c_5x_2^2 + c_6x_3^2 + c_7x_1x_2 + c_8x_1x_3 + c_9x_2x_3$
- Modified triquadratic = $c_0 + c_1x_1 + c_2x_2 + c_3x_3 + c_4x_1x_2 + c_5x_1x_3 + c_6x_2x_3$

where c_i is the regression coefficients.

4.1.4 Determine Polynomial Neuron (PN) Structure Using DE Design

In the DE approach similar to other evolutionary approaches, the initial population of solutions is randomized, which means that minimal heuristic knowledge is used. The appropriate number of input variable, order of polynomial, and which input variable are chosen are realized in a self-organizing manner and are tuned gradually throughout the DE iterations. Therefore, in the hybrid DE-GMDH design procedure described in this chapter, key issues are how to encode the number of *input variables*, *order of polynomial*, optimum *input variables* as a vector of solution (called individual of the population), and how to define a criterion to compute the fitness of each individual of the population. The following sub-sections describe the detailed representation of encoding strategy and the choice of fitness function. In designing an evolutionary-based GMDH, the most important consideration is the representation strategy. Therefore, in the DE-GMDH design, the most important consideration is how to encode the key factors into the vector of solution (called individual of the population). While in GA-PNN implementation, a binary coding has been used [21], [22], [23], [24] we employ a combinatorial DE coding approach [14][16][17] and a sequence of integer numbers is used in each vector of solution. In our DE-GMDH design, there are three system parameters (P_1, P_2, and P_3) which are now described. The first system parameter $P_1 \in [1,3]$ is randomly generated and represents the order of polynomial. The second system parameter $P_2 \in [1,r]$ is randomly generated and represents the number of input variables (where $r = \min(D, 5)$; D is the width of the input dataset; the default upper bound is $r = 2$). The designer must determine the maximum number (r) in consideration of the characteristic of system, design specification, and some prior knowledge of model. With this method the problem of conflict between over-fitting and generalization on one hand, and computation time on the other hand can be solved [24]. A consequence of complexity is the over-fitting problem and poor generalization. To avoid, this optimal number of inputs available to the model, the number of input variables and type or order in each PD need to be determined. In most cases, these parameters are determined by trial and error method leading to a heavy computational burden and low efficiency. Therefore, any modeling architecture that can alleviate these problems (as in the proposed DE-GMDH) would be preferred. The third parameter $P_3 \in [1,D]$ is the sequence of integers for each solution vector of the population of solutions, and it represents the entire candidates in the current layer of the network. The relationship between *vector of solution* and information on PD is shown in Figure 8, while the PD corresponding to the vector of solution is shown in Figure 9.

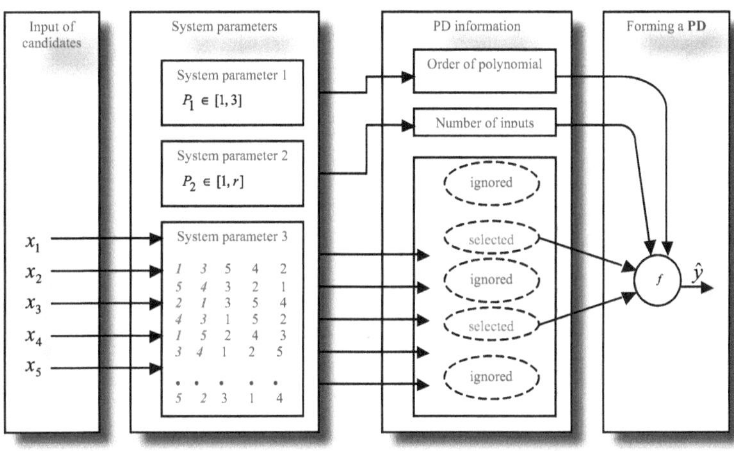

Fig. 8 Relationship between *vector of solution* and information on PD

Figure 8 shows the information for a PD for a case where the width (number of columns) of the dataset is 5. Therefore, a population of initial vectors of solutions is randomly generated for initialization. For $P_1 = 2$ and $P_2 = 2$, these correspond to the polynomial order of Type 2 (quadratic form), and there are two input variables to the node. This means that only the first two columns of the population of solutions will be considered, corresponding to the following pair-wise combinations of the input dataset being selected as input variables: 1 3, 5 4, 2 1, 4 3, 1 5, 3 4,..,5 2. For the fifth pair of selected input variables 1 5, the output of this PD is

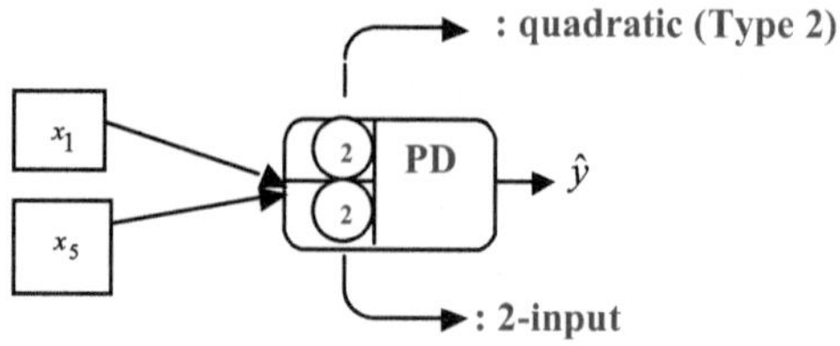

Fig. 9 Node with PD corresponding to *vector of solution*

$$\hat{y} = f(x_1, x_5) = c_1 + c_2x_1 + c_3x_5 + c_4x_1^2 + c_5x_5^2 + c_6x_1x_5 \tag{27}$$

where the coefficients $(c_1, c_2, c_3, c_4, c_5, c_6)$ are evaluated using the training dataset by means of the *least square error* (LSE), the *pseudo-inverse* method or where singular values occur, using the *singular value decomposition* (SVD) method. The LSE method is the simplest and where non-singular values occur, pseudo-inverse method is appropriate. The SVD non-trivial method is the optimum method to apply. The polynomial function, PD, is automatically formed according to the encoded information of the system parameters (P_1, P_2, and P_3).

4.1.5 Select Nodes (PNs) with the Best Predictive Capability, and Construct Their Corresponding Layer

In the DE-GMDH architecture, the selection of nodes (PNs) with the best predictive capability is decided by the DE optimization framework and subsequently the network construction with the corresponding layers are realized based on the search results. For each layer the best node is found based on the objective function (which is simply the external criterion used for solving the problem at hand). The nodes in the preceding layer connected to the best node in the current layer are marked for realizing the network as search progresses from layer to layer as shown in Figure 10.

4.2 Parametric Optimization: Coefficient Estimation of the Polynomial Corresponding to the Selected Node (PN)

The system of equations that need to be solved at each polynomial neuron earlier on described by equation 16 needs to be solved. To recap, these can be written as:

$$\sum_{n=1}^{N} aX = \sum_{n=1}^{N} b \tag{28}$$

The coefficients $(a_1, a_2, a_3, \ldots, a_N)$ are evaluated using the training dataset by means of the *least square error* (LSE) method, the *pseudo-inverse* (PI) method, or the *singular value decomposition* (SVD) method. There are a number of other known methods for parametric optimization but these three mentioned here are popular. The LSE method is simple but does not have the capability to deal with complex scenario where singularity occurs. On the other hand the SVD method has the

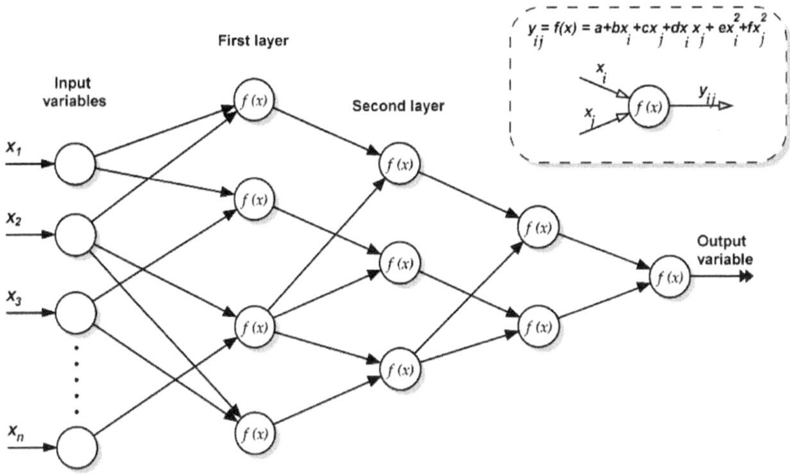

Fig. 10 GMDH-type network structure

capability to deal with complex scenario where singularity occurs but it is computationally expensive. The PI method is a compromise between these two extreme cases and it finds reasonably good results for real-life applications.

4.2.1 Regression Analysis Technique for Parametric Optimization

Consider the following problem of linear regression using least squares. A linear model in the form of $Y = X.a$ is expected, where vector a is the vector of regression coefficients, X is a matrix of input data with rows corresponding to observations and columns to predictor variables and Y is a vector of time series observations. A commonly used approach to get the unknown vector a utilizes pseudo inverse matrix:

$$a = \left(X^T X\right)^{-1} X^T Y \tag{29}$$

which works well in most of the cases and it determines the vector of the best coefficients of the quadratic equation 29 for the whole set of $\boldsymbol{M}$ data triples.. Problems arise, when the values of columns in X get close to each other. As a result the condition number of matrix X gets huge as well as the values of $\boldsymbol{a}$ produced by regression. The condition number is a measure of stability or sensitivity of a matrix - or the linear system it represents - to numerical operations (stated in other words, it is not wise to trust the results of computations on an ill-conditioned matrix).

This is exactly what happens when selecting only the best nodes on each layer according to some fitness function based on error criterion only. As the nodes with output closest to expected value are preferred the output of higher layers (starting often as low as on layer 3) gets almost unique which leads to ill conditioned matrices, that are constructed for regression from these outputs on higher layers. It should be noted that this procedure is repeated for each neuron of the next hidden layer according to the connectivity topology of the network. However, such solution directly from solving normal equations using regression analysis (RA) technique is rather susceptible to round off error and, more importantly, to the singularity of these equations.

4.2.2 Ill-Formed Problems

As already mentioned, utilizing regression analysis technique for solution of the coefficients is rather susceptible to round off error and, more importantly, to the singularity of these equations. Singularity occurs when a system of linear equations are ill-formed. Therefore, alternative methods need to be explored in order to obtain more stable solutions. We will discuss a very efficient technique for solving ill-formed (singular) problems.

4.2.3 Singular Value Decomposition for Parametric Optimization

Singular value Decomposition (SVD) is the method for solving most linear least squares problems that some singularities may exist in the normal equations. The SVD of a matrix, $X \in \Re^{M\times 6}$ is a factorisation of the matrix into the product of three matrices, column-orthogonal matrix $U \in \Re^{M\times M}$, diagonal matrix $W \in \Re^{M\times 6}$ with non-negative elements (singular values), and orthogonal matrix $V \in \Re^{6\times 6}$ such that

$$X = UWV^T \tag{30}$$

The most popular technique for computing the SVD was originally proposed in [25]. The problem of optimal selection of vector of the coefficients in equations 2 14 is firstly reduced to finding the modified inversion of diagonal matrix W [26] in which the reciprocals of zero or near zero singulars (according to a threshold) are set to zero. Then, such optimal a are calculated using the following relation

$$a = V\ [diag\,(1/W_j)]^T\, U^T Y \tag{31}$$

Such procedure of SVD approach of finding the optimal coefficients of quadratic polynomials,$\boldsymbol{a}$, improves the performance of self-organizing GMDH type algorithms that is employed to build networks based on input-output observation data triples.

However, such parametric identification problem is part of the general problem of modelling when structure identification is considered together with the parametric identification problem simultaneously. In this work, a new encoding scheme is presented in an evolutionary approach for simultaneous determination of structure and parametric identification of CI-GMDH networks for the modelling of complex problems.

4.2.4 The Moore-Penrose Pseudo Inverse

The Moore-Penrose pseudo-inverse is a general way to find the solution to the following system of linear equations:

$$\mathbf{b} = A\mathbf{y} \quad \mathbf{b} \in \Re^{m;} \quad \mathbf{y} \in \Re^{n;} \quad A \in \Re^{m \times n;}$$

Moore and Penrose showed that there is a general solution to these equations (which we will term the Moore-Penrose solution) of the form $\tilde{\mathrm{y}} = \mathrm{A}^{\dagger}\tilde{\mathrm{b}}$ The matrix $\mathrm{A}^{\dagger}$ is the Moore-Penrose "pseudo-inverse," and they proved that this matrix is the unique matrix that satisfies the following properties:

1. $AA^{\dagger}A = A$
2. $A^{\dagger}AA^{\dagger} = A^{\dagger}$
3. $\left(AA^{\dagger}\right)^{T} = AA^{\dagger}$
4. $\left(A^{\dagger}A\right)^{T} = A^{\dagger}A$

The Moore-Penrose pseudo-inverse and solution has the following properties. When:

- $m = n, A^{\dagger} = A^{-1}$ if A is full rank. The pseudo-inverse for the case where A is not full rank will be considered below.
- $m > n$ (which corresponds to a kinematically insufficient manipulator), the solution is the one that minimizes the quantity.

$$\|\mathbf{b} - A\mathbf{y}\|$$

  That is, in this case there are more constraining equations than there are free variables $\mathbf{y}$. Hence, it is not generally possible find a solution to these equations. The pseudo-inverse gives the solution $\mathbf{y}$ such that $A^{\dagger}\mathbf{y}$ is "closest" (in a least-squared sense) to the desired solution vector $\mathbf{b}$.
- $m < n$ (which corresponds to a kinematically redundant manipulator), then the Moore-Penrose solution minimizes the 2-norm of $\mathbf{y} : \|\mathbf{y}\|$. In this case, there are generally an infinite number of solutions, and the Moore-Penrose solution is the particular solution whose vector 2-norm is minimal

For application to redundant robot manipulators, we are concerned with the case where m ¡ n. To understand the Moore-Penrose solution in more detail, first recall that the Null Space of a matrix A, denoted N(A), is defined as follows:

$$N(A) = \{\mathbf{v}\} \mid A\mathbf{v} = 0$$

If r is the rank of matrix A, then the null space is a linear vector space with dimension

$$\dim(N(A)) = \max\{0, (r - n)\}$$

The Row Space of A, denoted Row (A), is the linear span of its rows. Clearly, every element in N (A) is orthogonal to any element in Row (A), and hence we say that N(A) and Row(A) are orthogonal to each other. Thus, any vector $\mathbf{y} \in R^n$ can be uniquely split into its row and null space components: $\mathbf{y} = \mathbf{y}_{row} + \mathbf{y}_{null}$. Note that:

$$\mathbf{b} = A\mathbf{y} = A\left(\mathbf{y}_{row} + \mathbf{y}_{null}\right) = A\mathbf{y}_{row}$$

From the claim above that the Moore-Penrose solution is the minimum norm solution, it must be true that the Moore-Penrose solution is the particular solution that has no null space component.

When A is full rank, the Moore-Penrose pseudo-inverse can be directly calculated as follows:

- case $m < n : A^{\dagger} = A^T(AA^T)^{-1}$
- case $m > n : A^{\dagger} = (A^TA)^{-1}A^T$

However, when A is not full rank, then these formulas cannot be used. More generally, the pseudo-inverse is best computed using the Singular Value Decomposition reviewed below.

4.2.5 The Singular Value Decomposition

Let $A \in R^{m\times n}$. Then there exist orthogonal matrices $U \in R^{m\times m}$ and $V \in R^{n\times n}$ such that the matrix A can be decomposed as follows:

$$A = U\Sigma V^T \tag{32}$$

where Σ is an m n diagonal matrix having the form:

$$\Sigma = \begin{bmatrix} \sigma_1 & 0 & 0 & \ldots & 0 & 0 \\ 0 & \sigma_2 & 0 & \ldots & 0 & 0 \\ 0 & 0 & \sigma_3 & \ldots & 0 & 0 \\ . & . & . & \ldots & . & . \\ 0 & 0 & 0 & \ldots & \sigma_p & 0 \end{bmatrix} \tag{33}$$

And

$$\sigma_1 \geq \sigma_2 \geq \ldots \sigma_p \geq 0 \qquad p = \min\{m, n\}$$

The σ_i are termed the singular values of the matrix A. The columns of U are termed the left singular vectors, while the columns of V are termed the right singular vectors. The decomposition described in 32 is called the “Singular Value Decomposition,” which is conveniently abbreviated as SVD.

Geometrically, the singular values of A are the lengths of the semi-axes of the hyperellipsoid E defined by

$$E = \{\mathbf{z}\ |\mathbf{z} = A\mathbf{x};\ \ \|\mathbf{x}\| = 1\} \tag{34}$$

Using the SVD, the pseudo-inverse of a matrix can be easily computed as follows. Let A be decomposed as in Equation 32. Then

$$A^{\dagger} = V \sum^{\dagger} UT \tag{35}$$

where the matrix $\sum^{\dagger}$ takes the form:

$$\Sigma = \begin{bmatrix} \frac{1}{\sigma_1} & 0 & 0 & \dots & 0 & 0 \\ 0 & \frac{1}{\sigma_2} & 0 & \dots & 0 & 0 \\ 0 & 0 & \frac{1}{\sigma_3} & \dots & 0 & 0 \\ . & . & . & \dots & . & . \\ 0 & 0 & 0 & \dots & \frac{1}{\sigma_p} & 0 \end{bmatrix} \tag{36}$$

for all of the non-zero singular values. If any of the σ_i is zero, then a zero is placed in corresponding entry of $\sum^{\dagger}$. If the matrix A, is rank deficient, then one or more of its singular values will be zero. Hence, the SVD provides a means to compute the pseudo-inverse of a singular matrix.

The computation of the SVD is a non-trivial issue. It suffices to know that all respectable software packages for doing mathematics (such as maple, MATLAB, or mathematical contain functions for computing the SVD. For our purposes, the existence of these procedures and the minimal facts outlined above should suffice.

4.3 Framework of the Design Procedure of the DE-GMDH Hybrid System

Step 1 *Determine system's input variables.*
Define the input variables of the system as x_i $(i = 1, 2, ..., n)$ related to output variable y. If required, normalization of input data can be completed as well.

Step 2 *Form training and testing data.*
The input-output data set $(x_i, y_i) = (x_{1i}, x_{2i}, ..., x_{ni}, y_i)$, $i = 1, 2, ..., n$ (n: the total number of data) is divided into two parts, that is, a training and a testing dataset. Their sizes are denoted by n_{tr} and n_{te} respectively. Obviously we have $n = n_{tr} + n_{te}$. The training data set is used to construct the *DE-GMDH* model. Next, the testing data set is used to evaluate the quality of the model.

Step 3 *Determine initial information for constructing the DE-GMDH structure.*
We determine initial information for the *DE-GMDH* structure in the following manner:

1. The termination method exploited here the maximum number of generations predetermined by the designer to achieve a balance between model accuracy and its complexity.
2. The maximum number of input variables used at each node in the corresponding layer.
3. The value of the weighting factor of the aggregate objective function.

Step 4 *Determine polynomial neuron (PN) structure using DE design.*
Determining the polynomial neuron (PN), is concerned with the selection of the number of input variables, the polynomial order, and the input variables to be assigned in each node of the corresponding layer. The PN structure is determined using DE design. The DE design available in a PN structure by using a solution vector of DE is the one illustrated in Fig. 5 in which the design of optimal parameters available within the PN (viz. the number of input variables, the order of the polynomials, and input variables) at last leads to a structurally and parametrically optimized network, which is more flexible as well as simpler in architecture than the conventional *DE-GMDH*. Each sub-step of the DE design procedure of three kinds of parameters available within the PN has already been discussed. The polynomials differ according to the number of input variables and the polynomial order.

Step 5 *Coefficient estimation of the polynomial corresponding to the selected node (PN).*
The vector of the coefficients of the PDs is determined using a standard mean squared error by minimizing the following index:

$$E_r = \frac{1}{n_{tr}} \sum_{i=1}^{n_{tr}} (y_i - z_{ki})^2, \qquad k = 1, 2, ..., r \tag{37}$$

where z_{ki} denotes the output of the k-th node with respect to the i-th data, r is the value in the second system parameter $P_2 \in [1, r]$ and n_{tr} is the number of training data subsets. This step is completed repeatedly for all the nodes in the current layer. Evidently, the coefficients of the PN of nodes in each layer are determined by the standard least square method. This procedure is implemented repeatedly for all nodes of the layer and also for all *DE-GMDH* layers starting from the input layer and moving to the output layer.

Step 6 *Select nodes (PNs) with the best predictive capability, and construct their corresponding layer.*
As shown in Fig. 2, all nodes of the corresponding layer of *DE-GMDH* architecture are constructed by DE optimization. The generation process of PNs in the corresponding layer is described in detail as the design procedure of 4 sub-steps. A sequence of the sub-steps is as follows:

Sub-step 1 We determine initial DE information for generation of the DE-GMDH architecture. That is, the number of generations and populations, mutation rate, crossover rate, and the length of a solution vector.

Sub-step 2 The nodes (PNs) are generated by DE design as many as the number of populations in the 1st generation. Where, one population takes the same role as one node (PN) in the *DE-GMDH* architecture and each population is operated by DE as shown in Fig. 2. That is, the number of input variables, the order of the polynomials, and the input variables as one individual (population) are selected by DE. The polynomial parameters are produced by the standard least squares method.

Sub-step 3 To evaluate the performance of PNs (nodes) in each population, we use an aggregate objective function that takes into account a sound balance between approximation and prediction capabilities of the one as shown in 12. And then, from the performance index obtained in 12, we calculate the fitness function of 13 as already discussed.

Sub-step 4 To produce the next generation, we carry out mutation, crossover, and selection operations using DE initial information and the fitness values obtained from **sub-step 3**. Generally, after these DE operations, the overall fitness of the population improves. We choose several PNs characterized by the best fitness values. Here, we select the node that has the highest fitness value for optimal operation of the next iteration in the *DE-GMDH* algorithm. The outputs of the retained nodes (PNs) serve as inputs in the subsequent layer of the network. The iterative process generates the optimal nodes of a layer in the *DE-GMDH* model.

Step 7 *Termination criterion.*
The termination method exploited here the maximum number of generations predetermined by the designer to achieve a balance between model accuracy and its complexity.

The *DE-GMDH* algorithm is carried out by repeating steps 4-6 consecutively. After the iteration process, the final generation of population consists of highly fit solution vectors that provide optimum solutions. After the termination condition is satisfied, one solution vector (PD) with the best performance in the final generation of population is selected as the output PD. All remaining other solution vectors are discarded and all the nodes that do not have influence on this output PD in the previous layers are also removed. By doing this, the *DE-GMDH* model is realized.

4.4 The Hybrid DE-GMDH Algorithm

The overall hybrid DE-GMDH flow or algorithm is described as follows. The inputs are the problem dimension (h), maximum number of generation or iteration ($G_{\max}$), population size (Np), mutation parameter (F), crossover parameter (CR) and lower bound and higher bound for permutative values $\mathbf{x}^{(lo)}, \mathbf{x}^{(hi)}$ respectively. The in-process parameters are initial population (P), forward transformed population (P_f), etc. The algorithm has been represented in Table 2 and the routines of Table 2 have been described in the following section.

Table 2 Hybrid DE-GMDH Algorithm

Input: h, $G_{\max}, Np \geq 4$, $F \in (0,+1), CR \in [0,1], \mathbf{x}^{(lo)}, \mathbf{x}^{(hi)}$
Output: x, the best solution ever found
$G_{\max} = 0$
$P(G) = initialize\left(\mathbf{x}^{(lo)}, \mathbf{x}^{(hi)}, Np, h\right)$
$F(G) = objective\,(P(G), Np, h)$
$best\_so\_far = \min(F(G))$
$while\ (G < G_{\max})$do
  $\{$*forward transformation*$\}$
  $P_f(G) \leftarrow forwardTransform\,(P(G), Np, h)$
  $\{$*DE strategies*$\}$
  $P_s(G) \leftarrow strategy\,(P_f(G), Np, h, strategy)$
  $\{$*backward transformation*$\}$
  $P_b(G) \leftarrow backwardTransform\,(P_s(G), Np, h)$
  $\{$*repair strategy: (i) front; (ii) back; (iii) random*$\}$
  $P''(G) \leftarrow relative\_mutate\,(P_b(G))$
  $\{$*improvement strategy: (i) mutation; (ii) insertion*$\}$
  $P_m(G) \leftarrow improvement\,(P''(G), Np, h, mutation\_type)$
  $\{$*selection*$\}$
  $G = G + 1$
end
  $\{$*local search*$\}$
  $P(G) \leftarrow local\_search\,(P(G), Np, h)$
$return(x)$;

$\{$***initialize population***$\}$
$initialize\left(\mathbf{x}^{(lo)}, \mathbf{x}^{(hi)}, Np, h\right)$
$P(G) = (\text{int})\left(rand_j[0,1] \cdot \left(\mathbf{x}^{(hi)} + 1 - \mathbf{x}^{(lo)}\right) + \mathbf{x}^{(lo)}\right),$
$\forall i \leq Np \wedge \forall j \leq h\ | x_{j,i} \notin \left\{x^0_{0,i}, \ldots, x^0_{j-1,i}\right\}$
$return\,(P(G))$

$\{$***objective function of the population of size Np and dimension h***$\}$
$objective\,(P(G), Np)$
$array \leftarrow P_{i1}, P_{i2}\ ; \forall i \leq Np$
$GMDH\_routine(array); \forall i \leq Np$
$return\,(error)$

$\{$***forward transformation***$\}$
$forwardTransformation\,(P, Np, h)$
α *is small positive constant*
$P_{i,j} = -1 + P_{i,j} * \alpha; \forall i \leq Np \wedge \forall j \leq h$
$return\,(P)$

{***backward transformation***}
$backwardTransformation\,(P,Np,h)$
α *is small positive constant*
$P_{i,j} = (1 + P_{i,j})\,/\,\alpha;\ \forall i \leq Np \wedge \forall j \leq h$
$return\ (P)$

{***DE strategy***}
$strategy\,(P,Np,h)$
for i $\leftarrow$ 1 to Np do
$r_1, r_2, r_3 \in \{1,2,...,Np\}$ randomly selected, except: $r_1 \neq r_2 \neq r_3 \neq i$
$j_{rand} \in \{1,2,...,h\}$
for j $\leftarrow$ 1 to h do

$$P_{j,i}^{G+1} = P_{j,r_3}^{G} + F * \left(P_{j,r_1}^{G} - P_{j,r_2}^{G}\right)$$

end
$array \leftarrow P_{i1}, P_{i2}\ ;\ \forall i \leq Np$
$GMDH\_routine(array);\ \forall i \leq Np$
end
$return\ (P)$

{***Repair strategies***}
$repair\,(P,Np,h,strategy)$
$case\ strategy\ (1)$
$front\_mutate\,(P,Np)$
$case\ strategy\ (2)$
$back\_mutate\,(P,Np)$
$case\ strategy\ (3)$
$random\_mutate\,(P,Np)$
$array \leftarrow P_{i1}, P_{i2}\ ;\ \forall i \leq Np$
$e_i = GMDH\_routine(array);\ \forall i \leq Np$
$P_i^* = \arg\min\,(e_i)\,;$
$return\ (P_i^*)$

{***Improvement strategies***}
$improvement\,(P,Np,h,mutation\_type)$
$case\ mutation\_type\ (1)$
$mutate\,(P,Np)$
$case\ mutation\_type\ (2)$
$insert\,(P,Np)$
$array \leftarrow P_{i1}, P_{i2}\ ;\ \forall i \leq Np$
$e_i = GMDH\_routine(array);\ \forall i \leq Np$
$P_i^* = \arg\min\,(e_i)\,;$
$return\ (P_i^*)$

{***Selection***}
$selection\,(P_m,P,Np,h)$

$best\_select = objective\,(P_m, Np, h)$
$if\ (best\_select < best\_so\_far)$
$\quad P(G+1) \leftarrow P_m\,(G);$
$\quad best\_so\_far \leftarrow best\_select;$
$else$
$\quad P(G+1) \leftarrow P\,(G);$
end
$return\,(P(G+1))$

{***local search***}
local_search (P, Np)
{*apply 2-opt local search approach*}
$array \leftarrow P_{i1}, P_{i2}\ ;\ \forall i \le Np$
$e_i = GMDH\_routine(array);\ \forall i \le Np$
$P_i^* = \arg\min\,(e_i)\,;$
$return\,(P_i^*)$

{***access GMDH routine***}
GMDH_routine $(array)$
{*generate data for regression*}
$(Zx, Zy) = regression(array, Y);$
{*where Y is the set of labels; Zx is the 6x6 transformation matrix; and Zy is the (6x1 matrix) corresponding labels of Zx.*}
{*pseudo-inverse for coefficients*}
$a = pseudo\_inverse\,(Zx, Zy)\,;$
$\hat{y} = a * Zx;$
$error \leftarrow mse\,(\hat{y}, y)\,;$
$return\,(error)$

5 DE-GMDH Mechanics Illustrated

In this section, this worked example outlines how the enhanced DE described in Section 3.9 integrates with GMDH to solve the modelling and prediction problem.

Initialization Phase

Step 1: Population Generation: An initial number of discrete trial solutions are generated for the initial population.

For the case of illustration, the operating parameter of *NP* and $G_{\max}$ are kept at a minimum. The other parameters $\mathbf{x}^{(lo)}$, $\mathbf{x}^{(hi)}$ and *D* are problem dependent. Initially the operating parameters are outlined: NP=10; D= 4; $G_{\max}$=1. This Step (1) initializes the population to the required number of solutions. Since NP is initialised to 10, only 10 permutative solutions are generated. Table 3 shows an initial population of solutions that are normally randomly generated. Traditional DE does not commence from this step; it normally general floating point solutions and not discrete, permutative solutions as we do in the forward/backward transformation-based approach DE scheme. This difference is significant in the two approaches because the classical or traditional DE cannot be used to solve the modelling/prediction problem that we set to solve using an integration of DE and GMDH. In the first layer or generation, we would like to answer the question: which parameters are inputs to each polynomial neuron? In subsequent layers or generations, we would also like to answer the question: which subsequent nodes are inputs to the current polynomial neurons? Answers to these questions are only available using the forward/backward transformation-based approach DE scheme.

In the example considered, there are four inputs and one output. Therefore, initialization would involve generating numbers between 1 and the number of inputs 1, 4 to obtain population of solutions as shown in Table 4.2. These labels are then used to determine the initial parameters connected to each polynomial neuron as discussed in Section 4.1.4 (determine polynomial neuron (PN) structure using DE design). Although each row of population of solutions has four entries, only the first N entries are considered in the DE structural or network design. In our case, $N = 2$ and so only the first two columns would be needed. Although in this illustration, the population size has been taken to be 10, in practice, a population size of 50 is recommended from experimentation.

Table 3 Initial population of solutions

Solution	1	2	3	4
1	4	2	3	1
2	3	1	2	4
3	2	4	1	3
4	1	2	4	3
5	4	1	2	3
6	4	1	3	2
7	4	1	2	3
8	4	2	3	1
9	1	2	4	3
10	4	3	1	2

Structural Optimization

Structural optimization takes place within the DE module. There is division of labour here. DE is responsible for structural optimization, while GMDH responsible for parametric optimization. In the DE-GMDH hybrid, DE and GMDH do not run in parallel; they are integrated and responsibilities are shared between them.

The first row of Table 3 shows that the elements 4, 2 are connected to a neuron. Consequently, other elements connected to other neurons are 3, 1, 1, 2, 4, 1, and 4, 3 respectively. The elements that not connected to a neuron are 2, 3. But as already mentioned, if the population size is large, all elements would be easily represented even for a very large number of elements or parameters.

This set of information is therefore used to connect parameters that define the problem to be solved to neurons in the first layer or generation. In the classical GMDH notation, "layers" are referred to, but in the hybrid DE-GMDH notation, "generations" are referred to. The inputs 4, 2, 3, 1, 1, 2, 4, 1, and 4, 3 are therefore used for connections as shown in Figure 11. This constitutes the first generation connections of the GMDH network being grown from generation to generation.

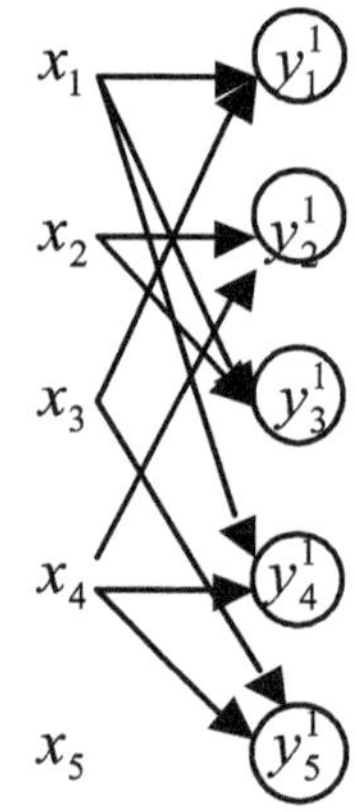

Fig. 11 First generation connections of the GMDH network being

Parametric Optimization

The next procedure is to calculate the objective function of each solution in the population. Parametric optimization includes determination of the fitness of each neuron, invoking GMDH external criterion, and tracking the connections to a neuron. The fitness of each solution is the basis for finding the best polynomial neuron. This responsibility is attributed to GMDH. A bit of explanation is necessary for understanding Figure 11 for parametric optimization. Elements 1, 3 are connected to neuron y_1^1; although these elements are represented as x_1 and x_3 they really refer to columns 1 and 3 respectively of input data. So these could be referred to as vectors x_1 and x_3 for the entire length of data. These are shown in Table 4.3 (highlighted) which is part of a real-life problem solved.

As has already alluded to, DE cannot work with these pieces of information yet because they are not in the form that DE can accept. So, some transformations are needed to transform the data in DE-data structure. What is DE-data structure? It simply means that the data should be in floating points, which when operated upon by DE mechanisms such as crossover and mutation, we should be able to transform back to permutative combinatorial for further processing by the GMDH component.

Table 4 Vectors x_1 and x_3 (highlighted)

x_1	x_2	x_3	x_4	y
19	0.5	65	320	0.213733
19	0.5	65	410	0.194167
19	0.5	127	600	0.2026
19	0.5	127	865	0.3444
19	1	264	320	0.549233
19	1	264	410	0.3256
19	1	500	600	0.182533
19	1	500	865	0.552633
38	1.5	65	320	0.5355
38	1.5	65	410	0.330833
38	1.5	127	600	1.5268
38	1.5	127	865	2.058133

Without these operators (crossover and mutation), DE cannot find competitive solutions in the global solution space. This is where the power of hybridization shows up. We need a stochastic approach to move the GMDH search out of local optima during the solution search. This is where DE comes in. DE is a stochastic optimization approach which has the mechanisms for pulling out a search out from local optima and moves towards the global optima within a reasonable computation time while realizing very good optimal solution.

The objective function is the error between the measured output and estimated output; this error is the external criterion. The measured output is the last column of Table 4. The estimated output is the one obtained using equation 16 or equation 28 in which the coefficients $(a_1, a_2, a_3, ..., a_N)$ are evaluated using the training dataset by means of the *least square error* (LSE) method, the *pseudo-inverse* (PI) method, or the *singular value decomposition* (SVD) method.

Conversion Phase

Step 2: Discrete to Floating Conversion: This conversion scheme transforms the parent solution into the required continuous solution. Table 5 gives the table with values in real numbers. Each value has been formulated with equation 24 and the results are presented in 3 decimal point format.

Step 3: DE Strategy: The DE strategy transforms the parent solution into the child solution using its inbuilt crossover and mutation schemas. In Step (3), DE strategies of Section 3.2 are applied to the real population in order to find better solutions.

Step 4: Floating to Discrete Conversion: This conversion schema transforms the continuous child solution into a discrete solution. This step is referred to as backward transformation which is applied to each solution. The results are given in Table 6.

Table 5 Solutions in real number format

Solution	1	2	3	4
1	1.002	0.001	0.501	-0.499
2	0.501	-0.499	0.001	1.002
3	0.001	1.002	-0.499	0.501
4	-0.499	0.001	1.002	0.501
5	1.002	-0.499	0.001	0.501
6	1.002	-0.499	0.501	0.001
7	1.002	-0.499	0.001	0.501
8	1.002	0.001	0.501	-0.499
9	-0.499	0.001	1.002	0.501
10	1.002	0.501	-0.499	0.001

Table 6 Backward transformation solutions

Solution	1	2	3	4
1	3	**5**	2	1
2	3	**5**	4	2
3	2	**5**	1	3
4	1	4	3	**5**
5	3	**5**	4	1
6	1	3	**5**	4
7	2	**5**	1	4
8	3	**5**	2	4
9	2	3	**5**	1
10	**5**	1	2	3

Mutation Phase

Step 5: Relative Mutation Schema Formulates the child solution into the discrete solution of unique values. Recursive mutation is applied in Step (5). For this illustration, the *random mutation schema* is used as this was the most potent and also the most complicated. The first routine is to drag all "bound offending values" to the offending boundary. The boundary constraints are given as $\mathbf{x}^{(lo)} = 1$ and $\mathbf{x}^{(hi)} = 4$ which is lower and upper bound of the problem. Table 7 gives the "bounded" solution.

In *random mutation*, initially all the duplicated values are isolated. The next step is to find out the missing values in each solution. Table 8 gives the missing values per solution. In the first, third, ninth, and tenth solutions, there is no missing value. In the second and eight solutions, there is no value of 1. In the fourth to sixth solutions, there is no value of 2. In the seventh and ninth solutions, there is no value of 3.

What is now needed, are the *positional indexes* which are randomly generated. A positional index indicates where the missing value will be inserted in the solution. First, we need to identify replications. For example, in Solution 2, the value

Table 7 Bounded solutions

	1	2	3	4
1	3	4	2	1
2	3	4	4	2
3	2	4	1	3
4	1	4	3	4
5	3	4	4	1
6	1	3	4	4
7	2	4	1	4
8	3	4	2	4
9	2	3	4	1
10	4	1	2	3

Table 8 Missing values

Solution	1
1	-
2	1
3	-
4	2
5	2
6	2
7	3
8	1
9	-
10	-

4 is replicated in 2 positions; these are columns 2 and 3. So a random number is generated between 2 and 3 to select the default value of where to retain the value 4. Let us assume that index 3 is selected. In this respect, value 4 is retained in position 3. This routine is applied to the entire population, solution piecewise in order to set the default values. A possible representation can be given as in Table 9.

There are two replications each for solutions 2, 4, 5, 6, 7, and 8; therefore there will be two labels from which a random number is to be drawn per solution (see Table 10). The positional index which is a random number drawn from the set of replicates in the second column is in the last column.

The positional index of Table 10 is therefore used in conjunction with the missing values of Table 8 to 'repair' the solutions of Table 7. Table 11 shows that the missing values will be placed in the *first* replicated index value of solutions 2, 4, 7, and 8, while the missing values will be placed in the second replicated index value of solutions 5 and 6. The solutions are now permutative. The fitness for each solution is then calculated.

Table 9 Replicated values

Solution	1	2	3	4
1				
2		4	4	
3				
4		4		4
5		4	4	
6			4	4
7		4		4
8		4		4
9				
10				

Table 10 Positional index

Solution	Set of replicates	Positional index
1	-	-
2	{1, 2}	1
3	-	-
4	{1, 2}	1
5	{1, 2}	2
6	{1, 2}	2
7	{1, 2}	1
8	{1, 2}	1
9	-	-
10	-	-

Table 11 Final placement of missing values

	1	2	3	4
1	3	4	2	1
2	3	1	4	2
3	2	4	1	3
4	1	2	3	4
5	3	4	2	1
6	1	3	4	2
7	2	3	1	4
8	3	1	2	4
9	2	3	4	1
10	4	1	2	3

Improvement Strategy Phase

Step 6: Mutation Standard mutation is applied to obtain a better solution. Step (6) describes the *Standard Mutation* schema. In standard mutation, a single value swap occurs. Assume that a list of random indexes are generated which show which values are to be swapped. It can be seen from Table 12, that the values indexed by 1 and 3 are to be swapped in Solution 1 and so forth for all the other solutions. The new "possible" solutions are given in Table 13; their fitness values are calculated. The highlighted values are the mutated values.

Step 7: Insertion Uses a two-point cascade to obtain a better solution. Step (7), Insertion also requires the generation of random indexes for cascading of the solutions. A new set of random numbers is shown in Table 14.

In Table 14 the values are presented in ascending order. Taking solution 1, the first process is to remove the value indexed by the first lower index (2) as shown.

The second process is to move all the values from the upper index (4) to the lower index.

Table 12 Random index

	1	2
1	1	3
2	2	1
3	1	4
4	2	3
5	3	1
6	2	4
7	2	1
8	4	1
9	1	2
10	3	4

Table 13 New "mutated" population

	1	2	3	4
1	**2**	4	**3**	1
2	**1**	**3**	4	2
3	**3**	4	1	**2**
4	1	**3**	2	4
5	**2**	4	3	1
6	1	**2**	4	**3**
7	**3**	**2**	1	4
8	**4**	1	2	**3**
9	**3**	**2**	4	1
10	4	1	**3**	**2**

Table 14 Random index

Solution	Index
1	{1, 3}
2	{2, 3}
3	{1, 2}
4	{3, 4}
5	{1, 2}
6	{2, 4}
7	{3, 4}
8	{1, 3}
9	{2, 4}
10	{2, 3}

Index	1	2	3	4
Solution 1		4	3	1

Index	1	2	3	4
Solution 1	4	3		1

Index	1	2	3	4
Solution 1	4	3	2	1

The last part is to insert the first removed value from the lower index into the place of the now vacant upper index.

Likewise, all the solutions are "cascaded" in the population and their new fitness calculated. Insertion leads to better solutions being found. These solutions replace the older solution in the population. The final population is shown in Table 15.

DE mechanics postulate that each "current" solution replaces it direct "preceding" solution in the population if it has better fitness. Comparing the final population with the initial population in Table 3, better solutions with fitness than the solutions in the old population are produced. Thus these "current" solutions replace the "preceding" solutions in the population for the next generation. Since we specified the $G_{max}=1$, only 1 iteration of the routine will take place.

Using the above outlined DE process, it is possible to formulate the basis for most permutative problems. Before termination, the two following steps are accessed.

Step 8: Repeat Execute steps 2-7 until reaching a specified cutoff limit on the total number of iterations.

Local Search Phase

Step 9: Local Search: Local search tries to find a better solution in the neighborhood.

Table 15 Final population after "insertion"

	1	2	3	4
1	4	3	2	1
2	1	4	3	2
3	4	3	1	2
4	1	3	4	2
5	4	2	3	1
6	1	4	3	2
7	3	2	4	1
8	1	2	4	3
9	3	4	1	2
10	4	3	1	2

6 Applications of the DE-GMDH Hybrid System

6.1 DE-GMDH for Modeling the Tool-Wear Problem

The end-milling experiment was carried out on the Seiki Hitachi milling machine and reported in [10][11]. A 16mm Co-high speed (HSS) Kobelco Hi Cut brand new end mill cutter was used to machine the work-piece. The end mill cutter had four flutes and the flute geometry was 30 degrees spiral. The overall length of the cutter was 77mm and the flute length was 26.5 mm. The work-piece machined was mid steel blocks which had a constant length of 100 mm for each trial. The machining was done under dry conditions. The milling experiment was conducted as designed. The work-piece used was mild steel which had a Brinell hardness number of 128.

Tool wear for turning was monitored using the Mitutoyo toolmakers microscope. The 55 degrees carbide insert with a positive rake angle of 7 degrees was removed from the tool holder and measured in the toolmakers microscope. A reference had to be made such that the distance from the reference line to the tip of the insert could be taken. It was very difficult to make a permanent reference line on the insert thus an insert holder was prepared. The insert holder was designed in such a way that the insert fitted inside the hole perfectly. The height of the surface of the insert was equal to height of the insert holder. Two reference lines were made on the insert holder, one at right angles to the tip of the insert which took into account the wear taking place on the nose of the insert and the other reference line was made parallel to the side of the insert which took into account the flank wear of the insert. A brand new end mill tool was measured on the toolmakers microscope from the reference line to the cutting edge. The tool was then used to machine a block of mild steel on the conditions of trial number 1. After machining the end mill was removed from the milling machine and the amount of wear on the end mill cutter was measured. The difference of the average of the first reading and the average of the current reading gave the extent of tool wear. Four readings were taken from each reference line and average of the readings was done.

The machining parameters were set during experimentation. This data set constituted the input to the self-organizing network and consisted of three inputs and one output. All inputs and were considered candidates to the causality relationship. For the specific application it was found that five replications were sufficient to yield a good approximation. The range of speed, feed, and depth of cut chosen for the experiments are respectively $v \in \{27, 39, 49\}$, $f \in \{0.0188, 0.0590, 0.1684\}$, $d_t \in \{0.5,\ 1.0,\ 1.5\}$.

The *DE-GMDH* self-organizing network was used to mine the causal relationships between the key input and output variables of the end-milling machining. The key process input variables were spindle speed, feed, and depth-of-cut, while the key output variable is tool-wear. With little knowledge of the cause-effect relationships at the outset, it is essential to determine firstly *which* sensor variables effect the key process output variables and secondly, establish a plausible quantitative relationship between the two, thereby establishing the desired causality. For ease of reference and clarity to readers, we highlight the main design steps discussed in the previous section.

Step 1 *Configuration of input variables*: The system input variables shown in Table 2 are the speed, feed of the machine, and depth-of-cut chosen for the experiments which are respectively $v \in \{27, 39, 49\}$, $f \in \{0.0188, 0.0590, 0.1684\}$, and $d_t \in \{0.5,\ 1.0,\ 1.5\}$.

Step 2 *From training and testing data*: Half of the dataset was used in designing the training data while the remaining half was used in designing the testing data.

Step 3 *Decision of initial information for constructing the DE-GMDH structure*: The number of generations, crossover and mutation rates are 100, 0.30 and 0.10 respectively.

Step 4 *Determine polynomial neuron (PN) structure using DE design*: The number of input variables of two, the polynomial order of Type 2, and the input variables were assigned to each node of the corresponding layer.

Step 5 *Coefficient estimation of the polynomial corresponding to the selected node (PN)*: The vector of the coefficients of the PDs is determined using a standard mean squared error using the number of training data subsets.

Step 6 *Select nodes (PNs) with the best predictive capability, and construct their corresponding layer*: All nodes of the corresponding layer of DE-GMDH architecture are constructed by DE optimization as already explained.

Step 7 *Termination criterion*: After the iteration process, the final generation of population consists of highly fit solution vectors that provide optimum solutions.

All the twenty seven trials were conducted using the same end mill cutter and each time after milling the measurements for wear was taken. The number of trials was kept at twenty seven due to the time it takes to carry out tool wear experiments. The average of the present measurement was subtracted from the previous one and the difference in the measurements gave the amount of wear. The results obtained for the twenty seven trials using mild steel blocks as work piece is shown in Table 16.

Table 16 Results of End Milling Experiment

Trial #	Speed (m/min)	Feed (mm/rev)	Depth-of-cut (mm)	Tool Wear (μm)
1	27	0.0188	0.5	8
2	27	0.0188	1	2.3
3	27	0.0188	1.5	2.6
4	27	0.059	0.5	1.7
5	27	0.059	1	2.45
6	27	0.059	1.5	2.7
7	27	0.1684	0.5	1.95
8	27	0.1684	1	2.55
9	27	0.1684	1.5	2.85
10	36	0.0188	0.5	2.9
11	36	0.0188	1	3.35
12	36	0.0188	1.5	4.35
13	36	0.059	0.5	3.105
14	36	0.059	1	3.55
15	36	0.059	1.5	4.5
16	36	0.1684	0.5	3.196
17	36	0.1684	1	3.95
18	36	0.1684	1.5	4.65
19	49	0.0188	0.5	4.95
20	49	0.0188	1	5.85
21	49	0.0188	1.5	7.7
22	49	0.059	0.5	5.2
23	49	0.059	1	6.25
24	49	0.059	1.5	10.2
25	49	0.1684	0.5	5.45
26	49	0.1684	1	6.75
27	49	0.1684	1.5	19.51

6.1.1 The End-Mill Tool Wear Model

The tool wear can be modeled as

$$VB = c_1 + c_2x_1 + c_3x_2 + c_4x_1x_2 + c_5x_1^2 + c_6x_2^2 \tag{38}$$

where x_1 = feed, x_2 = depth of cut, and x_3 = spindle speed. The methodology presented in Section 4.5.3 was applied to the output of the DE-GMDH used for the work reported in this article to determine model coefficients 4.7999, 0.0262678, -12.1041, 0.400388, -0.00216996, -0.00289143 for the quadratic equation given in equation (15) for the milling operation, leading to a predictive model given as

$$VB = 4.7999 + 0.0262678x_1 - 12.1041x_2 + 0.400388x_1^2 - 0.00216996x_2^2 - 0.00289143x_1x_2 \tag{39}$$

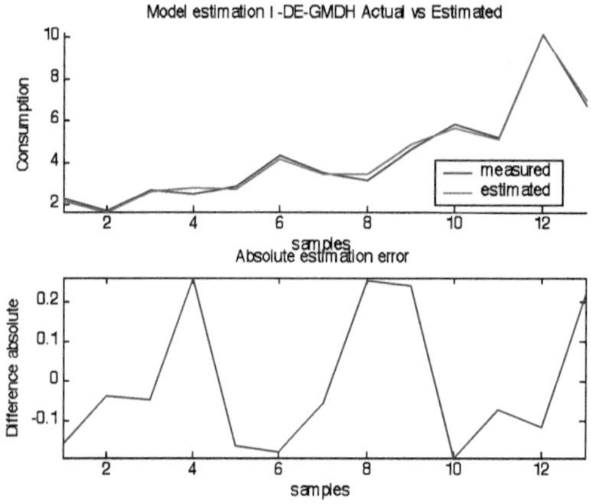

Fig. 12 The DE-GMDH actual & estimated and absolute difference (testing) for the tool wear problem

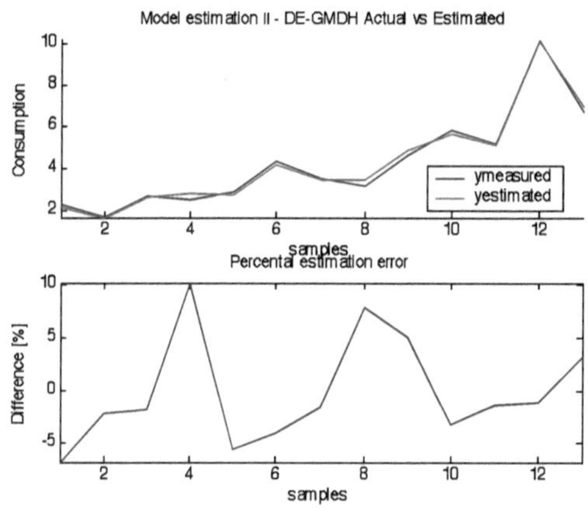

Fig. 13 The DE-GMDH actual & estimated and percentage error (testing) for the tool wear problem

The evaluation criterion is the mean square error based on equation 37. The training error, $PI = 0.0000833$ and the testing error, and $EPI = 0.0000743$. Realizing the model for our tool wear in milling operation is a useful and practical tool in industrial applications. Our model in equation 39 holds for any value of the inputs and becomes a vital tool to predict tool-wear for any input conditions. We can predict the wear level of the milling tool once we have an idea of the spindle speed and feed as well as the material depth-of-cut during operation.

The modeling approach that we have presented in this paper has the following advantages over models realized from the standard GMDH:

- model is based on quadratic regression polynomial in each layer (including the output layer);

- model realized is in quadratic form irrespective of how complex a problem is, which is easy for the user to understand (as in equation 38);

Figures 12 shows the actual & estimated, and absolute difference figures for the tool wear problem. Figure 13 shows the actual & estimated and percentage difference figures for the tool wear problem.

6.1.2 Comparative Study of the DE-GMDH Model for the Tool Wear Problem

In order to validate the efficacy of the proposed DE-GMDH modeling approach, the results obtained for the tool wear problem was benchmarked with the results obtained using polynomial neural network (PNN), and an enhanced-GMDH as shown in Table 17. The results show that the proposed hybrid DE-GMDH performs better than PNN and the e-GMDH.

Table 17 Performance index of identified model

Model	Polynomial type	*PI*	*EPI*
PNN	Type II: Quadratic	4.345	3.694
e-GMDH	Type II: Quadratic	0.033419	0.154649

6.2 Exchange Rates Forecasting Using the DE-GMDH Paradigms

In our experimentation, we used three different datasets (Euros, Great Britain Pound and Japanese Yen) in our forecast performance analysis. The data used are daily Forex exchange rates obtained from the Pacific Exchange Rate Service [27]. The data comprises of the US dollar exchange rate against Euros, Great Britain Pound (GBP) and Japanese Yen (JPY). The length of the data is 1 January 2000 to 31 December 2002 (partial data sets excluding holidays). Half of the data set was used as training data set, and half as evaluation test set or out-of-sample datasets, which are used to evaluate the good or bad performance of the predictions, based on evaluation measurements.

The forecasting evaluation criteria used is the normalized mean squared error (Variation Accuracy Criterion or Ivachnenko's δ^2)

$$\delta^2 = \frac{\sum_{i=1}^{n} (y_i - \hat{y}_i)^2}{\sum_{i=1}^{n} (y_i - \bar{y}_i)^2} = \frac{1}{\sigma^2}\frac{1}{N}\sum_{i=1}^{n}(y_i - \hat{y}_i)^2 \tag{40}$$

where y_i and $\hat{y}_i$ are the actual and predicted values, σ is the estimated variance of the data and $\bar{y}_i$ the mean. The ability to forecast movement direction or turning points can be measured by a statistic developed by Yao and Tan [28]. Directional change statistics (D_{stat}) can be expressed as

$$D_{stat} = \frac{1}{N}\sum_{t=1}^{n} a_t \times 100\% \,, \tag{41}$$

where $a_t = 1$ if $a_t = (y_{t+1} - y_t)\,(\hat{y}_{t+1} - \hat{y}) \geq 0$, and $a_t = 0$ otherwise.

For simulation, the five-day-ahead data sets are prepared for constructing DE-GMDH models. A *DE-GMDH* model was constructed using the training data and then the model was used on the test data set. The actual daily exchange rates and the estimated ones for three major internationally traded currencies are shown in Figures 14, 15, 16, 17, 18 and 19.

6.2.1 Analysis for EURO

The minimum training error is 0.0171448, while the minimum testing error is 0.0159927. The coefficient of determination (r-squared value), $r^2 = 0.995994$.

Figure 14 shows the DE-GMDH prediction and absolute difference error for the EURO exchange rate problem. The absolute difference error, is found be within the range of ± 0.01. Here, there is an excellent match between the measured and predicted values, showing that the proposed DE-GMDH model can be used as a feasible solution for exchange rate forecasting. From Figure 14 the absolute difference error is found be within the range of ± 1. In Figures 14 and 15, the predictor is closely following the actual values.

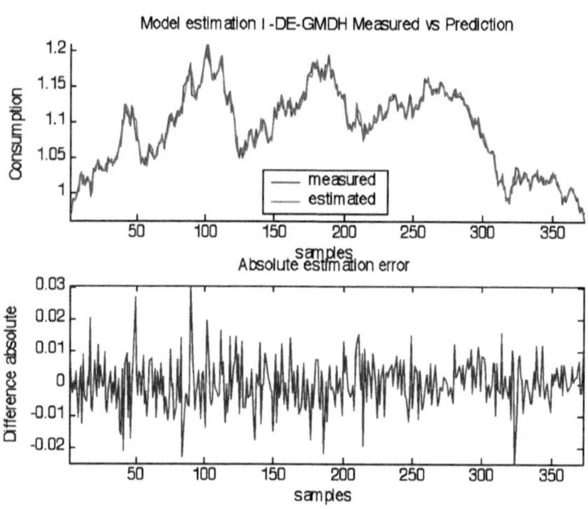

Fig. 14 The DE-GMDH actual & estimated and absolute difference error for the EURO exchange rate problem

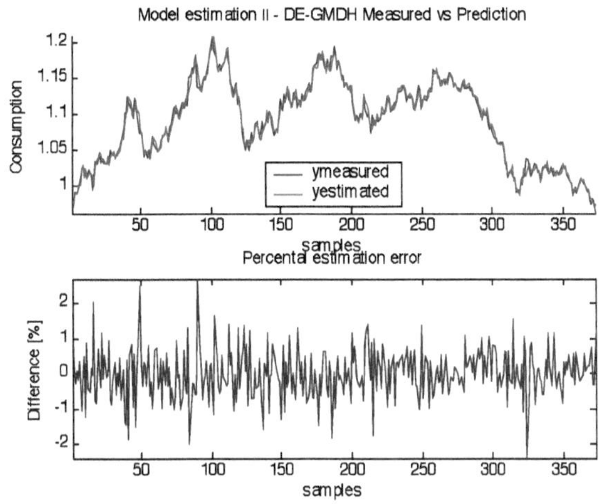

Fig. 15 The DE-GMDH actual & estimated and percentile error for the EURO exchange rate problem

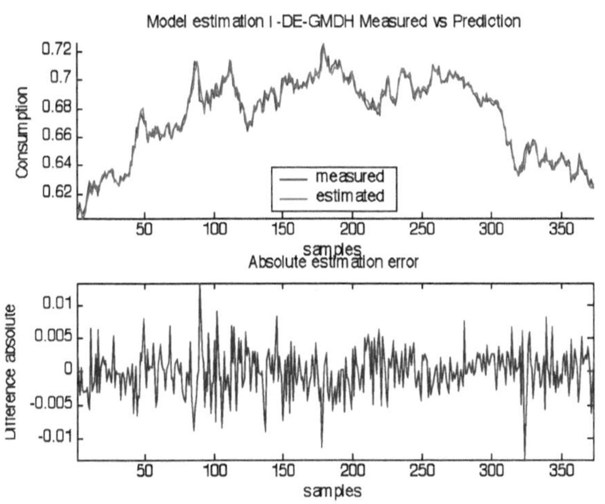

Fig. 16 The DE-GMDH actual & estimated and absolute difference error for the GBP exchange rate problem

6.2.2 Analysis for GBP

The minimum training error is 0.0146048, while the minimum testing error is 0.0139706. The coefficient of determination (r-squared value), $r^2 = 0.99729$.

Figure 16 shows the DE-GMDH prediction and absolute difference error for the GBP exchange rate problem. The absolute difference error is found be within the range of ± 0.01. Here, there is an excellent match between the measured and predicted values (see Figure 15), showing that the proposed DE-GMDH model can be

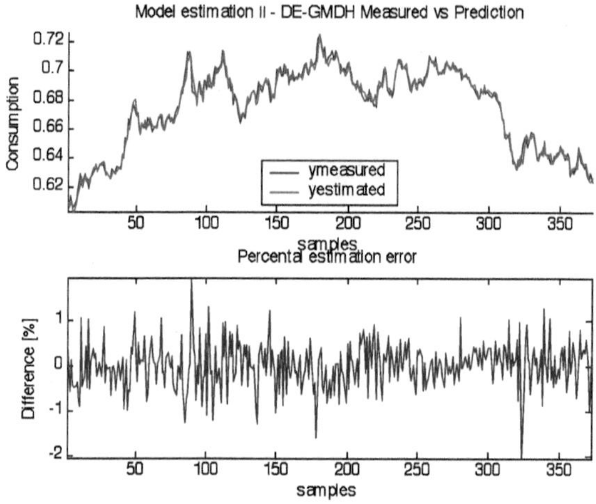

Fig. 17 The DE-GMDH actual & estimated and percentile error for the GBP exchange rate problem

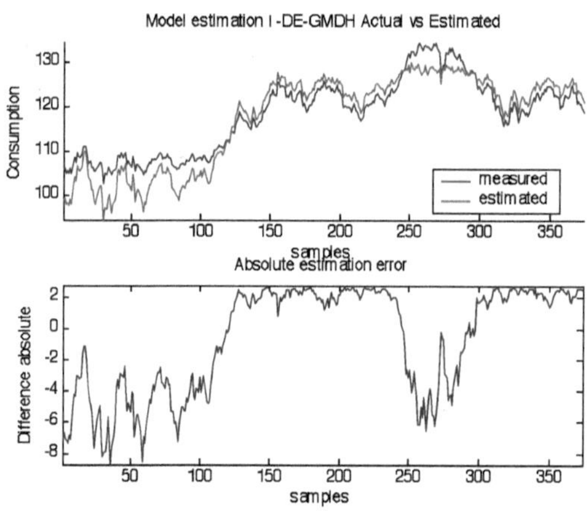

Fig. 18 The DE-GMDH actual & estimated and absolute difference error for the YEN exchange rate problem

used as a feasible solution for exchange rate forecasting. From Figure 16 the absolute percentage error is found be within the range of ± 1. In Figures 15 and 16, the predictor is closely following the actual values.

6.2.3 Analysis for YEN

The minimum training error is 0.186101, while the minimum testing error is 0.183277. The coefficient of determination (r-squared value), $r^2 = 0.954328$.

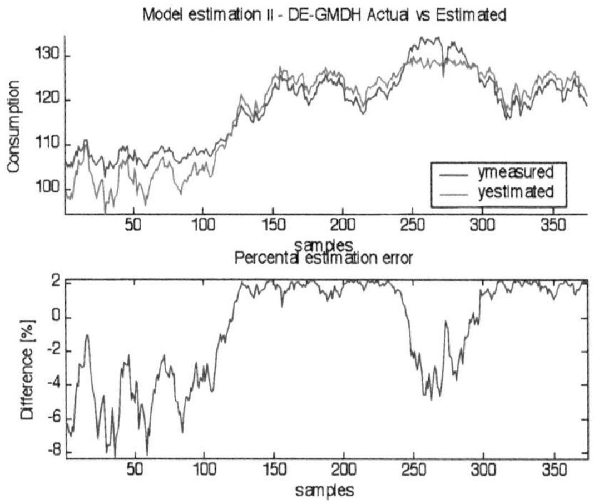

Fig. 19 The DE-GMDH actual & estimated and percentile error for the YEN exchange rate problem

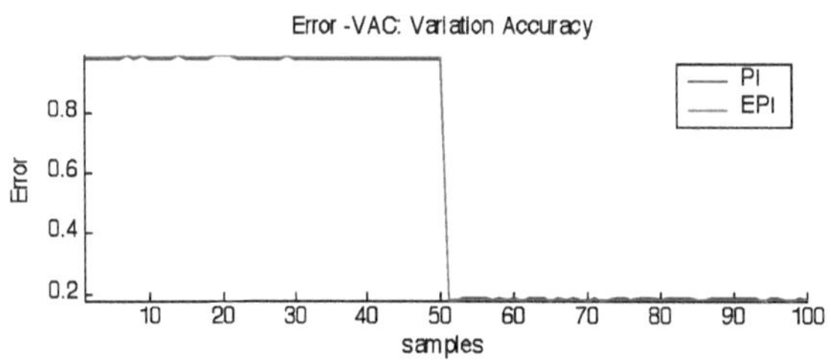

Fig. 20 Training error (PI) and testing error (EPI) against number of generations for the DE-GMDH for the YEN exchange rate problem

Figure 18 shows the DE-GMDH prediction and absolute difference error for the YEN exchange rate problem. The absolute difference error is found be within the range of ± 0.01. Here, there is not a very good match between the measured and predicted values, although it shows that the proposed DE-GMDH model can be used as a feasible solution for exchange rate forecasting. From Figure 19 the absolute difference error is found be within the range of ± 1. The training and testing errors for the DE-GMDH is shown in Figure 20.

6.2.4 Comparative Study of the DE-GMDH Model for the Exchange Rate Problem

For comparison purpose, the forecast performances of a traditional multilayer feed-forward network (MLFN) model and an adaptive smoothing neural network (ASNN) [29] model are also shown in Table 18. From Table 18, using NMSE performance index, it is observed that the proposed DE-GMDH forecasting models are significantly

Table 18 Forecast performance evaluation for the three exchange rates (NMSE for testing)

Exchange rate	Euros	British pounds	Japanese yen
MLFN [29]	0.5534	0.2137	0.2737
ASSN [29]	0.1254	0.0896	0.1328
FNT [44]	0.018	0.0142	0.0084
e-GMDH [45]	**0.0156**	0.0147	**0.0077**
DE-GMDH (This chapter)	0.0159	**0.0139**	0.0096

better than other neural networks models and competes with FNT and enhanced GMDH (e-GMDH) for the three major internationally traded currencies studied except for Japanese Yen. The e-GMDH has extra features than the standard GMDH.

6.3 Gas Furnace Experimentation Using the DE-GMDH Learning Network

In this section we illustrate the performance of the DE-GMDH network by experimenting with data of the gas furnace process which have been intensively studied as a benchmark problem in the previous literature [30] - [43].

For the design of experiment, the delayed term of the observed gas furnace process data, $y(t)$ is used as system input variables made up of six terms given as follows: $u\,(t-3)\,, u\,(t-2)\,, u\,(t-1)\,, y\,(t-3)\,, y\,(t-2)$ and $y\,(t-1)$. The processed data $y\,(t)$ (which is the output) resulted in 293 rows and 6 columns input variables of nodes in the first generation of the DE-GMDH structure.

The criterion used was the MSE, Mean Square Error. The gas furnace process data is compared to the value estimated by the self-organizing DE-GMDH network as shown in Figures 21 and 22. The time-series predictions in both Figures are seen to be very close to the gathered original data. Figure 21 shows that the error based on the difference between the measured and estimated (predicted) values cluster within the range of which is very reasonable. Figure 22 shows that the error based on the difference expressed in percentage between the measured and estimated (predicted) values again cluster close to , which again is very reasonable. The predicted values are well within acceptable measurement errors. The correlation coefficient of determination (r-squared value) is $r^2 = 0.99327$.

6.3.1 Gas Furnace Experimentation Using the DE-GMDH Learning Network

Table 19 shows the contrasts between the performances of the proposed DE-GMDH-type network with other models studied in the literature. The experimental results clearly demonstrates that the proposed DE-GMDH-type network outperforms the existing models both in terms of better approximations capabilities (lower values of

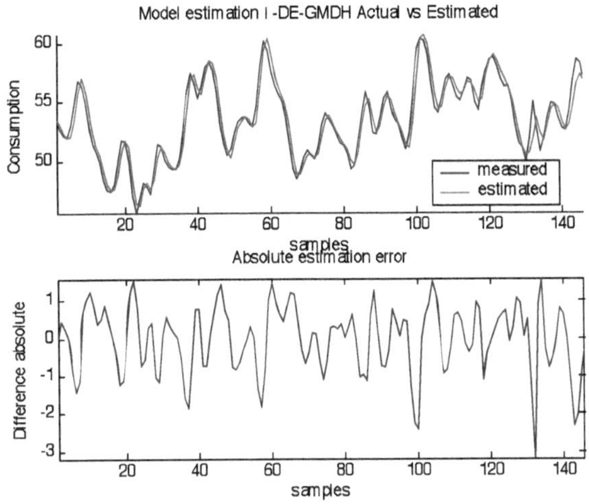

Fig. 21 The DE-GMDH actual & estimated and absolute difference error for the gas furnace problem

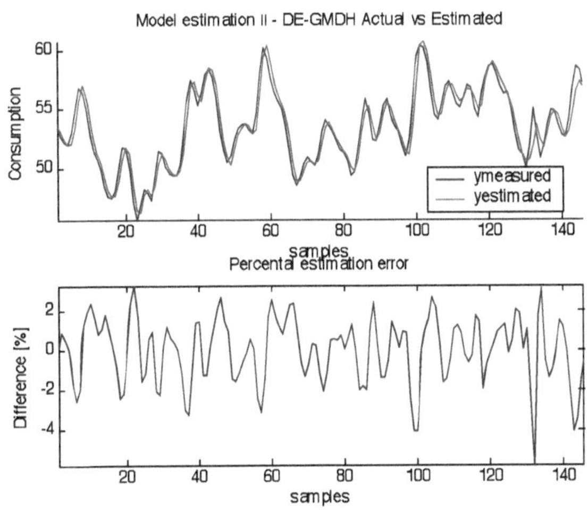

Fig. 22 The DE-GMDH actual & estimated and percentage error for the gas furnace problem

the performance index on the training data, PIs) as well as generalization abilities (lower values of the performance index on the testing data, EPIs).

6.4 CPU Time Cost of the DE-GMDH Algorithm

Although DE is a global optimization algorithm that is computationally intensive in terms of CPU time cost, an enhanced DE version was implemented [16][18]

Table 19 Comparison of identification error with other models

Model		Mean squared error		
		PI	PIs	EPIs
Box and Jenkin's model [30]		0.71		
Tong's model [31]		0.469		
Sugeno and Yasukawa's model [32]		0.355		
Sugeno and Yasukawa's model [33]		0.19		
Xu and Zailu's model [34]		0.328		
Pedrycz's model [35]		0.32		
Chen's model [39]		0.268		
Gomez-Skarmeta's model [41]		0.157		
Oh and Pedrycz's model [36]		0.123	0.02	0.271
Kim et al's model [37]		0.055		
Kim et al's model [38]		0.034	0.244	
Leksi and Czogala's model [40]		0.047		
Lin and Cunningham's model [42]			0.071	0.261
Oh and Pedrycz's model [43]	Type I Basic Case 1	0.057	0.017	0.148
	PNN Case 2	0.057	0.017	0.147
	Modified Case 2	0.046	0.015	0.103
	PNN Case 2	0.045	0.016	0.111
	Type II Basic Case 1	0.029	0.012	0.085
	PNN Case 2	0.027	0.021	0.085
	Modified Case 2	0.035	0.017	0.095
	PNN Case 2	0.039	0.017	0.101
DE-GMDH-type network [10][11]			0.00058	0.00053

PI-performance index over entire training data; PIs-performance index on the training data; EPIs-performance index on the testing data.

which is appreciably fast and competes with most other optimization techniques. The average CPU time cost for the tool wear problem is approximately 2 seconds while for the exchange rate problems is approximately 9 seconds for each of the three international currencies; similar CPU time cost is incurred for the Box-Jenkins gas furnace process problem. The CPU time cost for the DE used for the proposed compares favourably well with other optimization techniques implemented on the same platform for the same problem domain.

7 Conclusions

In this chapter, newly proposed design methodology of the hybrid *GMDH* and *DE* (which we refer to as *DE-GMDH*) is described. The architecture of model is not predefined, but can be self-organized automatically during the design process. In our approach, we first present a methodology for modeling, and then develop predictive model(s) of the problem being solved in form of second-order equations based on the input data and coefficients realized. The studies of the experiment carried out helped with the comparison of the *DE-GMDH* network and the standard GMDH network and PNN for this class of modeling problem and it was found that the *DE-GMDH* network appears to perform better than the standard GMDH algorithm and the PNN model. We have applied the *DE-GMDH* approach to the problem of developing predictive model for tool-wear in turning operations. Using the turning input parameters (speed, feed, and tool diameter) and the response (tool wear), a predictive model based on *DE-GMDH* approach is realized which gives a reasonably good solution. For the tool wear problem results presented show that the proposed *DE-GMDH* algorithm appears to perform better than the standard *GMDH* algorithm and its variants as well as the polynomial neural network (PNN) model. For the exchange rate problem, the results of the proposed DE-GMDH algorithm are competitive with all other approaches except in one case. For the Box-Jenkins gas furnace data, the experimental results clearly demonstrates that the proposed *DE-GMDH*-type network outperforms the existing models both in terms of better approximations capabilities (lower values of the performance index on the training data, PIs) as well as generalization abilities (lower values of the performance index on the testing data, EPIs).

The selection procedure of proposed inductive modeling approach has three main advantages over the standard selection method.

- Firstly, it allows unfit individuals from early layers to be incorporated at an advanced layer where they generate fitter solutions;
- Secondly, it also allows those unfit individuals to survive the selection process if their combinations with one or more of the other individuals produce new fit individuals, and;
- Thirdly, it allows more implicit non-linearity by allowing multi-layer variable interaction.

Although there are other population-based optimization techniques (Ant Colony Optimization [ACO], Scatter Search [SS], etc.) that are not yet investigated for hybridization with GMDH, these are fresh grounds for active research.

References

1. Ivakhnenko, A.G.: The Group Method of Data Handling-A rival of the Method of Stochastic Approximation. Soviet Automatic Control, vol 13 c/c of avtomatika 1(3), 43–55 (1968)
2. Ivakhnenko, A.G.: Polynomial theory of complex systems. IEEE Trans. on Systems, Man and Cybernetics SMC-1, 364–378 (1971)
3. Ivakhnenko, A.G., Ivakhnenko, G.A., Muller, J.A.: Self-organization of neural networks with active neurons. Pattern Recognition and Image Analysis 4(2), 185–196 (1994)
4. Farlow, S.J. (ed.): Self-organizing Methods in Modeling. GMDH Type Algorithms. Marcel Dekker, New York (1984)
5. Madala, H.R., Ivakhnenko, A.G.: Inductive Learning Algorithms for Complex Systems Modelling. CRC Press Inc., Boca Raton (1994)

6. Mueller, J.-A., Lemke, F.: Self-Organizing Data Mining: An Intelligent Approach to Extract Knowledge From Data, Dresden, Berlin (1999)
7. Howland, J.C., Voss, M.S.: Natural gas prediction using the group method of data handling. In: ASC 2003: Seventh IASTED International Conference on Artificial Intelligence and Soft Computing, Banff, Alberta (2003)
8. Iba, H., de Garis, H., Sato, T.: Genetic programming using a minimum description length priniciple. In: Kinnear Jr., K.E. (ed.) Advances in Genetic Programming, pp. 265–284. MIT, Cambridge (1994)
9. Nariman-Zadeh, N., Darvizeh, A., Ahmad-Zadeh, G.R.: Hybrid genetic design of GMDH-type neural networks using singular value decomposition for modelling and prediction of the explosive cutting process. Proc. Inst. Mech. Engrs, Part B: Journal of Engineering Manufacture 217, 779–790 (2003)
10. Onwubolu, G.C.: Design of Hybrid Differential Evolution and Group Method of Data Handling for Inductive Modeling. In: Proceedings of International Workshop on Inductive Modeling, Prague, Czech, pp. 87–95 (2007)
11. Onwubolu, G.C.: Design of Hybrid Differential Evolution and Group Method of Data Handling for Modeling and Prediction. Information Sciences (accepted, 2008) (in press)
12. Storn, R.M., Price, K.V.: Differential evolution - a simple evolution strategy for global optimization over continuous space. Journal of Global Optimization 11, 341–359 (1997)
13. Price, K.V., Storn, R.M.: Differential evolution homepage (Web site of Price and Storm) as at (2001), `http://www.ICSI.Berkeley.edu/~storn/code.html`
14. Onwubolu, G.C.: Optimization using differential evolution, Institute of Applied Science Technical Report, TR-2001/05 (2001)
15. Storn, R.M., Price, K.V., Lampinene, J.A.: Differential Evolution: A Practical Approach to Global Optimization. Springer, Berlin (2005)
16. Davendra, D.: Hybrid Differential Evolution and Scatter Search for Discrete Domain Problems, MSc Thesis, The University of the South Pacific (2003)
17. Onwubolu, G.C., Davendra, D.: Scheduling flow shops using differential evolution algorithm. European Journal of Operational Research 171, 674–692 (2006)
18. Davendra, D., Onwubolu, G.C.: Scheduling flow shops using enhanced differential evolution algorithm. In: European Conference on Modeling and Simulation (ECMS), Prague, Czech (2007)
19. Price, K.V.: An introduction to differential evolution. In: Corne, D., Dorigo, M., Glover, F. (eds.) New Ideas in Optimization, pp. 79–108. McGraw Hill, UK (1999)
20. Onwubolu, G., Kumalo, T.: Optimization of multipass tuning operations with genetic algorithms. International Journal of Production Research 39(16), 3727–3745 (2001)
21. Hiassat, M., Abbod, M., Mort, N.: Using Genetic Programming to Improve the GMDH in Time Series Prediction. In: Bozdogan, H. (ed.) Statistical Data Mining and Knowledge Discovery, pp. 257–268. Chapman & Hall CRC (2003)
22. Oh, S.-K., Park, B.-J., Kim, H.-K.: Genetically optimized hybrid fuzzy neural networks based on linear fuzzy inference rules. International Journal of Control, Automation, and Systems 3(2), 183–194 (2005)
23. Park, H.-S., Park, B.-J., Kim, H.-K., Oh, S.-K.: Self-organizing polynomial neural networks based on genetically optimized multi-layer perceptron architecture. International Journal of Control, Automation, and Systems 2(4), 423–434 (2004)
24. Kim, D., Park, G.-T.: GMDH-type neural network modeling in evolutionary optimization. In: Ali, M., Esposito, F. (eds.) IEA/AIE 2005. LNCS, vol. 3533, pp. 563–570. Springer, Heidelberg (2005)
25. Golub, G.G., Reincsh, C.: Singular value decomposition and least square solutions. Numer. Math. 14(5), 403–420 (1970)

26. Press, W.H., Teukolsky, S.A., Vetterling, W.T., Flannery, B.P.: Numerical Recipes in FORTRAN: The Art of Scientific Computing, 2nd edn. Cambridge University Press, Cambridge (1992)
27. `http://fx.sauder.ubc.ca/`
28. Yao, J.T., Tan, C.L.: A case study on using neural networks to perform technical forecasting of forex. Neurocomputing 34, 79–98 (2000)
29. Yu, L., Wang, S., Lai, K.K.: Adaptive smoothing neural networks in foreign exchange. In: Sunderam, V.S., van Albada, G.D., Sloot, P.M.A., Dongarra, J. (eds.) ICCS 2005. LNCS, vol. 3516, pp. 523–530. Springer, Heidelberg (2005)
30. Box, G.E.P., Jenkins, F.M.: Time Series Analysis: Forecasting and Control, 2nd edn. Holden-Day, San Francisco (1976)
31. Tong, R.M.: The evaluation of fuzzy models derived from experimental data. Fuzzy Sets Syst. 13, 1–12 (1980)
32. Sugeno, M., Yasukawa, T.: Linguistic modeling based on numerical data. In: IFSA 1991, Brussels, Computer, Management & Systems Science, pp. 264–267 (1991)
33. Sugeno, M., Yasukawa, T.: A fuzzy-logic-based approach to qualitative modeling. IEEE Trans. Fuzzy Syst. (1), 7–31 (1993)
34. Xu, C.W., Zailu, Y.: Fuzzy model identification self-learning for dynamic system. IEEE Trans Syst., Man Cybern. SMC 17(4), 683–689 (1987)
35. Pedrycz, W.: An identification algorithm in fuzzy relational system. Fuzzy Sets Syst. 13, 153–167 (1984)
36. Oh, S.K., Pedrycz, W.: Identification of fuzzy systems by means of an auto-tuning algorithm and its application to nonlinear systems. Fuzzy Sets Syst. 115(2), 205–230 (2000)
37. Kim, E., Park, M.K., Ji, S.H., Park, M.: A new approach to fuzzy modeling. IEEE Trans Fuzzy Syst. 5(3), 328–337 (1997)
38. Kim, E., Lee, H., Park, M., Park, M.: A simple identified Sugeno-type fuzzy model via double clustering. Inf. Sci. 110, 25–39 (1998)
39. Chen, J.Q., Xi, Y.G., Zhang, Y.G.: A clustering algorithm for fuzzy model identification. Fuzzy Sets Syst. 98, 319–329 (1998)
40. Leski, J., Czogala, E.: A new artificial neural networks based fuzzy inference system with moving consequents in if-then rules and selected applications. Fuzzy Sets Syst. 108, 289–297 (1999)
41. Gomez-Skarmeta, A.F., Delgado, M., Vila, M.A.: About the use of fuzzy clustering techniques for fuzzy model identification. Fuzzy Sets Syst. 106, 179–188 (1999)
42. Lin, Y., Cunningham, G.A.: A new approach to fuzzy-neural modeling. IEEE Trans Fuzzy Syst. 3(2), 190–197 (1995)
43. Oh, S.K., Pedrycz, W.: The design of self-organizing polynomial neural networks. Inf. Sci. 141, 237–258 (2002)
44. Chen, Y., Yang, B., Abraham, A.: Flexible neural trees essemble for stock index modelling. Neurocomputing 70(4-6), 697–703 (2007)
45. Buryan, P., Onwubolu, G.C.: Design of enhance MIA-GMDH learning networks. International Journal of Systems Science (accepted) (in press) (2008)

Hybrid Particle Swarm Optimization and GMDH System

Anurag Sharma and Godfrey Onwubolu

Abstract. This chapter describes a new design methodology which is based on hybrid of particle swarm optimization (PSO) and group method of data handling (GMDH). The PSO and GMDH are two well-known nonlinear methods of mathematical modeling. This novel method constructs a GMDH network model of a population of promising PSO solutions. The new PSO-GMDH hybrid implementation is then applied to modeling and prediction of practical datasets and its results are compared with the results obtained by GMDH-related algorithms. Results presented show that the proposed algorithm appears to perform reasonably well and hence can be applied to real-life prediction and modeling problems.

1 Introduction

The GMDH is a heuristic self-organizing modeling method which Ivakhnenko [22] has developed for modeling purpose as a rival method of stochastic approximation. GMDH is ideal for complex, unstructured systems where the investigator is only interested in obtaining a high-order input-output relationship [1]. GMDH algorithm can be applied to given data set of a system where it tries to find relation between input data and output data without much interference/involvement of an investigator. Hence this can be treated as a good data mining tool where data is transformed into knowledge for decision making. Data mining is specifically used on those data sets where no priori knowledge is available by applying any appropriate data mining tool to extract some hidden knowledge. GMDH works well for its purpose but it is infected with some shortcomings. Generally we use data mining on very large multi-dimensional datasets. GMDH struggles to find good model solution where

Anurag Sharma
School of School of Computing, Information System, Mathematical Sciences and Statistics, Faculty of Science & Technology, The University of the South Pacific, Private Bag, Suva, Fiji
e-mail: sharma_au@usp.ac.fj

Godfrey Onwubolu
Knowledge Management & Mining, Richmond Hill, Canada
e-mail: onwubolu_g@dsgm.ca

G.C. Onwubolu (Ed.): Hybrid Self-Organizing Modeling Systems, SCI 211, pp. 193–231.
springerlink.com

dimension is very big because of it combinatorial behavior to solve the problem. A novel algorithm has been proposed here for modeling and predic-tion purpose that tries to overcome the existing discrepancies of traditional GMDH algorithm. This novel algorithm is actually hybridization of GMDH with an adaptive heuristic Particle Swarm Optimization algorithm. This proposed novel algorithm is named PSO-GMDH which indicates the hybridization of two separate heuristic algorithms. Heuristic optimization algorithms are normally applied to problems where no specific algorithm of a problem exists or a known specific problem has very high time complexity that does not work for large size problems. Since, GMDH is unable to deal with large size problems the ideas of PSO heuristic algorithm has been combined with GMDH algorithm. Specifically, the selection process of individual variables (nodes) in traditional GMDH has been replaced with heuristic selection process of PSO algorithm which also provides the termination criteria of the algorithm.

2 The Group Method of Data Handling (GMDH)

2.1 Overview of Traditional GMDH

This section describes the original GMDH modeling that was proposed by A. G. Ivakhnenko in 1960s. This method is particularly useful in solving the problem of modeling multi-input to single-output data. It just requires the data set of a particular area of application for training and testing in order to realize a mathematical model.

The details of original Group Method of Data handling (GMDH) modeling has been described in an article by S. J. Farlow [1]:

GMDH algorithm is basically works on *interpolation* that is used to find the approximate values of a complex function using some other easier function.

The simplest form of interpolation uses a straight line as if it were the given function *f(x)* whose values are need to be approximated. For this *linear interpolation* the particular straight line was chosen to pass through the two end points of the interval, where we knew the values of the function as shown in Fig. 1 [2].

Function *f(x)* can be approximated by a straight line through the two points. The line is given by

$$y(x) = f(x_1) + \frac{f(x_2) - f(x_1)}{x_2 - x_1}(x - x_1) \tag{1}$$

Eq. 1 is the desired formula for estimating the value of *f(x)* from the given value of *x*.

The values that we are given of a function are sometimes spaced so far apart that linear interpolation is not sufficiently accurate for our purposes. In such cases we use nth order polynomials.

$$P_n(x) = a_0 + a_1 x + a_2 x^2 + \ldots + a_n x^n \tag{2}$$

and chooses a polynomial which passes through selected samples i.e. given data set of the function as shown in Fig. 2.

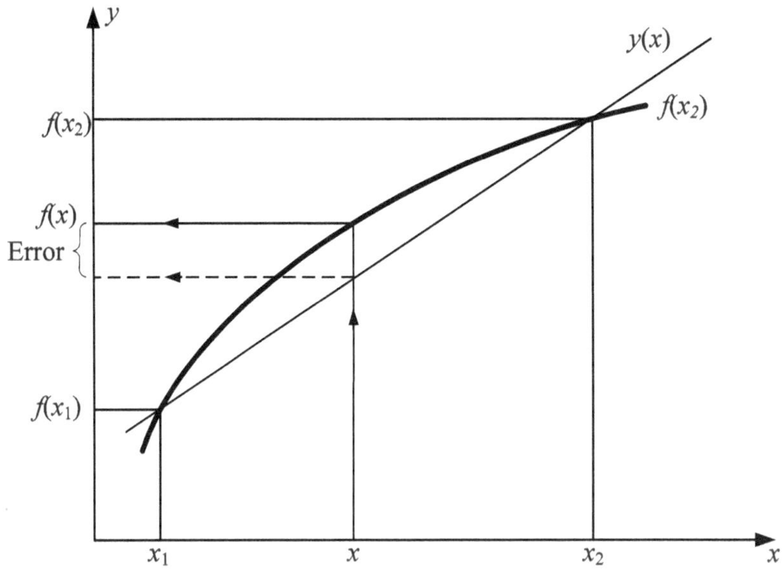

Fig. 1 Linear Interpolation

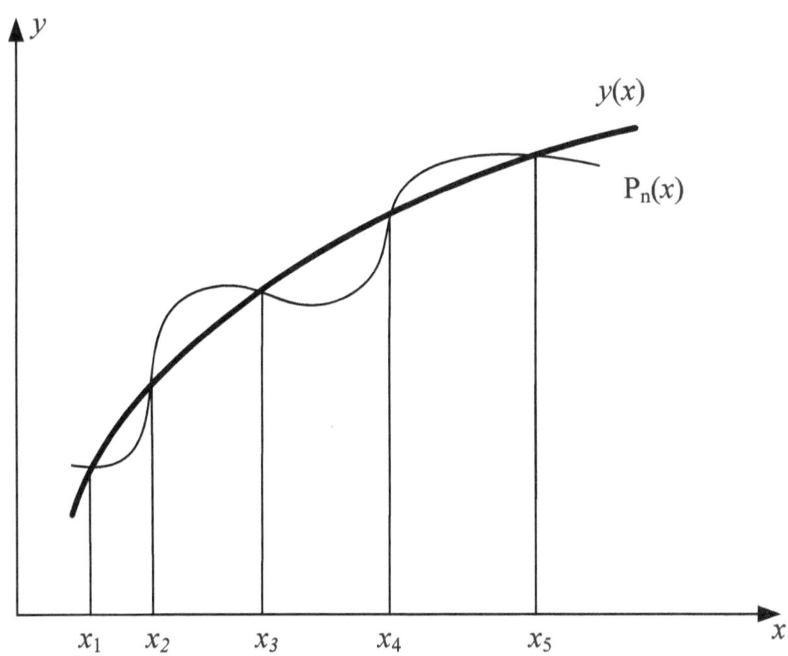

Fig. 2 Polynomial Interpolation

In the particular cases of polynomial interpolation in a table of values i.e. data set of a function *y(x)*, the condition that the polynomial pass exactly through the point (x_i, y_i) is that

$$P_n(x_i) = y_i = a_0 + a_1 x_i + a_2 x_i^2 + \ldots + a_n x_i^n \tag{3}$$

Hence the approximation of y(x) is a polynomial $P_n(x)$ instead of a straight line. With a given set of values the coefficients can be found using mathematical methods which are not given here. Interested readers can find more about interpolation and its applications.

Traditional GMDH uses similar regression methods to find a model from the given data set. It uses a high-order polynomial of the form shown below [1]:

$$y = a + \sum_{i=1}^{m} b_i x_i + \sum_{i=1}^{m}\sum_{j=1}^{m} c_{ij} x_i x_j + \sum_{i=1}^{m}\sum_{j=1}^{m}\sum_{k=1}^{m} d_{ijk} x_i x_j x_k + \ldots \tag{4}$$

Which relates m input variables $x_1, x_2, x_3, .., x_m$ to a single output variable y.

Preamble: collect regression-type data of n-observations and divide the data into n_{tr} training and n_{te} testing sets as shown in Table 1 [5].

Table 1 Data Set Format

n_{tr}	x_{11}	x_{12}	x_{13}	...	x_{1m}	y_1
	x_{21}	x_{22}	x_{23}	...	x_{2m}	y_2
	x_{31}	x_{32}	x_{33}	...	x_{3m}	y_3
	...	...	...	...	...	...
	x_{nt}	$x_{nt,2}$	$x_{nt,3}$	...	$x_{nt,m}$	y_{nt}
n_{te}	...	...	...	...	...	...
	x_{n1}	x_{n2}	x_{n3}	...	x_{nm}	y_n

Mathematically we have set X of input data set and set Y of output data set.

Step 1: Construct $^mC_2$ new variables $z_1, z_2, z_3, \ldots, z_{\binom{m}{2}}$ in the training dataset for all independent variables (columns of X), two at a time $\left(x_{i,k-1}, x_{i,k} : i \in [1, m]\, and\, k \in \left[2, \binom{m}{2}\right]\right)$ and construct the regression polynomial:

$$z_1 = A + Bx_1 + Cx_2 + Dx_1^2 + Ex_2^2 + Fx_1x_2 \text{ at points } (x_{11}, x_{12}) \tag{5}$$

$$z_k = A + Bx_{k-1} + Cx_k + Dx_{k-1}^2 + Ex_k^2 + Fx_{k-1}x_k \text{ at points } \left(x_{i,k-1}, x_{i,k}\right) \tag{6}$$

Step 2: For each of these regression surfaces, evaluate the polynomial at all n data points (i.e. *A, B, C, D, E,* and *F* obtained from $x_{i,k-1}, x_{i,k}; y_i$ for training). The coefficients for the polynomial are found by least square fitting as given in [20], or singular value decomposition (SVD) for singular-value problems as given in [21] using the data in the training set.

Step 3: Eliminate the least effective variables: replace the columns of X (old variables) by those columns of Z (new variables) that best estimate the dependent variables y in the testing dataset such that

$$d_k^2 = \sum_{i=nt+1}^{n} \left(y_i - z_{i,k}\right)^2 \text{ where } k \in \left[1,2,....,\binom{m}{2}\right] \tag{7}$$

Order Z according to the least square error $d_k| \left\|d_j\right\| < R$ where R is some prescribed number chosen a priori. Replace columns of X with the best Z's ($Z_{<R}$); in other words $X_{<R} \leftarrow Z_{<R}$.

Step 4: Test for convergence. Let $DMIN = d_l$ where l = number of iterations. If $DMIN_l = DMIN_{l-1}$ go to Step 1, else stop the process.

2.2 *Drawbacks of Traditional GMDH*

The original GMDH is entailed with some discrepancies such as it generates quite complex polynomial as it progresses through the layers. Every addition of a layer increases the order of polynomial exponentially. Hence it becomes very difficult to keep extending the layers. To avoid the complete explosion of this algorithm GMDH has taken *greedy approach* where it chops off unfit solutions in each layer and only allows predefined number of fit solutions to move to next layer. However, greedy approach itself is inclined to its ineffectiveness in finding global best solution and most of the time it stuck into local best solution. Its usage of greedy selection approach to get the best polynomial makes it bias towards those individual variables that are unfit at early stage but might become fit on later stage. Another drawback can be pointed out on its termination criteria of algo-rithm where it terminates the algorithm as soon as it receives solution of poor quality. It assumes good solution in past will always give good solution in future which is again *greedy approach* that is not always true.

Besides suffering from few shortcomings, GMDH is still an effective algorithm to determine a model from a given data set. Diminishing its shortcomings might improve its performance. As indicated above that it has two major drawbacks. First is its combinatorial behavior of algorithm which is unlikely to be solved but its impact can be minimized. Generic optimization algorithms are known to be quite effective in solving combinatorial problems where it searches for solution by taking all possible combinations of fit and unfit solutions from one generation/layer/iteration to another generation/layer/iteration. Generic algorithms are based on heuristic approach where it can solve any optimization problem. In the case of GMDH, the selection of nodes from one layer to another to get the optimum model solution is actually an optimization problem where an optimum solution is need to be found from infinite possible solution. Particle Swarm Optimi-zation is one of the generic optimization algorithms which could be applied in GMDH to replace its greedy approach based selection process with heuristic selection approach. The combination of these two separate algorithms into one could be named hybrid PSO-GMDH algorithm.

3 Particle Swarm Optimization Algorithm

Particle Swarm Optimization is a heuristic algorithm that solves continuous and discrete optimizing problem of a large domain. It is a generic algorithm; that means it can be used for any discrete/combinatorial problem for which good specialized algorithms do not exist. This algorithm is inspired by the social behavior of bird flocking or fish schooling which is described with its mathematical functions in details in this section.

The idea of optimization plays a major role in engineering, computer science, system theory, economics and other areas of science [6]. Optimization principles are of increasing importance in modern design and system operation in various areas [6]. The contents of this chapter are mostly from our published work in [18].

To solve a particular optimizing problem, algorithms are developed. These algorithms solve only specific problems. Computer scientists often make use of generic algorithms to work on a wide range of problem domain, e.g. tabu search algorithms [25], simulated annealing [26], genetic algorithms [24], ant colony optimization [27] and particle swarm algorithms [7].

Particle swarm optimization (PSO) was originally designed by R. C. Eberhart, and J. Kennedy [7], which solves many kinds of continuous and binary problems of large domain. It is certainly not as powerful as some specific algorithms, but, on the other hand, it can easily be modified for any discrete/combinatorial problem for which good specialized algorithms do not exist [8]. To grasp the abstract phenomenon behind this powerful algorithm, a simple analogy is given.

The algorithm is inspired by the social behavior of bird flocking or fish schooling. The analogy involves simulating social behavior among individuals (particles) "flying" through a multidimensional search space, each particle representing a single intersection of all search dimensions. The particles evaluate their positions relative to a goal (fitness) at every iteration; particles in a local neighborhood share memories of their "best" positions, then use those memories to adjust their own velocities, and thus subsequent positions [9].

The original PSO formulae define each particle as potential solution to a problem in D-dimensional space. The position of particle i is represented as

$$X_i = (x_{i1}, x_{i2}, \ldots, x_{iD}) \tag{8}$$

Each particle also maintains a memory of its previous best position, represented as

$$P_i = (p_{i1}, p_{i2}, \ldots, p_{iD}) \tag{9}$$

A particle in a swarm is moving; hence, it has a velocity, which can be represented as

$$V_i = (v_{i1}, v_{i2}, \ldots, v_{iD}) \tag{10}$$

At each iteration, the velocity of each particle is adjusted so that it can move towards the neighborhoods best position known as lbest (P_i) and global best position known as gbest (P_g) attained by any particle present in the swarm [9].

After finding the two best values, each particle updates its velocity and positions according to Eq. 11 and Eq. 12 weighted by a random number *c1* and *c2* whose upper limit is a constant parameter of the system, usually set to value of 2.0 [10]. *c1* is known as cognitive influence factor and *c2* is social influence factor.

$$V_i(t+1) = V_i(t) + c_1 \cdot (P_i - X(t)) + c_2 \cdot (P_g - X_i(t)) \tag{11}$$

$$X_i(t+1) = X_i(t) + V_i(t+1) \tag{12}$$

All swarm particles tend to move towards better positions; hence, the best position (i.e. optimum solution) can eventually be obtained through the combined effort of the whole population.

The methodology of obtaining the velocity vector $v_i(t+1)$ is illustrated in Fig. 3, and it has been observed that particle *i* has moved towards the neighborhood's best position and its own best position.

Fig. 3 illustrates the operators' modification until the finest form is not ob-tained. A few points can be noted during the modification: (1) the new position is converging towards the best position region; (2) the velocity could explode towards infinity. Hence new position on that situation is undefined.

All swarm particles tend to move towards better positions; hence, the best position (i.e. optimum solution) can eventually be obtained through the combined effort of the whole population.

The following section describes how the movement of a particle is restrained through mathematical function to search near the specified space where the solution is likely to be found rather than diverse it away from possible solution space.

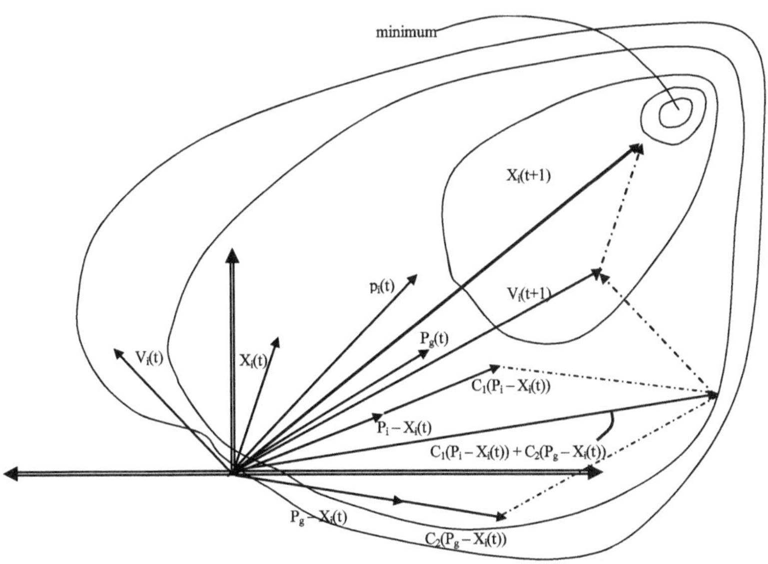

Fig. 3 Contour Lines and the Process for Generating New Locations in PSO Scheme

3.1 Explosion Control

Constriction parameters have a great impact on the operations. Varying these parameters has the effect of varying the strength of the pull "towards" the two best position, which could be verified by observing Fig. 3. If accidentally the coefficients *c1* and *c2* exceeds the value 4.0, both the velocities and positions explode towards infinity. Thus almost all implementation of the particle Swarm limit each of the two coefficient *c1* and *c2* to 2.0 [11]. To control the explosion of the system a new constriction coefficient is used, which is called inertia weight. A large inertial weight facilitates global exploration while a small inertial weight tends to fa-cilitate local exploration to fine tune the current search area [12]. Hence, equation 4 can be modified to have a place for new constriction constant, inertial weight (α) [13].

$$V_i(t+1) = \alpha \cdot V_i(t) + c_1 \cdot (P_i - X(t)) + c_2 \cdot (P_g - X_i(t)) \quad (13)$$

where inertial weight $(\alpha) = \frac{k}{abs\left[\frac{1-\frac{a}{2}-\sqrt{|a^2-4a|}}{2}\right]}$.

$k \in (0, 1]$ and $a = c1 + c2$ such that $a > 4$.

Equation 13 is generic operator for all types of objective functions. This is to tune up these coefficients for any particular kind of problem domain.

3.2 Particle Swarm Optimization Operators

This section describes all predefined operators that are used to move a particle from one position to another.

3.2.1 Position Minus Position: Subtraction

$$(position, position) \rightarrow velocity$$

Particles move from one place to another in search for a better position. To move a particle it must have some velocity to move in specified direction (towards better position).

Let say particle p_1 has position x_1 and particle p_2 has position x_2. To move the particle p_1 from position x_1 to x_2 velocity v is given.

$$v = x_2 \Theta x_1 \quad (14)$$

Suppose position x_1 and x_2 are represented in the form of array of dimension 4.

$$x_1 = \begin{bmatrix} 0 \\ 1 \\ 2 \\ 3 \end{bmatrix}, x_2 = \begin{bmatrix} 0 \\ 2 \\ 3 \\ 1 \end{bmatrix}$$

Then the velocity v will be:

$$v = \begin{bmatrix} 0 \\ 2 \\ 3 \\ 1 \end{bmatrix} \Theta \begin{bmatrix} 0 \\ 1 \\ 2 \\ 3 \end{bmatrix} = \begin{bmatrix} 1 \leftrightarrow 2 \\ 1 \leftrightarrow 3 \end{bmatrix}$$

Meaning: on x_1, exchange 1 and 2, then exchange 1 and 3. The final result would be x_2. The mathematical idea behind this procedure is to memorize the transformation the velocity will eventually do when applied to position (see "Position plus velocity"). Another important operator is described next.

3.2.2 Coefficient Times Velocity

This operator is stochastic, and defined only for a coefficient between 0 and 1. For a coefficient greater than 1, say *coeff* = $k + c$, with k is integer part and c is decimal part, then simply k times velocity_plus_velocity and one times coefficient_times _velocity is used [13].

Suppose a velocity v and coefficient c is given, then $c\ x\ v$ can be computed for a coefficient c given as

$$\begin{cases} c \in [0,1], c' = random(0,1) \\ c' \leq c \to (i \leftrightarrow j) \to (i \leftrightarrow i) \\ c' > c \to (i \leftrightarrow j) \to (i \leftrightarrow j) \end{cases} \tag{15}$$

The following example shows how this operator works for a given v, $0.5v$ ($c < 1 = 0.5$) is given as:

$$0.5 \otimes \begin{bmatrix} 0\ 1 \\ 2\ 3 \\ 3\ 0 \\ 4\ 4 \\ 0\ 2 \\ 1\ 0 \end{bmatrix} = \begin{bmatrix} 0\ 1 \\ 2\ 2 \\ 3\ 3 \\ 4\ 4 \\ 0\ 2 \\ 1\ 1 \end{bmatrix} = \begin{bmatrix} 0 \leftrightarrow 1 \\ 0 \leftrightarrow 2 \end{bmatrix}$$

This operator is mainly used to change the velocity. The size of velocity is also changed when this operator is imposed. The operator is helpful when velocity of a particle needs to be changed. Either move away or move closer to another particle. For the previous sections the velocity of the particle was deduced. Now particle has a velocity to move into new position. One other important operator is velocity plus velocity which is discussed below.

3.2.3 Velocity Plus Velocity: Addition

This operator is required only when coefficient is greater than one, suppose a particle p has two velocities v_1 and v_2 , then the new $velocity_{vnew}$ will be:

$$v_1 \oplus v_2 = v_{new} \tag{16}$$

The technique of addition is as follows. The sequence of transpositions describing v_2 is simply "added"to the one describing v_1. An example of this operator is shown below.

$$\begin{bmatrix} 0 \leftrightarrow 1 \\ 2 \leftrightarrow 3 \end{bmatrix} \oplus \begin{bmatrix} 0 \leftrightarrow 2 \\ 3 \leftrightarrow 4 \\ 3 \leftrightarrow 1 \end{bmatrix} = \begin{bmatrix} 0 \leftrightarrow 1 \\ 2 \leftrightarrow 3 \\ 0 \leftrightarrow 2 \\ 3 \leftrightarrow 4 \\ 3 \leftrightarrow 1 \end{bmatrix} = \begin{bmatrix} 0 \leftrightarrow 3 \\ 1 \leftrightarrow 2 \\ 1 \leftrightarrow 4 \end{bmatrix}$$

This operator is not commutative, we usually do not have $v_1 \oplus v_2 = v_2 \oplus v_1$.

3.2.4 Position Plus Velocity

Particles can be moved according to their current velocity. Suppose particle p has position x and velocity v then the new position will be

$$x_{new} = x + v \tag{17}$$

This function is used to obtain global best particle of the swarm. Let say particle p has position x, velocity v and new position x_{new}. Suppose postion x has some arbitrary values of independent variables $(a, b, c, d, e, f\,)$ and velocity v has values (b, d) in first component and values (a, f) in second component. Then the x_new is shown as $x\ v\ x_{new}$.

$$\begin{bmatrix} a \\ b \\ c \\ d \\ e \\ f \end{bmatrix} \oplus \begin{bmatrix} b \rightarrow d \\ a \rightarrow f \end{bmatrix} = \begin{bmatrix} f \\ d \\ c \\ b \\ e \\ a \end{bmatrix}$$

The technique of transformation is that each component of the velocity (that is a transposition) is successively applied, first to x, then to the position obtained.

As discussed above PSO algorithm is based on social psychological metaphor where swarm particles interact with each other in their neighborhood and their whole society. Section 3.3 describes the different ways of neighborhood defined in PSO algorithm.

3.3 Particle Swarm Optimization Neighborhood

The particle swarm algorithm is an adaptive algorithm based on a social-psychological metaphor. Each particle is influenced by a success of their topological neighbors [17]. This external function provides a particle its neighbor of a given type. Also, there are many ways to define a "neighborhood" [14], but we can distinguish three classes:

3.3.1 Social Neighborhood

The social neighborhood, just takes *relationship* into account. In practice, for each particle, its neighborhood is defined as a list of particles at the very begin-ning, and does not change. Note that, when the process converges, a social neigh-borhood becomes a physical one. Al social neighborhood is a type when a particle chooses k nearest particles according to its location. Mathematically, social neighborhood is defined for a particle to simply get $\frac{k}{2}$ particles on each side of 1-D array.

3.3.2 Physical Neighborhood

The physical neighborhood, takes distances into account. In practice distances are recomputed at each time step, which is quite costly, but some clustering techniques need this information in this type of neighborhood, a particle "chooses" k best particles from the entire swarm; the distance between a particle and k particles in the globe is normally calculated. Normally the social neighborhood be-comes a physical neighborhood during the process of algorithm.

3.3.3 Queen

Instead of using these two types of neighborhood. An extra particle can be used, which "summarizes" the neighborhood. This method is combination of the (unique) queen method defined in [14] and of the multi-clustering method described in [16]. For each neighborhood, we iteratively build a gravity center and take it as best neighbor. This method needs some mathematical computations. The coefficient is obtained as follows:

$$(c_i) = \sum_j \frac{(f_0 + 1)}{(f_0 + 1) + (f_j + 1)} \tag{18}$$

Since PSO is an adaptive algorithm based on social-psychological metaphor. The population of individuals adapt by returning stochastically towards previous successful regions in the search space. Move towards is the most important external function in PSO, because it explains the movement of the particles [17].

Sometimes PSO algorithm is unable to output the desired or expected value, then the swarm is known to be in *no-hope* state. Like other generic optimization techniques PSO has some strategies to avoid being stuck in local maximum or minimum. PSO has a well-defined procedure to move out from the *no-hope* state which is called *re-hope*. PSO uses *no-hope* and *re-hope* processes to improve search performances. The following section covers procedures to determine when PSO is in *no-hope* state and the details of *re-hope* process.

3.3.4 Particle Swarm Optimization Improvement Strategies

As in other generic optimization techniques such as genetic algorithm, differen-tial evolution, ant colony optimization, etc., PSO has some strategies to avoid being stuck in local maximum or minimum. For PSO, in particular, the *No-Hope/Re-Hope* process is used in order to improve search performances; in other words, for the search to be adaptive.

3.3.5 No-Hope Tests

For any objective function of any type, the decision has to be made about the optimum value. The PSO algorithm has a task to decide whether the optimum value is achievable or not, and if achievable than how good is it.

If the PSO algorithm is unable to output the desired or expected value, then the swarm is in *no hope* state. There are some reasons behind this status. Firstly, if not single particles are moving then there is no question of movement of swarm, hence no better position is expected. Secondly, no effective movement is occurring i.e. either the swarm has reduced very much or the movement is extremely low. Finally when the algorithm is producing the same best value greater than the *threshold* times, then the better solution than the same best value is unexpected and the swarm is in *no hope* state. When the PSO algorithm gets into a *no-hope* state, the only way out is to either accept the result of the current situation or *re-hope* for better result. The following criteria are useful for the *no-hope* test [15].

Criterion 0

If a particle has to move towards another one, which is at distance 1, wither it does not move at all or it goes exactly to the same position as this other one, de-pending on the social/confidence coefficients. It may be possible that all moves computed according to equation 16 are null. In this case there is absolutely no hope to improve the current best solution.

Criterion 1

The No-Hope test defined in [15] is *swarm too small.* In this test, the swarm diameter is computed at each time step, which is costly. But in a discrete case, as soon as the distance between two particles tends to become too small, the two particles become identical, usually first by positions and then by velocities. Hence, at each time step, a *reduced* swarm is computed in which all particles are different, which is not very expensive, and the No-Hope test becomes *swarm too reduced*, say by half the original size.

Criterion 2

Another criterion has been added to the *no-hope* test criterion is the *swarm too slow*. This criterion compares the velocities of all particles to a threshold, either individually or globally. In one version of the algorithm, this threshold is in fact modified at each time step, according to the best result obtained so far and to the statistical distribution of arc values.

Criterion 3

Another very simple criterion that is defined is the *no improvement for too many times*. In practice, it appears that criteria 1 and 2 are sufficient.

3.3.6 Re-Hope Tests

PSO has a well-defined procedure to move out from the no-hope state is called Re-hope. As soon as there is no hope, the swarm is reexpanded. The idea behind this method is to check if there is still hope to reach the better solution. If there is no hope, then swarm is moved [15]. The particles are moving slowly and continuously to get better position. The movement of particles is categorized in four: (i) lazy descent method, (ii) energetic descent method, (iii) local iterative leveling, and (iv) Adaptive re-hope method. There are a number of re-hope strategies defined for PSO. The re-hope strategies in [15] and [19] inspire the first two meth-ods described here.

Lazy Descent Method (LDM)

Each particle goes back to its pervious best position and, from there, moves randomly and slowly (i.e. size of velocity is 1) and stop as soon as it finds a better position or when a maximum number of moves (problem size) is reached. If the current swarm is smaller than the initial one, it is completed by a new set of particles randomly chosen.

Energetic descent method (EDM)

Each particle goes back to its previous best position and, from there moves slowly (i.e. velocity size is 1) as long as it finds a better position in at most maxi-mum number of move (problem size). If the current swarm is smaller than the initial one, it is completed by a new set of particles randomly chosen. The only drawback of this method is it is more expensive than LDM.

Local Iterative Leveling (LIL)

This method is more expensive and more powerful. This method is used when EDM fails to find a better position. For each immediate physical neighbor y (at distance 1) of the particle p, a temporary objective function value f(y) is computed by using the following algorithm:

- Find all neighbors at a distance 1;
- Find the best distance, i.e. y_{min};
- Assign y to the temporary objective function value $f(y) = \frac{f(y_{\min})+f(x)}{2}$
- Move p towards its best neighbor;

Usually this algorithm's big O order ranges from polynomial to exponential hence this procedure is only used when the swarm is in no-hope state [15].

Adaptive Re-Hope Method (ARM)

The three above methods can be automatically used in an adaptive way, accord-ing t how long (number of times steps) the best solution has not been improved. An example of adaptive re-hope strategy is shown in Table 2.

Parallel and Sequential Versions

The algorithm can run either in (simulated) parallel mode or in sequential mode. In the parallel mode, at each time step, new positions are computed for all particles

Table 2 Data Set Format

Same best[a]	Re-Hope type
0	No Re-Hope
1	Lazy Descent Method (type = 0)
2	Energetic Descent method (type = 1)
≥ 3	Local Iterative Leveling (type = 2)

[a] Number of time steps without improvement.

and then the swarm is globally moved. In sequential mode, each particle is moved at a time on a cycling way. So, in particular, the best neighbor used at time t+1 may be not anymore the same as the best neighbor at time t, even if the iteration is not complete. Eq. 12 implicitly supposes a parallel mode, but in practice there is no clear difference in performances, and the sequential method is a bit less expensive.

4 The Proposed Hybrid PSO-GMDH Algorithm

4.1 Overview

Particle Swarm Optimization is used to solve any optimizing problems whereas GMDH itself is a self-organizing modeling method which heuristically determines the optimum model of a given problem. The reason for bringing the idea of hybridization was to overcome the existing shortcomings of traditional GMDH which is mainly of its combinatorial behavior of progressing through layers to find the optimum model of a given problem. Firstly, it makes the assumption that good approximation quality in the past guarantees the good approximation in the immediate future which is a greedy approach [3]. As discussed in section 2.1 GMDH only picks pre allocated number of best solutions of the current layer to move to next layer. It ignores all other solutions which are unfit in early stages but might generate very fit solution in later stages. Hence choice of solution is always locally best [4]. That means traditional GMDH has high chances of being trapped into local best solution. Secondly, the termination condition of traditional GMDH process depends on the quality of output value in layer by layer approach of nodes selection. It also uses a greedy approach that keeps track of local best solutions in the current layer. The iteration process is stopped as soon as new layer generates poorer solution than previous layer. GMDH tries to refine the model in each layer until the best estimation of the model that predicts the output with least error is obtained. The process of termination condition is shown in Fig. 4. The Y axis shows the best solution in the corresponding layer. Initially the curve declines then starts inclining after iteration 5 which indicates GMDH was getting better solutions up to iteration 5 then the poor solution is obtained in iteration 6. Hence the iteration has to be stopped in iteration 5.

The principal approach of modeling of PSO-GMDH and traditional GMDH is more or less same. In PSO the whole population (of constant size) of swarm particles

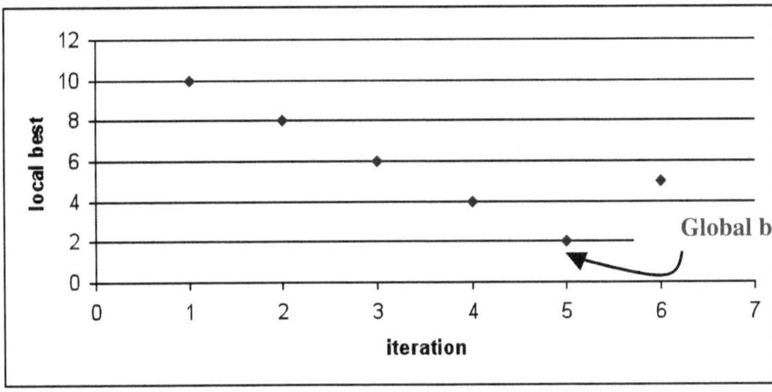

Fig. 4 Termination Criteria of Traditional GMDH

progresses iteratively until the optimum solution is found. Iterative process of PSO can be compared with layered approach of GMDH where swarm particles search for better position in each iteration as GMDH nodes look for better solution in each layer.

The proposed hybrid PSO-GMDH uses a heuristic search process which makes it more attractive for efficiently searching for large and complex search spaces. Fig. 3 shows how PSO can achieve global best solution by the combined effort of the whole population. It is likely that solution found by traditional GMDH is trapped into local minimum whereas PSO's domain of search space is infinitely large and it has its internal mechanism to avoid being trapped into local minimum. The tendency of PSO is that each particle keeps searching for better solution. Un-like traditional GMDH PSO doesn't stop the search process if the best solution of the next layer is not better than previous layer. Fig. 5 shows that PSO-GMDH uses its own termination conditional which doest not follow *greedy approach*. Unlike GMDH which moves only predefined number of fit solutions to the next layer, PSO's selection process is based on heuristic approach where combination of fit and unfit solutions are moved

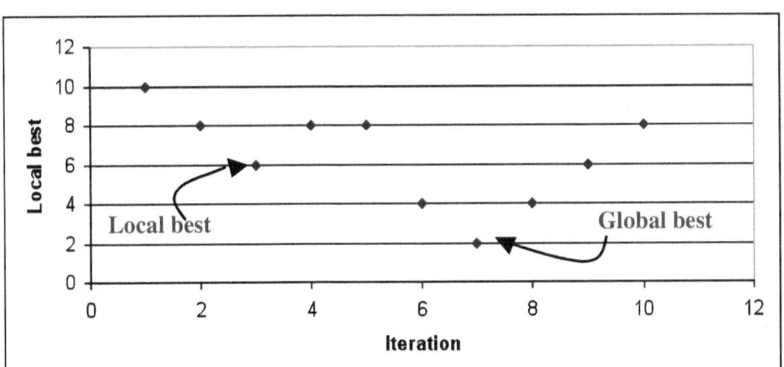

Fig. 5 Termination Criteria of Hybrid PSO-GMDH Algorithm

to next layer. In every iteration it keeps looking for better solution and doesn't stop the search process even after getting unfit solution because these unfit solutions at initial stage can make fitter solution at the later stage.

Now the question that is raised here is in which part of GMDH process could PSO be applied? In GMDH the model is created through making polynomials of various orders and combinations of features which approximate the given output. Creation of polynomials is part of an iterative process where some given combinations of nodes from current layer make the polynomials for next layer. These polynomials are candidates for model functions which are then used to calculate the output values that are used as input nodes for next layer. So we have the iterative approach where generated outputs for current layer work as input nodes for next layer of GMDH network. Traditional GMDH selects all possible combinations of nodes from previous layer to make polynomials for next layer. Out of which only best M (predefined number) is selected to stay in the next layer. This step is necessary to avoid the explosion of combinatorial behavior of the algorithm. This selection process is the first drawback of traditional GMDH. Suppose a given dataset has very large dimension D and GMDH algorithm tends to move D nodes from one layer to another. Suppose it combines d nodes from previous layer to make a polynomial for current layer then it will end up in making total of ${}^{D}C_{d}$ polynomials from which best D will be selected. In every iteration total of ${}^{D}C_{d}$ combinations would prove very expensive in terms of processing time and memory. For example if D is 50 and d is 4 for each layer then there are total of ${}^{50}C_{4}$ i.e. 230,300 polynomials will have to be made from different combinations of 4 nodes for each layer which will then be sorted according to their fitness value and from which only best 50 will be picked to move to next layer. On the other hand PSO-GMDH would heuristically determine which combinations of nodes will be moved to next layer. If PSO generates k particles then each particle will specify its set of nodes from which a polynomial will be made. After generating total of k polynomial best 50 polynomials (or new nodes) will be moved to next layer. Hence in each layer we will have same number of nodes. Suppose we have 100 particles in the above case then we have maximum of 100 possible combination of nodes/layer from which we will choose 50 nodes (combination of fit and unfit nodes) which is far less than 230,300 nodes generated in traditional GMDH. Hence PSO-GMDH diminishes the curse of dimensionality phenomenon.

To solve this problem and additionally provide greater variations in selection of nodes, PSO comes into picture. The detail selection process is described in section 4.2 but in brief it provides the list of nodes from current layers to be combined to make polynomials in the next layer. The selection process doesn't have to take every possible combination of previous layer's data as traditional GMDH does. PSO-GMDH generates a limited workable amount of particles that remain constant throughout the modeling process. Besides, PSO-GMDH offers a new selection process where PSO stochastically determines how many nodes need to be combined and which nodes need to be combined to form the polynomials. The selection is based on heuristic manner hence combinations of fit and unfit individuals from previous layers are selected. Unfit individuals can survive the selection process for next layer if they produce fit solutions. Only D polynomials (not necessarily best ones) will be retained for next layer as traditional GMDH does. Then rest of modeling process is same as mentioned above that it will take these calculated polynomials as the input nodes for next layer. So even after applying PSO to manage the selection of nodes the core of GMDH process remains same.

Traditional GMDH makes the assumption about the stopping criteria that good approximation quality in the past guarantees the good approximation in the immediate future [3]. It looks for better approximation in each iteration until it gets poorer solution which indicates no better solution is possible in the coming layers. Hence the search for the solution stops as soon as poorer solution compared to previous layer is computed. This kind of approach is known as greedy approach where an algorithm concentrates on arriving at a solution by making a sequence of choices, each of which simply looks the best at that iteration/layer. Hence choice of solution is always locally best [4]. On the other hand PSO-GMDH uses different termination criterion. Its termination criterion totally depends on PSO's internal heuristic approach of finding the global best solution. Its optimization process determines when to stop the search for solutions. The search process can also be terminated after reaching maximum number of iterations predetermined by the user or if the system reaches *no-hope* stage where no further better solution is possible.

4.2 Technical View

PSO-GMDH is simply a GMDH process where selection of nodes to move to next layer is determined by PSO. Technically speaking PSO-GMDH runs two algorithms simultaneously and interacting with each other. PSO tells GMDH which variables (nodes) to select for next layer then GMDH uses those selected nodes to make polynomials for its modeling process. GMDH then returns the computed output value of polynomials back to PSO which it uses as fitness function that is used to determine the quality of polynomial (model function). Progress in each iteration of PSO makes GMDH to progress one layer.

This section describes the technical view of PSO-GMDH algorithms using the following notations:

P_T	Total number of particles used in PSO-GMDH algorithm
M	Total number of nodes to be moved from one layer to another in traditional GMDH algorithm.
X	Only input variables of data set.
Y	Only output variable of data set.
x_i	i^{th} feature of data set.
n_{tr}	size of training data set.
n_{te}	size of testing data set.
n	size of dataset.
P_1, P_2, P_3	System variables exists in each particle.
c_i	i_{th} coefficient of a polynomial function.
E_j	mean square error between actual output and output from j^{th} polynomial.

The PSO-GMDH algorithm starts with generation of swarm particles of size P_T which is normally greater than constant M the number of variable to move from one layer to another. Each particle is initialized with some random positions which are updated on each iteration by PSO. Now the question that is raised here is how every layer of GMDH process integrates with PSO (iterative process) process? The detailed framework of PSO-GMDH algorithm is given below which describes the step by step process:

4.2.1 System's Input Variables

Modeling of a system requires its corresponding data set. Modeling is a kind of knowledge discovery from existing data set where we need to have some predefined number of features and several combinations of different features making up the dataset. The dataset is consists of input variables (i.e. feature vector) and one output variable only. For a given input vector $X = (x_1, x_2, x_3, .. , x_n)$ there has to be one output variable Y. $x_1, x_2, x_3, .. , x_n$ indicate different feature or input variable of the dataset.

4.2.2 Forming Training and Testing Data

As mentioned above modeling required a dataset from which some knowledge has to be discovered. The quality of the derived model has to be tested with sepa-rate set of dataset which has not been used in derivation of a model. Hence the derivation of a model requires two separate sets of dataset of same feature vectors. One data set is used to make the models and one is used to test those models. The separate sets of datasets can be made simply by dividing the given dataset of size n into two sets of training and testing data sets whose sizes can be denoted by n_{tr} and n_{te} respectively, where total size $n = n_{tr} + n_{te}$. Training data set is used to construct various candidates of model functions whereas testing data set is used to evaluate the quality of those models at each iteration of PSO-GMDH.

4.2.3 Network Realization

The traditional GMDH is based on a heuristic approach where it selects some predefined number M of relatively fit individuals from one layer to another until the termination condition is met. It keeps track of local best solutions in each layer until it finds poorer solution than previous layer which is the termination criterion for the algorithm. As mentioned above that GMDH tries to refine the model in each layer until the best estimation of the model that predicts the output with least error is obtained. The principal approach of modeling of PSO-GMDH and traditional GMDH is more or less same. In PSO the whole population (of constant size) of swarm particles progresses iteratively until the optimum solution is found. Iterative process of PSO can be compared with layered approach of GMDH where swarm particles search for better position in each iteration as GMDH nodes look for better solution in each layer.

As mentioned above in section 3 PSO uses swarm particles as its unit search agent. The position (sequence of numbers) of each particle is used to determine which salient combinations of input variables of previous layer to be moved to next layer. Each particle must contain 3 system parameters (*P1, P2, P3*). $P_1 \in [1,3]$ is randomly generated and represents the order of polynomial that will be generated from previous layer, which can be either 2 or 3 but for simplicity we normally take 2 in each layer. The advantage of doing this is to generate more complex universe of data permutations [5]. Traditional GMDH normally just combines 2 or 3 variables from previous layers to next layers but PSO-GMDH varies with different combinations from one layer to another. $P_2 \in [1,r]$, $r = \min(D,5)$ is again randomly generated number which determines number of input variables to be taken from previous layer where D is the width of the input dataset; the default lower bound is r = 2. $P_3 = \{a \in Z^+ | 1 \leq a \leq D\}$ is a sequence of integers stored as position of a particle represents the entire can-didates in the current layer of the network. The determination of combinations of nodes to be moved to next layer is determined by all three parameters. Firstly P_1 determines the order of polynomial to be formed then P_2 determines the number of combinations of nodes forming the polynomial which in turn will be selected from P_3 as first P_2 nodes from the whole sequence. Fig. 6 depicts the usage of all 3 system parameters to determine the polynomial. The polynomial has the following form:

$$y = a + \sum_{i=1}^{m} b_i x_i + \sum_{i=1}^{m}\sum_{j=1}^{m} c_{ij} x_i x_j + \sum_{i=1}^{m}\sum_{j=1}^{m}\sum_{k=1}^{m} d_{ijk} x_i x_j x_k + \ldots$$

Using the above form the polynomial of order 2 can be written in the following simplified form which is normally used in GMDH:

$$y = A + Bu + Cv + Du^2 + Ev^2 + Fuv$$

Fig 6 shows PSO-GMDH process of each layer. Each particle consists of separate set of system parameters which are used to generate the polynomial. These polynomials are actually used as objective function for PSO. It is a candidate for model function which determines how promising the model is.

Generic Behavior of PSO: PSO is a generic optimizing algorithm which re-quires two external information of a given problem, its input dataset and a tailor made objective function of the problem. Fig 7 shows the generic behavior of PSO and how GMDH has fit-in into PSO to make it hybrid PSO-GMDH. To find the model of a given problem a dataset of that problem is fed into PSO-GMDH system where PSO interacts with GMDH which behaves as objective function of PSO to produce the solution. The solution is a polynomial function (or a model) that can predict very closely to the output of the given problem.

Parallel Process: The most important part of the hybrid PSO-GMDH is interac-tion between PSO algorithm and GMDH algorithm. Conceptually, both algorithms run

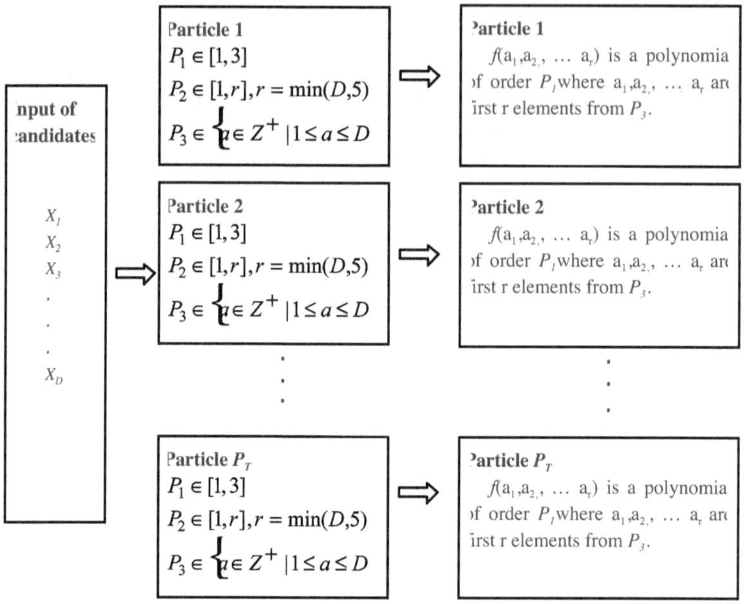

Fig. 6 Formation of Polynomials by Swarm Particles using 3 System Parameters P_1, P_2 and P_3

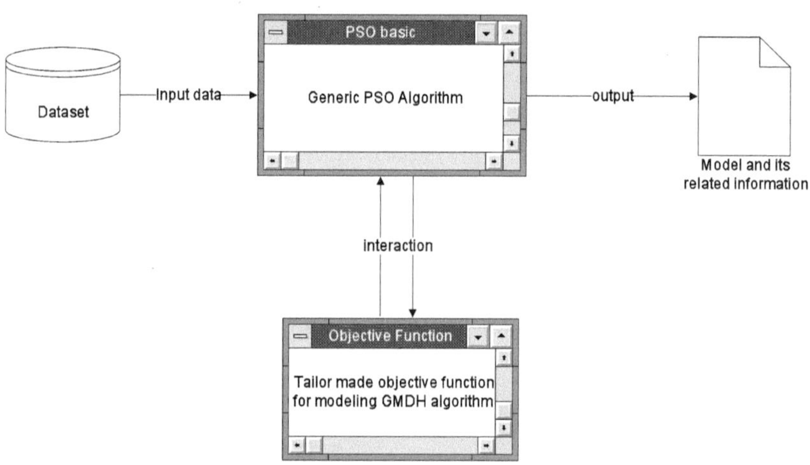

Fig. 7 Generic Behavior of PSO depicted using GMDH as a tailor made objective function

in parallel and interact with each other as shown in Fig. 8. It shows that each iteration of PSO corresponds to one layer of GMDH. The particles of PSO carry all 3 system parameters which determine which nodes of GMDH would make up the polynomial; therefore PSO's particles send this set of information to the nodes of GMDH where it picks only those combinations of nodes that have been indi-cated by PSO to make

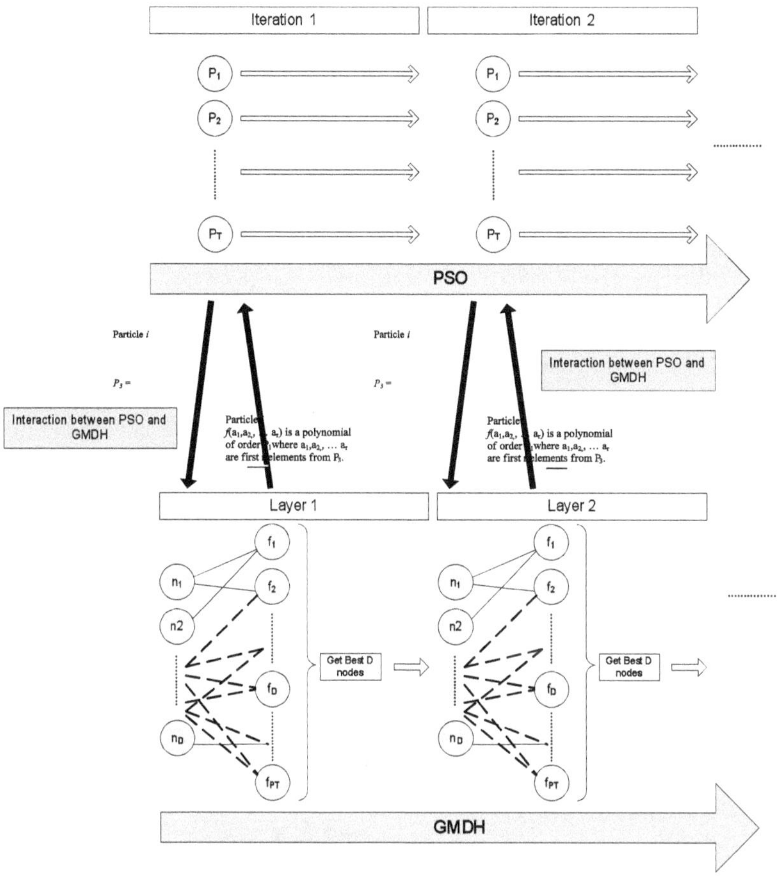

Fig. 8 Hybridization of GMDH and PSO shown as parallel process

up the polynomial. The functional value of these polynomials is then compared with the given output and then error value is returned back to particles of PSO which use them as their fitness value for objective function.

It should not be misunderstood that we have confluence of two separate unaltered algorithms PSO and GMDH that just interacts with each other in parallel where iteration *i* of PSO corresponds to layer *i* of GMDH. Actually the selection of nodes process from GMDH has been replaced with PSO algorithm that provides the information for selection of nodes to move to next layer. Even the stopping criterion of GMDH is replaced with generic stopping criteria of PSO. Those researchers who have experience with heuristic algorithms can view this novel hybridization as transformation of GMDH algorithm into objective function of PSO where PSO controls the termination criteria as well as the fitness function (here GMDH algorithm). Each call to fitness function results in polynomial formation of nodes that returns the fitness value (mean square error between polynomial functional value and actual output value) back to PSO algorithm.

Illustrative Example

To understand network realization let us take an example. Suppose we have input dataset of 5 dimensional vectors. Each vector has only one output. PSO-GMDH process would require number of particles to be specified by the user. We can take 10, 15, 20 or any other appropriate size of the swarm particles. Here we can take just 5 for simplicity. Take a scenario where a particle determines the type of polynomial for a current iteration using its 3 system parameters namely P_1, P_2 and P_3. Suppose value of P_1 is 2 which means polynomial will be of order 2. Value of P_2 is also 2 which will tell the particle to choose first 2 integers from the sequence of integers stored in parameter P_3 (recall that P_3 contains the position of a particle). Various possible values of P_3 for all 5 particles representing positions of particles at any given instance are given in Table 3 together with dummy Error values E_k i.e. mean square error between actual output and output from polynomial $f(x_i,x_j)$. PSO-GMDH uses these error values for its fitness value, hence the polynomial with least error value is the optimum solution.

Table 3 Attributes of Particles in first layer

Particles	Parameter P_3 (or position of corresponding particles)	Selected first P_2 i.e. 2 nodes to make a polynomial	E_k = MSE btw $f(x_i,x_j)$ & actual output
Particle 1	1, 3, 5, 4, 2	1, 3	15.2
Particle 2	5, 4, 3, 2, 1	5, 4	4.5
Particle 3	2, 1, 3, 5, 4	2, 1	8.9
Particle 4	4, 3, 1, 5, 2	4, 3	5.2
Particle 5	1, 5, 2, 4, 3	1, 5	7.2

For instance, the generated polynomial for particle 3 in the first iteration would be: $f(x_2,x_1) = c_1 + c_2x_2 + c_3x_1 + c_4x_2x_1 + c_5x_2^2 + c_6x_1^2$ where c1, c2, .. c6 are the constants evaluated using training dataset.

Hence Particle 3 has following attributes:

- $P_1 = 2$
- $P_2 = 2$
- $P_3 = \{2, 1, 3, 5, 4\}$ $f(x_2,x_1) = c_1 + c_2x_2 + c_3x_1 + c_4x_2x_1 + c_5x_2^2 + c_6x_1^2$, E_3 = mean square error between actual output and the output from polynomial $f(x_2,x_1)$. Lets say the error value is 8.9.

In the same manner other particles also assign the values for their attributes.

The step by step process of PSO-GMDH using this example is described next. PSO-GMDH is also compared alongside with traditional GMDH.

Layer 1 of the process

All possible combinations of nodes and their corresponding dummy error val-ues of the first layer are shown in Table 4.

Table 4 All combinations of nodes

Order No	x_i	x_j	E_k val for $f(x_i,x_j)$
1	1	2	8.9
2	1	3	15.2
3	1	4	6.6
4	1	5	7.2
5	2	3	6
6	2	4	7
7	2	5	6
8	3	4	5.2
9*	3	5	4
10	4	5	4.5

Traditional GMDH uses the greedy approach to pick the best 5 nodes which is shown in Table 5 with grey color whereas PSO heuristically determine which 5 nodes to be picked. Each swarm particle contains 3 main system parameters where parameter P_3 contains information about which nodes will be picked for next itera-tion as shown in Table 3. Table Table 6 shows picked nodes from all available nodes in grey color. It could be noted that nodes combination of 3 and 5 of order No 9 is the best polynomial which is picked by traditional GMDH algorithm but not by PSO-GMDH. Nodes combination of 4 and 5 is the best polynomial for PSO-GMDH at this stage. It is not necessary that PSO will pick the best 5 nodes of current layer. It

Table 5 Traditional GMDH Approach

Order No	x_i	x_j	E_k val for $f(x_i,x_j)$ in sorted order
9*	3	5	4
10	4	5	4.5
8	3	4	5.2
5	2	3	6
7	2	5	6
3	1	4	6.6
6	2	4	7
4	1	5	7.2
1	**1**	**2**	**8.9**
2	1	3	15.2

* best value at this stage

can pick non promising solutions initially which might become very promising later on. Observe a node of order No 1 which has been rejected by traditional GMDH but is taken by PSO-GMDH. On later stage this rejected (or taken) node becomes the best node. Table 6 shows that PSO-GMDH has picked a node of order 1 combined through (x_1, x_2) which is not prom-ising at this stage.

Table 6 PSO-GMDH Approach

Order No	x_i	x_j	E_k val for $f(x_i, x_j)$ in sorted order
9	3	5	4
10*	4	5	4.5
8	3	4	5.2
5	2	3	6
7	2	5	6
3	1	4	6.6
6	2	4	7
4	1	5	7.2
1	**1**	**2**	**8.9**
2	1	3	15.2

* best value at this stage

Layer 2 of the process

For second layer PSO-GMDH would use the nodes shown in Table 7 derived from Table 6. Observe the new order numbers after sorting the selected nodes. Now the new order No of node 1 is 4.

Table 7 PSO-GMDH Approach: Chosen candidates for next iteration

Prev Order No	New Order No	$x_i or x_j$
10*	1	4.5
8	2	5.2
4	3	7.2
1	4	8.9
2	5	15.2

* best value at this stage

The next iteration of PSO will move its particles from one position to another with calculated velocities in a search space therefore the value of parameter P_3 for each particle will change (recall P_3 stores the current position of a particle). The new attributes for the particles are shown in Table 8. The non-promising node of order

Table 8 Attributes of Particles in second layer

Particles	Parameter P_3 (or position of corresponding particles)	Selected first P_2 i.e. 2 nodes to make a polynomial	E_k = MSE btw $f(x_i,x_j)$ & actual output
Particle 1	1, 3, 4, 5, 2	1, 3	3.5
Particle 2	2, 4, 3, 5, 1	2, 4	5.9
Particle 3	2, 3, 1, 5, 4	2, 3	4.4
Particle 4	5, 3, 1, 4, 2	5, 3	5.8
Particle 5	5, 1, 2, 4, 3	5, 1	8.5

Table 9 All combinations of nodes

Order No	x_i	x_j	E_k val for $f(x_i,x_j)$
1	1	2	5.8
2	1	3	3.5
3	1	4	8.6
4	1	5	8.5
5	2	3	4.4
6	2	4	5.9
7*	2	5	3.5
8	3	4	11
9	3	5	5.8
10	4	5	4.9

best value at this stage

no. 4 has combined with node 2 and has given a promising node. This new node is shown with bold letters in Table 8.

All possible combinations of nodes and their corresponding dummy error val-ues in the second layer are shown in Table 9. Note that the non-promising node is now promising with order no 6 but it is still not the best solution.

All new combinations of nodes provided by P3 parameter will again be selected for next layer. Table 10 shows the selected combinations with grey background color. Table 11 shows the 5 selected nodes to be processed for next layer.

Layer 3 of the process

The third layer will pick the nodes from second layer which is shown in Table 11. The new attributes for the particles are shown in Table 12. The non-promising node has now become the most promising node captured by particle 1 and is shown with bold in Table 12.

All possible combinations of nodes from layer 2 are now shown in Table 13 and selected selected nodes for next layer is shown in Table 14. Note very carefully that the best polynomial is of order no 3 which is constructed from that same node which was very non promising at layer 1. Traditional GMDH would have rejected

Table 10 PSO-GMDH Approach

Order No	x_i	x_j	E_k val for $f(x_i,x_j)$ in sorted order
2*	1	3	3.5
7	2	5	3.6
5	2	3	4.4
10	4	5	4.9
1	1	2	5.8
9	3	5	5.8
6	2	4	5.9
4	1	5	8.5
3	1	4	8.6
8	3	4	11

best value at this stage

Table 11 PSO-GMDH Approach: Chosen candidates for next iteration

Prev Order No	New Order No	x_i or x_j
2*	1	3.5
5	2	4.4
10	3	4.9
6	4	5.9
4	5	8.5

best value at this stage

Table 12 Attributes of Particles in third layer

Particles	Parameter P_3 (or position of corresponding particles)	Selected first P_2 i.e. 2 nodes to make a polynomial	E_k = MSE btw $f(x_i,x_j)$ & actual output
Particle 1	**1, 4, 3, 5, 2**	**1, 4**	**3.4**
Particle 2	2, 4, 5, 3, 1	2, 4	6.9
Particle 3	2, 1, 3, 5, 4	2, 3	4.4
Particle 4	4, 3, 1, 5, 2	5, 3	5.7
Particle 5	5, 1, 3, 4, 2	5, 1	8.3

that node at initial stage but PSO has used that same node which was initially non promising but at the layer 3 it has become the most promising and the best node. This node is made from combination of node 1 and 4 of layer 2 where node 4 of layer 2 was made from nodes 2 and 4 of layer 1 that was initially made from input data 1 and 2. Node transformation from non-promising to very promising is shown in Fig. 9.

Table 13 All combinations of nodes

Order No	x_i	x_j	E_k val for $f(x_i,x_j)$
1	1	2	5.8
2	1	3	3.5
3	**1**	**4**	**3.4**
4	1	5	8.3
5	2	3	4.4
6	2	4	6.9
7*	2	5	3.5
8	3	4	11
9	3	5	5.7
10	4	5	4.9

best value at this stage

Table 14 PSO-GMDH Approach: Chosen candidates for next iteration

Prev Order No	New Order No	x_i or x_j
3*	**1**	**3.4**
1	2	5.8
6	3	6.9
4	4	8.3
8	5	11

best value at this stage

This search for better node will continue until the predefined maximum number of iteration is reached.

4.2.4 Coefficient Estimation of Polynomial Corresponding to the Selected Particle

The estimation of coefficient of each particle is done exactly as traditional GMDH using regression of matrix multiplication [23]. Aim is to find the best possible model such that square of difference between the actual output and the predicted output is minimized.

$$E_j = \sum_{i=1}^{n_{tr}} (y_i - z_{ij})^2 \quad j = 1,2,....P_T \tag{19}$$

where z_{ij} denotes the output of the $j-th$ particle with respect to the $i-th$ data. P_T is the total number of particles in the swarm and n_{tr} is the size of training set. To use the matrix regression first the polynomial function has to be converted to matrix form. For example polynomial of order 2 can be written in the following form:

$$Y = A + Bu + Cv + Du^2 + Ev^2 + Fuv$$

Note that only training dataset is considered for coefficient estimation.

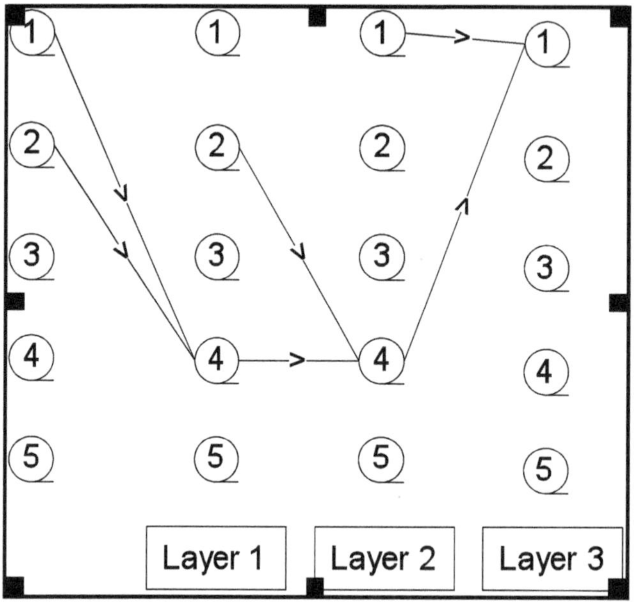

Fig. 9 PSO-GMDH selection process of non-promising node 4 of layer 1 which combines with other nodes and becomes promising node in layer 3

The above function can be transformed into the following matrix form:

$$[\mathrm{A} + \mathrm{Bu} + \mathrm{Cv} + \mathrm{Du}^2 + \mathrm{Ev}^2 + \mathrm{Fuv}] = [\mathrm{Y}] \tag{20}$$

$$\begin{bmatrix} 1 & u & v & u^2 & v^2 & uv \end{bmatrix}_{[nt \times 6]} \begin{bmatrix} A \\ B \\ C \\ D \\ E \\ F \end{bmatrix}_{[6 \times 1]} = [Y] \tag{21}$$

$$\text{or simply, } [\mathrm{X}][\mathrm{coeffs}] = [\mathrm{Y}] \tag{22}$$

where [coeffs] represents $[ABCDEF]^T$, vector of coefficients.

The least square technique from multiple-regression analysis provides the formula to get the coefficients in the following form 23:

$$[\mathrm{coeffs}] = (X^T X)^{-1} X^T Y \tag{23}$$

4.2.5 Selection of Input Variables through Swarm Particles for Next Layer

At each iteration of PSO, swarm particles use their corresponding parameter P_3 which utilizes the position (sequence of integers) of particles, to determine which input variables are to be combined to make a polynomial. These polynomials will become candidates for the selection of nodes for next layer. The selection is based on fitness function of the model which is simply the mean square error:

$$E_r = \frac{1}{n_{te}} \sum_{i=1}^{n_{te}} (y_o - y_c) \tag{24}$$

The lower the error the better the fitness. As shown in the Eq. 24 mean square error is calculated using testing dataset only. This provides the independent testing of model function.

Total particles P_T (normally $P_T >$) in each iteration are supposed to provide the selection guidelines using steps 1, 2 3, and 4 for M nodes of corresponding layer of GMDH process to move to next layer until termination condition is met. PSO will keep P_T particles and GMDH will keep M nodes.

4.2.6 Termination Criterion

PSO-GMDH uses PSO's termination method which is predetermined by the user by specifying number of iterations required for the process. PSO also has auto termination method which analyses the system and determines if no better solution is possible then the execution is terminated.

5 Experimentation

Few experiments are conducted to determine the feasibility and efficiency of this novel algorithm. The following dataset were used.

5.1 Tool Wear Problem

5.1.1 Experimental Setup

The problem discussed here is the end-milling experiment which was carried out on the Seiki Hitachi milling machine and also reported in chapter 4. A 16mm Co-high speed (HSS) Kobelco Hi Cut brand new end mill cutter was used to machine the work-piece. The end mill cutter had four flutes and the flute geometry was 30 degrees spiral. The overall length of the cutter was 77mm and the flute length was 26.5 mm. The work-piece machined was mid steel blocks which had a constant length of 100 mm for each trial. The machining was done under dry conditions. The milling experiment was conducted as designed. The work-piece used was mild steel which had a Brinell hardness number of 128.

Monitoring of the tool wear of the end mill was conducted in the toolmakers microscope. The 16mm Co-high speed (HSS) Kobelco Hi Cut brand new end mill cutter having 4 teeth was measured in the toolmakers microscope. A tool holder was designed so that it could hold the tool on the toolmakers microscope table and readings could be taken.

Tool wear for turning was monitored in the toolmakers microscope. The 55 degrees carbide insert with a positive rake angle of 7 degrees was removed from the tool holder and measured in the toolmakers microscope. A reference had to be made such that the distance from the reference line to the tip of the insert could be taken. It was very difficult to make a permanent reference line on the insert thus an insert holder was prepared. The insert holder was designed in such a way that the insert fitted inside the hole perfectly. The height of the surface of the insert was equal to height of the insert holder. Two reference lines were made on the insert holder, one at right angles to the tip of the insert which took into account the wear taking place on the nose of the insert and the other reference line was made parallel to the side of the insert which took into account the flank wear of the insert. A brand new end mill tool was measured on the toolmakers microscope from the reference line to the cutting edge. The tool was then used to machine a block of mild steel on the conditions of trial number 1. After machining the end mill was removed from the milling machine and the amount of wear on the end mill cutter was measured. The difference of the average of the first reading and the average of the current reading gave the extent of tool wear. Four readings were taken from each reference line and average of the readings was done.

5.1.2 Design of Experiment

The machining parameters were set during experimentation. This data set constituted the input to the self-organizing network and consisted of three inputs and one output. All inputs and were considered candidates to the causality relation-ship. For the specific application it was found that five replications were sufficient to yield a good approximation. The range of speed, feed, and depth of cut chosen for the experiments are respectively $v \in \{27, 39, 49\}$, $f \in \{0.0188, 0.0590, 0.1684\}$, $d_t \in \{0.5,\ 1.0,\ 1.5\}$.

5.1.3 Experimental Results and Discussions

All the twenty seven trials were conducted using the same end mill cutter and each time after milling the measurements for wear was taken. The average of the present measurement was subtracted from the previous one and the difference in the measurements gave the amount of wear. The cutting conditions and results obtained for the twenty seven trials using Mild steel blocks as work piece is shown in Table 15.

The turning parameters (see Table 15) that were fed into the enhanced-GMDH network shown in Figure 10 as inputs are x_1 = speed (v); $X - 2$ = feed, (f), and x_3 = depth-of-cut (DOC, d_t). The targets for the tool wear are given in the last columns of Table 15. The outputs of the e-GMDH learning network reported in this paper were used to develop the mathematical model of the tool wear in Sections 4.

Table 15 Cutting conditions and measured values for the milling operation

Trial #	Speed, v (m/min)	Feed, f (mm/rev)	DOC, d_t (mm)	Wear, VB (μm)
1	27	0.0188	0.5	8
2	27	0.0188	1	2.3
3	27	0.0188	1.5	2.6
4	27	0.059	0.5	1.7
5	27	0.059	1	2.45
6	27	0.059	1.5	2.7
7	27	0.1684	0.5	1.95
8	27	0.1684	1	2.55
9	27	0.1684	1.5	2.85
10	36	0.0188	0.5	2.9
11	36	0.0188	1	3.35
12	36	0.0188	1.5	4.35
13	36	0.059	0.5	3.105
14	36	0.059	1	3.55
15	36	0.059	1.5	4.5
16	36	0.1684	0.5	3.196
17	36	0.1684	1	3.95
18	36	0.1684	1.5	4.65
19	49	0.0188	0.5	4.95
20	49	0.0188	1	5.85
21	49	0.0188	1.5	7.7
22	49	0.059	0.5	5.2
23	49	0.059	1	6.25
24	49	0.059	1.5	10.2
25	49	0.1684	0.5	5.45
26	49	0.1684	1	6.75
27	49	0.1684	1.5	19.52

Fig. 10 shows the measured and estimated outputs for the tool wear problem. Figure 11 shows the GMDH network for the tool wear problem. Table 16 shows the database of output results for the tool wear problem.

The interpretation of the network for each of the four layers is as follows:

$$\begin{aligned}
&y_1 = f(x_1, x_3);\; y_2 = f(x_1, x_3);\; y_3 = f(x_2, x_3)\\
&y_4 = f(y_1, y_2);\; y_5 = f(y_2, y_3);\; y_6 = f(y_2, y_3)\\
&y_7 = f(y_4, y_5);\; y_8 = f(y_5, y_6)\\
&y_9 = f(y_7, y_8)
\end{aligned}$$

However, these have to be corrected mapped to the database results shown in Table 16. For the database results, layer 1 outputs are first given followed by those for layer 2, and so on until the last layer is reached. In layer 1, y_1, y_2 and y_3 of the network correspond to y_3, y_2 and y_1 of the database respectively. In layer 2, y_4, y_5 and

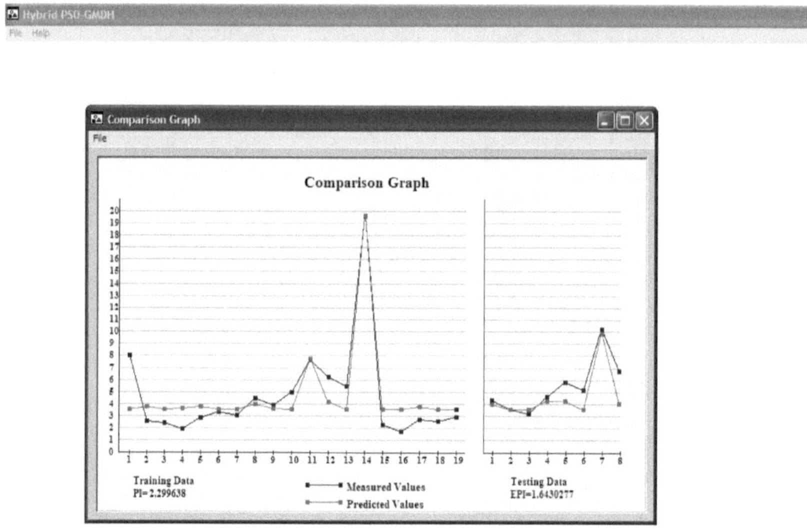

Fig. 10 Measured and estimated outputs for the tool wear problem

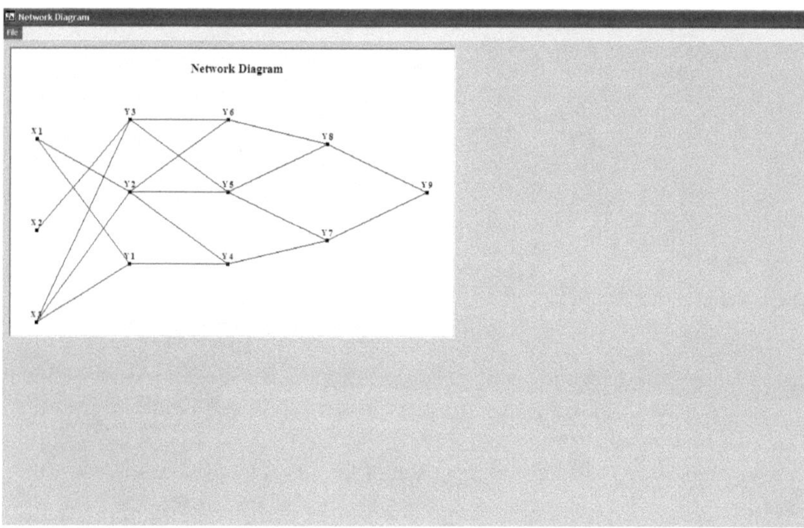

Fig. 11 GMDH network for the tool wear problem

y_6 of the network correspond to z_1, z_3 and z_2 of the database respectively. In layer 3, y_7 and y_8 of the network correspond to w_1 and w_2 of the database respectively. In layer 4, y_9 of the network correspond to v_1 of the database respectively.

The coefficients per node are also shown in Table 16. For example the coefficients of node y_3 in layer 1 are 3.53183, 3.41982, -0.43112, 0.091457, 2.60811, and

Table 16 Database of output results for the tool wear problem

PARTICLE SWARM OPTIMIZATION								
Processing time: 1.891 seconds								
Iterations: 4								
Evaluations: 84								
Optimum value :			1.64303					
Best Layer at:			3					
EPI:			1.64303					
PI:			2.29964					
Network of Layers								
3	3	1	3.53183	3.41982	-0.43112	0.091457	2.60811	0.009742
2	3	1	3.53183	3.41982	-0.43112	0.091457	2.60811	0.009742
1	3	2	3.79263	-2.15968	0.978243	1.54233	2.59583	0.210299
3	3	2	3.69111	-2.34E-06	-0.39864	0.033452	0.000891	9.06E-11
2	2	1	3.811	-0.36014	0.274229	-0.07148	0.049272	-6.18E-05
1	2	1	3.811	-0.36014	0.274229	-0.07148	0.049272	-6.18E-05
2	3	2	4.00562	-7.1202	6.50868	-1.27664	1.36893	1.30E-06
1	1	2	4.00176	-6.68E-06	-0.38546	0.083097	-4.28E-05	-3.64E-10
1	2	1	3.93398	-0.66593	0.30708	-0.02418	0.094024	1.66E-05
Training Data Set								
8			3.59999					
2.6			3.74768					
2.45			3.5648					
1.95			3.61073					
2.85			3.78751					
3.35			3.58753					
3.105			3.5448					
4.5			4.02922					
3.95			3.6568					
4.95			3.56642					
7.7			7.71615					
6.25			4.19115					
5.45			3.55488					
19.52			19.5956					
2.3			3.55567					
1.7			3.60276					
2.7			3.7572					
2.55			3.59692					
2.9			3.54588					
Testing/Checking Data Set								
4.35			3.96518					
3.55			3.60264					
3.196			3.54198					
4.65			4.25716					
5.85			4.25657					
5.2			3.56175					
10.2			9.84692					
6.75			4.0349					

The parameters used for the modeling of this problem are:

Swarm size	0
Maximum best column	0
Even row selection	1
Maximum tour	9000000
Intensity 'n' = no; 'y' = yes	n
Maximum evaluation	4
Convergence case	5

0.009742. It therefore becomes easy to determine the connections per layer until the final output as well as the equations connecting the nodes. The polynomial for a node is defined using these coefficients as:

$$f(x_i,x_j) = c_1 + c_2x_i + c_3x_j + c_4x_ix_j + c_5x_i^2 + c_6x_j^2$$

As could be observed, very nonlinear set of equations can occur and certainly the degree of the polynomials increases by the power of 2 from layer to layer (l^2) which means that for layer 2, the degree of polynomial is 4, etc.

5.2 *Gas Furnace Problem*

The problem solved is the "Box and Jenkins furnace data" [28] which is a benchmark problem often reported in the literature. There are originally 296 data points {y(t),u(t)}, from t=1 to t=296. In this problem, y(t) is the output CO2 con-centration and u(t) is the input gas flow rate. Here we are trying to predict y(t) based on

{y(t-1), y(t-2), y(t-3), y(t-4), u(t-1), u(t-2), u(t-3), u(t-4), u(t-5), u(t-6)}.

This reduces the number of effective data points to 290. Most methods find that the best set of input variables for predicting y(t) is {y(t-1),u(t-4)}. Sugeno and Yasukawa has found that the best set of input variables for predicting y(t) is {y(t-1), u(t-4), u(t-3)}.

output y(t);
input y(t-1) y(t-2) y(t-3) y(t-4) u(t-1) u(t-2) u(t-3) u(t-4) u(t-5) u(t-6)

5.2.1 PSO-GMDH Modeling

The number of swarm size is given by the user but when PSO-GMDH is re-quired to determine this automatically the value of '0' is given. This is the case for the maximum best column also. For the selection for the test data, three cases exist: random (0), even (1), and odd (2). When random selection is made, PSO-GMDH randomly selects data for testing but when 1 or 2 is selection even or odd rows of data are selected for testing. When a brute-force search is required, then intensity is 'yes' or 'y' otherwise, the choice is 'no' or 'n'. the maximum evaluation is set by the user or left to PSO-GMDH to determine if '0' is chosen. There are five convergences cases to choose from.

The results generated by the PSO-GMDH after each iteration are reported as:

After iteration 1 the best particle is particle [1]
position: 7 2 1 9 6 5 3 8 4 10 7
velocity: (size = 1) : magnitude [1 , 7]
objective functional value: 0.0801373
distance: 0
After iteration 2 the best particle is particle [7]
position: 1 6 3 4 2 5 7 8 9 10 1
velocity: (size = 6) : magnitude [9 , 5] [5 , 4] [2 , 6] [4 , 9] [4 , 2] [9 , 5]
objective functional value: 0.0714395
distance: 0
After iteration 3 the best particle is particle [7]
position: 1 6 3 4 2 9 7 8 5 10 1
velocity: (size = 6) : magnitude [9 , 5] [5 , 4] [2 , 6] [4 , 9] [4 , 2] [9 , 5]
objective functional value: 0.0687775
distance: 0
After iteration 4 the best particle is particle [7]
position: 1 6 3 4 2 9 7 8 5 10 1
velocity: (size = 6) : magnitude [9 , 5] [5 , 4] [2 , 6] [4 , 9] [4 , 2] [9 , 5]
objective functional value: 0.0687775

Figure 12 shows the measured and estimated outputs for the gas furnace problem. The training error of 0.126 and the testing error of 0.068 are displayed. These errors are base on mean square error (MSE). The testing error is often used in the literature to determine the efficacy of a modeling approach; this gives more information than the graph.

Figure 13 shows the GMDH network for the gas furnace problem. There are 10 inputs and one output, but we have only three inputs shown in the network. The explanation is simple. Only parameters 2, 3 and 10 are connected to the three best nodes in layer 1. This means that the objective functions of pair-wise combina-tions of parameters 1, 4, 5, 6, 7, 8, and 9 do not give competitive nodes, so those nodes are relegated to the background; in other words they are dropped.

Table 17 shows the database of output results for the gas furnace problem. This database is rich in information because it gives the network topology and the equations for each node. A bit of interpretation of the output data is necessary. Whereas the network shows that y_1 is connected to x_2 and x_3, y_2 is connected x_2 and x_{10} while y_3 is connected x_2 and x_{10} while the database gives the true interpre-tation that y_7 is connected to x_2 and x_3, y_2 is connected x_2 and x_{10} while y_1 is connected x_2 and x_{10}. The differences are merely due to the ways of numbering on the network and internally in the database. The database gives the exact connections but the network labels are sequential for the active neurons. Non-active neurons are not numbered on the network but internally the numbering is maintained for the database archiving.

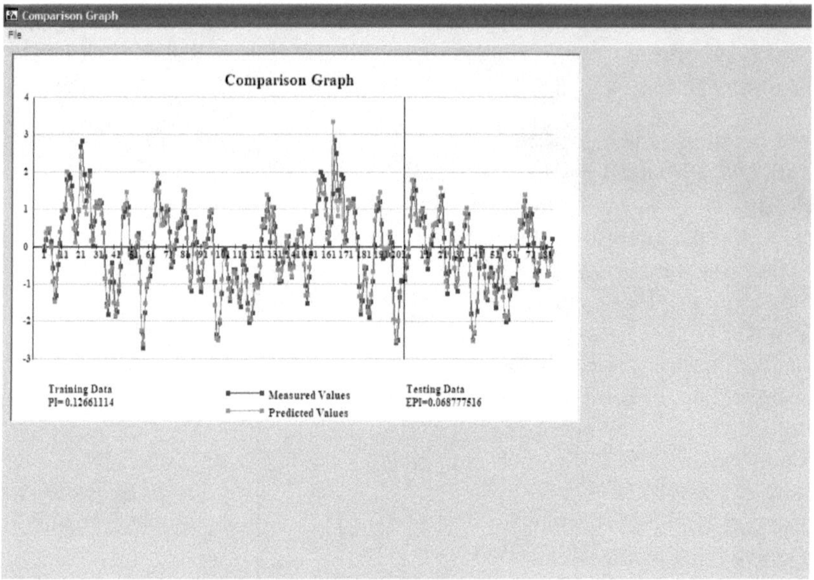

Fig. 12 Measured and estimated outputs for the gas furnace problem

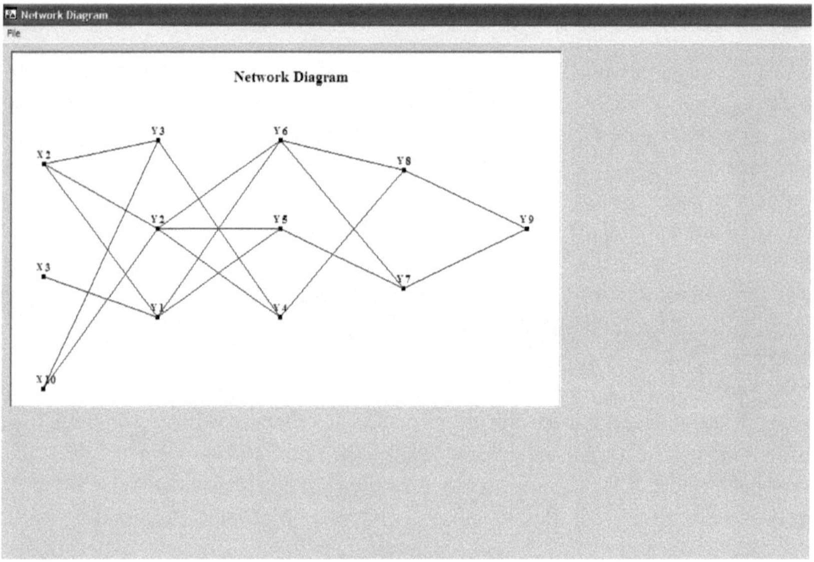

Fig. 13 GMDH network for the gas furnace problem

Table 17 Database of output results for the gas furnace problem

PARTICLE SWARM OPTIMIZATION								
Optimum value :			0.068778					
Best Layer at:			3					
EPI:			0.068778					
PI:			0.126611					
Network of Layers								
7	2	3	-0.99279	0.315822	0.365218	-0.00781	0.001328	6.38E-11
2	10	2	-0.99713	-0.0828	0.018472	0.018933	0.055631	1.32E-06
1	10	2	-0.99713	-0.0828	0.018472	0.018933	0.055631	1.32E-06
1	7	2	-0.9924	-1.67656	0.083294	0.063762	0.103198	0.002982
6	1	2	-2.6647	-3.86E-06	1.13451	1.03578	1.89E-05	-3.96E-08
2	7	2	-0.9924	-1.67656	0.083294	0.063762	0.103198	0.002982
1	1	6	-1.00367	1.35655	-0.34384	0.048684	0.282728	-0.00344
6	1	2	-2.4166	8.10E-06	1.26712	0.730777	2.12E-05	-5.85E-08
1	1	6	-1.00337	1.51282	-0.43421	-0.04207	0.375178	0.000249
Processing time: 8.188 seconds								
Iterations: 4								
Evaluations: 872								

6 Conclusion

This chapter has presented a hybrid PSO-GMDH for modeling and prediction of complex, real-life problems. The inherent setbacks found in the classical GMDH are resolved when a hybrid of PSO and GMDH is realized. The hybrid PSO-GMDH architecture realized is more flexible in determining network topol-ogy of the problem being solved. The hybrid PSO-GMDH learns much more easily and generalizes much better than the classical GMDH.

In this chapter, we have presented for the first time a hybrid PSO-GMDH framework because we are not aware of any literature that has reported this. Two types of problems have been solved, one is based on experimentation in a manu-facturing laboratory in which the controlling parameters were used to generate output response in the form of tool wear; the other is the time series problem. Although the results obtained are not as competitive as those obtained in chapter 4 for tool wear problem, they are quite promising.

The man-machine interface for the hybrid PSO-GMDH software developed and reported in this chapter is very conducive for utilizing as a data mining platform.

Acknowledgements

Amal Shankar, Ashwin Dayal, Deepak Bhartu, and Kenneth Katafono were responsible for integrating the various PSO and GMDH components and writing the GUI codes under the supervision of the two authors of this chapter. The authors of this chapter provided the full PSO and partial GMDH codes.

References

1. Farlow, S.J.: The GMDH Algorithm of Ivakhnenko. The American Statistician 35(4) (1981)
2. Hamming, R.W., Feigenbaum, E.A.: Interpolation and Roundoff Estimation. In: Introduction to Applied Numerical Analysis, pp. 143–148. McGraw-Hill, New York (1971)
3. Zaychenko, Y.P., Kebkal, A.G., Krachkovckii, V.F.: The Fuzzy Group Method of Data Handling and its Application to the Problems of the Macroeconomic Indexes Forecasting (2007), `http://www.gmdh.net/`
4. Neapolitan, R.E., Naimipour, K.: The greedy approach, Fondation of Algorithms using C++ Pseudocode. Jones and Bartlet publishers Inc. (2003)
5. Onwubolu, G.C.: Design of Hybrid Differential Evolution and Group Method of Data Handling for Inductive Modeling. In: International Workshop on Inductive Modeling, IWIM Prague, Czech, pp. 23–26 (2007)
6. Kreyszig, E.: Unconstrained optimization, linear programming. In: Advance Engineering Mathematics, 2nd edn. John Wiley, Inc., Chichester (1993)
7. Eberhart, R.C., Kennedy, J.: A new optimizer using particle swarm theory. In: Proc. Sixth International Symposium on Micro Machine and human science, Nagoya, Japan. IEEE Service Center, Piscataway (1995)
8. Clerc, M.: Discrete particle swarm optimization illustrated by the traveling salesman problem. In: New Optimization techniques in Engineering. Springer, Berlin (2004)
9. Carlistle, A., Dozier, G.: Adapting Particle Swarm Optimization to Dynamic Environments (1998), `http://www.CartistleA.edu`
10. Kennedy, J., Eberhart, R.C.: The particle swarm: social adaptation in information processing systems. In: Corne, D., Dorigo, M., Glover, F. (eds.) New Ideas in Optimization, pp. 379–387. McGraw-Hill, London (1999)
11. Kennedy, J., Eberhart, R.C.: A discrete binary version of the particle swarm algorithm. In: International Conference on Systems, Man, and Cybernetics (1997)
12. Kennedy, J.: The particle swarm: social adaptation of knowledge. In: IEEE international conference on Evolutionary computation, indianpolis, Indiana. IEEE Service Center, Piscataway (1997)
13. Clerc, M., Kennedy, J.: The particle swarm: explosion, stability, and convergence in a multidimensional complex space. IEEE transactions on Evolutionary Computation 6, 58–73 (2002)
14. Kennedy, J.: Small worlds and mega-minds: effects of neighborhood topology on particle swarm performance. In: Congress on Evolutionary computation, Washington D.C. IEEE, Los Alamitos (1999)
15. Clerc, M.: The Swarm and the queen: towards a deterministic and adaptive particle swarm optimization. In: Congress on Evolutionary Computation, Washington D.C, pp. 1951–1957. IEEE Service Center, Piscataway (1999)

16. Kennedy, J.: Stereotyping: Improving Particle Swarm Performance with Cluster Analysis. Presented at Congress on Evolutionary Computation (2000)
17. Kennedy, J., Spears, W.: Matching algorithms to problems: An experimental test of the particle swarm and some genetic algorithms on the multimodal problem generator. In: Proceedings of the IEEE Congress on Evolutionary Computation (CEC 1998), Anchorage, Alaska (1998)
18. Onwubolu, G.C., Sharma, A.: Particle Swarm Optimization for the assignment of facilities to locations. In: New Optimization Techniques in Engineering. Springer, Heidelberg (2004)
19. He, Z., Wei, C.: A new population-based incremental learning method for the traveling salesman problem. In: Congress on Evolutionary Computation, Washington D.C. IEEE, Los Alamitos (1999)
20. Press, W.H., Teukolsky, S.A., Vetterling, W.T., Flannery, B.P.: Numerical recipes in C: The art of scientific computing. Cambridge University Press, Cambridge (1992)
21. Nariman-Zadeh, N., Darvizeh, A., Ahmad-Zadeh, G.R.: Hybrid genetic design of GMDH-type neural networks using singular value decomposition for modeling and predicting of the explosive cutting process, Nariman-Zadeh. In: Proc. Instn Mech. Engrs Vol 217 Part B: Nariman-Zadeh, 779–790 (2003)
22. Ivakhnenko, A.G.: The Group Method of Data Handling-A rival of the Method of Stochastic Approximation. Soviet Automatic Control, vol 13 c/c of avtomatika 1(3), 43–55 (1968)
23. Larson, R., Edward, B.H., Falvo, D.C.: Application of Matrix Operations, 5th edn. Elementary Linear Algebra, pp. 107–110. Houghton Mifflin, New York (2004)
24. Holland, J.H.: Adaptation in Natural and Artificial Systems. University of Michigan Press, Ann Arbor (1975)
25. Glover, F.: Heuristics for interger programming using surrogate constraints. Decision Sciences 8, 156–166 (1977)
26. Kirkpatrick, S., Gelatt, C.D., Vecci, M.P.: Optimization by Simulated Annealing. Science 220(4598), 671–680 (1983)
27. Dorigo, M.: Optimization, Learning and Natural Algorithm, PhD thesis, Politecnico di Milano, Italy (1992)
28. Box, G.E.P., Jenkins, G.M.: Time Series Analysis, Forecasting and Control, pp. 532–533. Holden Day, San Francisco (1970)

GAME – Hybrid Self-Organizing Modeling System Based on GMDH

Pavel Kordík

Abstract. In this chapter, an algorithm to construct hybrid self-organizing neural network is proposed. It combines niching evolutionary strategies, nature inspired and gradient based optimization algorithms (Quasi-Newton, Conjugate Gradient, GA, PSO, ACO, etc.) to evolve neural network with optimal topology adapted to a data set. The GAME algorithm is something in between the GMDH algorithm and the NEAT algorithm. It is capable to handle irrelevant inputs, short and noisy data samples, but also complex data such as "two intertwined spirals" problem. The self-organization of the topology allows it to produce accurate models for various tasks (classification, prediction, regression, etc.). Bencharking with machine learning algorithms implemented in the Weka software showed that the accuracy of GAME models was superior for both regression and classification problems. The most successful configuration of the GAME algorithm is not changing with problem character, natural evolution selects all important parameters of the algorithm. This is a significant step towards the *automated data mining*.

1 Introduction

In this chapter, you will find a description of the recently introduced Adaptive Models Evolution (GAME) algorithm [24] with respect to its self-organizing properties and the hybrid nature of its building blocks. The GAME algorithm uses a data driven approach. Resulting models fully reflect the character of a data set used for training. For simple problems, it evolves simple models (in terms of topology and transfer functions) and for a complex relationship of variables, a complex model is evolved.

Pavel Kordík
Department of Computer Science and Engineering, FEE, Czech Technical University, Prague, Czech Republic
e-mail: `kordikp@fel.cvut.cz`

G.C. Onwubolu (Ed.): Hybrid Self-Organizing Modeling Systems, SCI 211, pp. 233–280.
springerlink.com

The GAME algorithm proceeds from the Multilayered Iterative Algorithm (MIA *GMDH*) [18].

GAME models are *self-organized* layer by layer by means of a special genetic algorithm preserving diverse solutions.

The optimization of transfer functions in neurons (or units or partial descriptions in the GMDH terminology) is solved independently. Several optimization methods compete to adjust the coefficients of transfer functions.

Neurons can be of several types – the polynomial transfer function can be good at fitting certain relationships, but often a different transfer function is needed (e.g. sigmoid for classification purposes). GAME models mostly consist of several different types of neurons that are optimized by different methods – they have a *hybrid* character. Also the ensemble of GAME models is often produced to get even better bias-variance trade-off and to be able to estimate the credibility of the output for any configuration of input variables.

The hybrid character of GAME models and their self-organizing ability give them an advantage over standard data mining models. Our experiments show that the performance of hybrid models is superior on a large scale of different data sets.

Below, you will find a detailed description of the GAME algorithm and the ideas behind it.

1.1 Self-Organizing Modelling

The Group Method of Data Handling (GMDH) was invented by A.G. Ivakhnenko in the late 1960s [18]. He was looking for computational instruments allowing him to model real world systems characterized by data with many inputs (dimensions) and few records. Such ill-posed problems could not be solved traditionally (ill-conditioned matrixes) and therefore a different approach was needed. Prof. Ivakhnenko proposed the GMDH method, which avoided the solution of ill-conditioned matrixes by decomposing them into submatrices of lower dimensionality that could be solved easily. The main idea behind the GMDH is the adaptive process of combining these submatrices back to the final solution, together with external data set validation preventing data overfitting. The original GMDH method is called Multilayered Iterative Algorithm (MIA GMDH). Many similar GMDH methods based on the principle of induction (problem decomposition and combination of partial results) have been developed since then.

The only possibility of modelling real world systems before the GMDH was to manually create a set of mathematical equations mimicking the behavior of a real world system. This involved a lot of time, domain expert knowledge and also experience with the synthesis of mathematical equations.

The GMDH allowed for the automatic generation of a set of these equations. A model of the real world system can also be created by Data Mining (DM) algorithms, particularly by artificial Neural Networks (NNs).

Some DM algorithms such as decision trees are simple to understand, whereas NNs often have so complex structure that they are necessarily treated as a black-box model. The MIA GMDH is something in between - it generates polynomial equations which are less comprehensible than a decision tree, but better interpretable than any NN model.

The main advantage of the GMDH over NNs is that the optimal topology of the network (number of layers and neurons, transfer functions) is determined automatically. Traditional neural networks such as MLP [32] require the user to experiment with the size of the network. Recently, some NNs also adopted the self-organizing principle and induct topology of models from data.

The MIA GMDH builds models layer by layer, while the accuracy of the model on the validation data set increases (as described in the Chapter 1. of this book). However, the accuracy of resulting models is not very high when applied to different benchmarking problems [27]. The reason is that it selects the optimal model from a very limited state space of possible topologies.

1.1.1 Limitations of MIA GMDH

The MIA GMDH (see Fig. 1) was invented 40 years ago and therefore incorporates several limitations that were necessary to make the computation feasible.

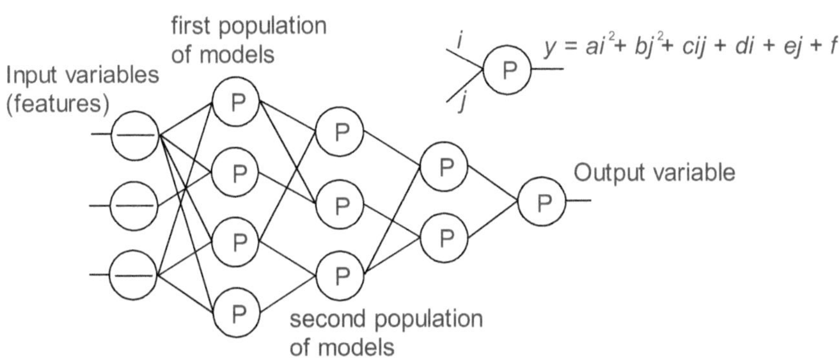

Fig. 1 An example of the inductive model produced by the GMDH MIA algorithm.

The limitations of the MIA GMDH are as follows:

- All neurons have the same transfer function
- Transfer function is simple polynomial
- Each polynomial neuron has exactly two inputs
- Inputs are chosen from the previous layer only

These structural limitations allow the MIA algorithm to check all possible combinations of a neuron's interconnections and choose the best. In other words, the algorithm searches the whole state space of possible MIA GMDH topologies, and returns the optimal topology for this state space.

The problem is that all other possible topologies of models (e.g. model containing a neuron with 4 inputs and sigmoid transfer function) are not examined, although they can provide us with a better solution.

If we decide to drop GMDH MIA limitations, our search space expands in both size and dimensionality. With new degrees of freedom (number of layers and neurons, interconnections of neurons, their type of transfer functions, values of coefficients, type of optimization method used etc.), the search space of all possible topologies becomes mind-bogglingly huge.

Advances in computer technology and the appearance of modern heuristic optimization methods allow an algorithm to navigate through the search space efficiently, obtaining almost the optimal solution in a reasonable time.

The experimental results show that the GAME algorithm, we have proposed for this purpose, outperforms GMDH MIA considerably.

1.1.2 Self-organizing Neural Networks

Self Organizing Map (SOM) [22] is a typical example of how *self-organization* is understood in the area of neural networks. This network has fixed topology and neurons self-organize during training by weights updates to reflect the density of data in hyperspace. In the Growing SOM [40] variant, the matrix of neurons increases in size (and dimension) from the minimal form – the topology is self-organized as well. These networks are unsupervised, whereas this book focuses mainly on supervised methods.

Supervised neural networks such as MLP have fixed topology and only weights and neuron biases are the subject of training. A suitable topology for given problem has to be determined by user, usually by an exhaustive trial and error strategy.

Some more recently introduced neural networks demonstrate self-organizing properties in terms of topology adaptation.

The Cascade Correlation algorithm [10] generates a feedforward neural network by adding neurons one by one from a minimal form. Once a neuron has been added to the network, its weights are frozen. This neuron then becomes a feature-detector in the network, producing outputs or creating other feature detectors. This is a very similar approach to the MIA GMDH as described in Chapter 1.

It has been shown [10] that cascade networks perform very well on the "two intertwined spirals" benchmarking problem (a network consisting of less than 20 hidden neurons was able to solve it) and the speed of training outperformed Backpropagation.

According to experiments on real-world data performed in [52], the algorithm has difficulties with avoiding premature convergence to complex topological structures. The main advantage of the Cascade Correlation algorithm is also its main disadvantage. It easily solves extremely difficult problems therefore it is likely to overfit the data.

In the next section, we introduce a robust algorithm that can generate feedforward neural networks with adaptive topology – learning, structure of network and transfer functions of neurons are tailored to fit the data set.

2 Group of Adaptive Models Evolution (GAME)

2.1 *The Concept of the Algorithm*

The Multilayered perceptron neural networks trained by the Backpropagation algorithm [41] are very popular even today, when many better methods exist. The success of this paradigm is mostly given by its robustness. It works reasonably well for a large scale of problems of different complexity, despite the fixed topology of the network and uniform transfer functions in neurons. The Group of Adaptive Models Evolution (GAME) algorithm, proposed in this chapter, has the ambition to be even more robust and more adaptive to individual problems. The topology of GAME models adapts to the nature of a data set supplied.

2.1.1 The Pseudocode of the GAME Algorithm

The GAME algorithm is a supervised method. For training, it requires a dataset with input variables (features) and the output variable (target).

The GAME algorithm, in summary, is described below:

1. Separate the validation set from the training data set (50% random subset)
2. Initialize first population of neurons – input connections, transfer functions and optimization methods chosen randomly
3. Optimize coefficients of neurons' transfer functions by assigned optimization method – error and gradients computed on the training data set
4. Compute fitness of neurons by summarizing their errors on the validation set
5. Apply Deterministic Crowding to generate new population (randomly select parents, competition based on fitness and distance to be copied into the next generation)
6. Go to 3), until diversity level is to low or the maximum number of epochs (generations) is reached
7. Select the best neuron from each niche – based on fitness and distance, freeze them to make up the layer and delete the remaining neurons
8. 8) Until the validation error of the best neuron is significantly lower than the best neuron from the previous layer, proceed with the next layer and go to 2)
9. Mark the neuron with the lowest validation error as the output of the model and delete all neurons not connected to the output

2.1.2 An Example of the GAME Algorithm on the Housing Dataset

We will demonstrate our algorithm on the Housing dataset that can be obtained from the UCI repository [2]. The dataset has 12 continuous input variables and one continuous output variable. In Fig. 2 you can find the description of the most important variables.

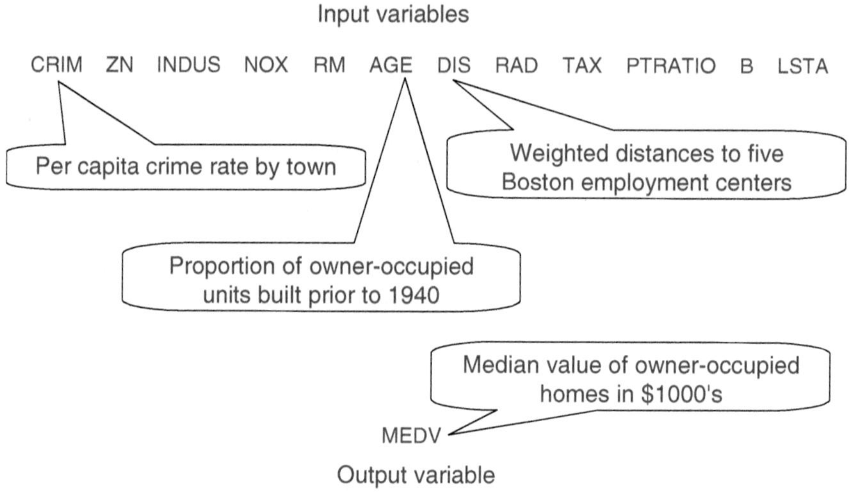

Fig. 2 Housing data set: the description of the most important variables.

Firstly, we split the data set into a subset used for training (A+B) and a test set to get an unbiased estimate of the model's error (see Fig. 3). Alternatively, we can perform k-fold crossvalidation [21].

Then we run the GAME algorithm, which separates out the validation set (B) for the fitness computation and the training set (A) for the optimization of coefficients.

In the second step, the GAME algorithm initializes the population of neurons (the default number is 15) in the first layer. For instant GAME models, the preferable option is *growing complexity* (number of input connections is limited to index of layer). Under this scheme, neurons in the first layer cannot have more than one input connection, as shown in Fig. 2.1.2. The type of the transfer function is assigned randomly to neurons, together with the type of method to be used to optimize coefficients of transfer function.

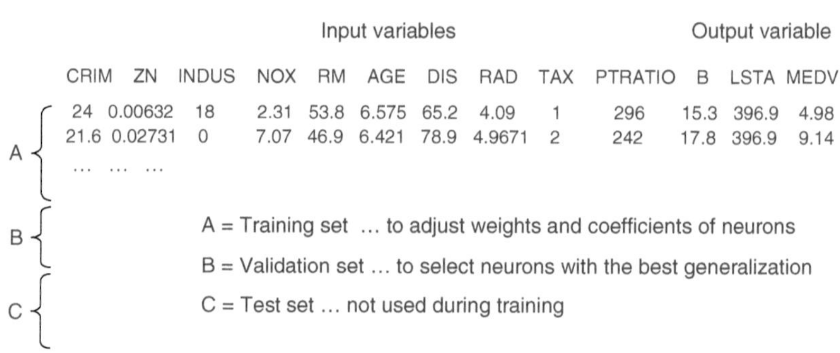

Fig. 3 Splitting the data set into the training and the test set; the validation set is separated from the training set automatically during GAME training.

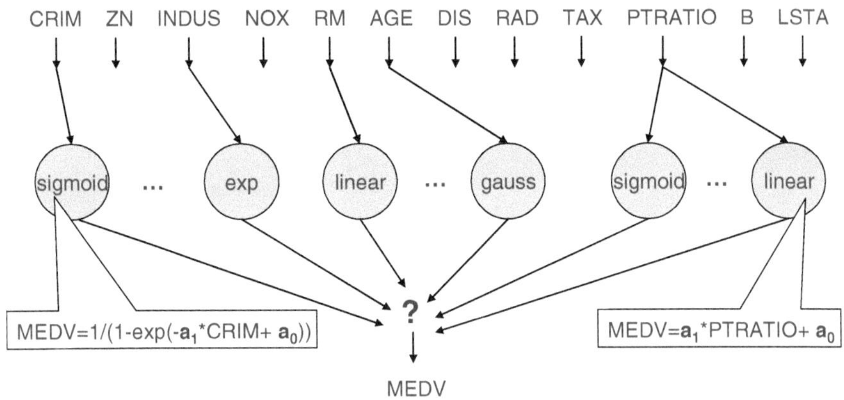

Fig. 4 Initial population of neurons in the first GAME layer.

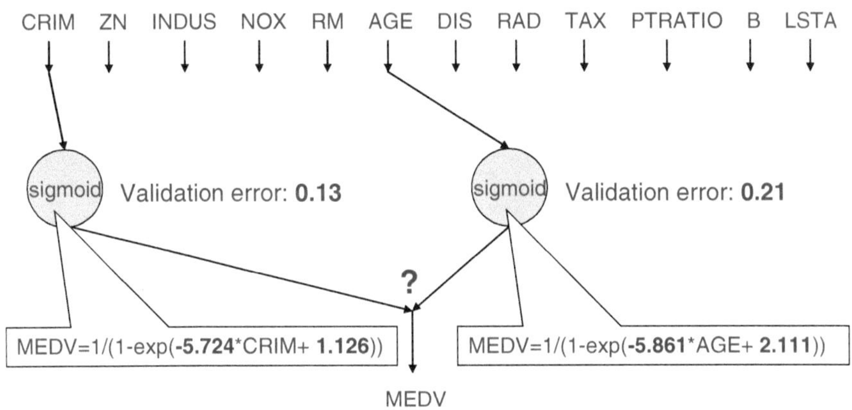

Fig. 5 Two individuals from different niches with coefficients optimized on set A and validated on set B. The neuron connected to the AGE feature has a much higher validation error than neurons connected to CRIM and survives thanks to *niching*.

The type of transfer function can be sigmoid, Gaussian, linear, exponential, sine and many others (a complete and up-to-date list of implemented transfer functions is available in the FAKEGAME application [5]), see Fig. 5. If the sigmoid transfer function is assigned to a neuron, the output of this neuron can be computed, for example as MEDV=1/(1-exp(-$\mathbf{a}_1$*CRIM+ $\mathbf{a}_0$)), where coefficients $\mathbf{a}_0$ and $\mathbf{a}_1$ are to be determined.

To determine these coefficients, an external optimization method is used. The optimization method is chosen from a list of available methods (Quasi-Newton, Differential Evolution, PSO, ACO, etc.).

The fitness of each individual (neuron) is computed as the inverse of its validation error. The genetic algorithm performs selection, recombination and mutation and the

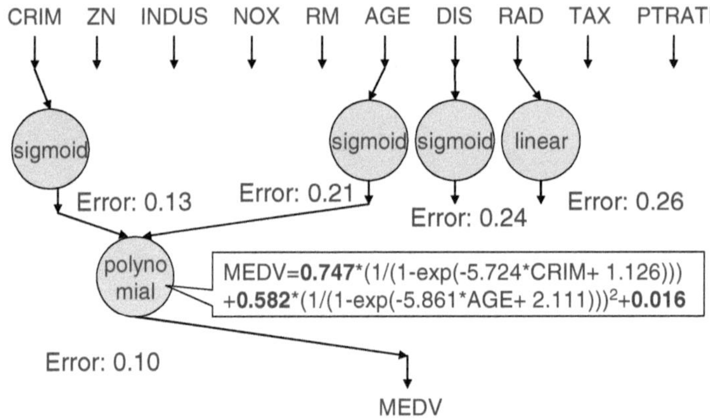

Fig. 6 In our example, the best individual evolved in the second layer combines the outputs of neurons frozen in the first layer (feature detectors)

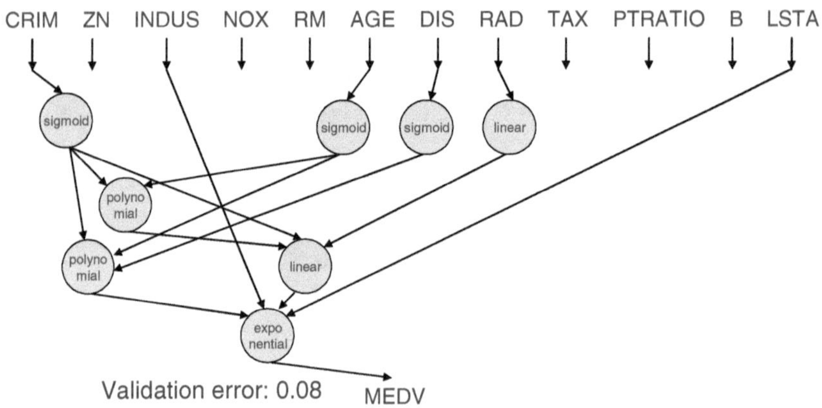

Fig. 7 The GAME model of the MEDV variable is finished when new layers do not decrease the validation error significantly.

next population is initialized. After several epochs, the genetic algorithm is stopped and the best neurons from individual niches are frozen in the first layer (Fig. 6).

Then the GAME algorithm proceeds with the second layer. Again, an initial population is generated with random chromosomes, evolved by means of the niching genetic algorithm, and then the *best and diverse* neurons are selected to be frozen in the second layer.

The algorithm creates layer by layer, until the validation error of the best individual decreases significantly. Fig. 2.1.2 shows the final model of the MEDV variable.

2.2 *Contributions of the GAME Algorithm*

The GAME algorithm proceeds from the MIA GMDH algorithm. In this section, we summarize improvements to the Multilayered GMDH as described in Chapter 1.

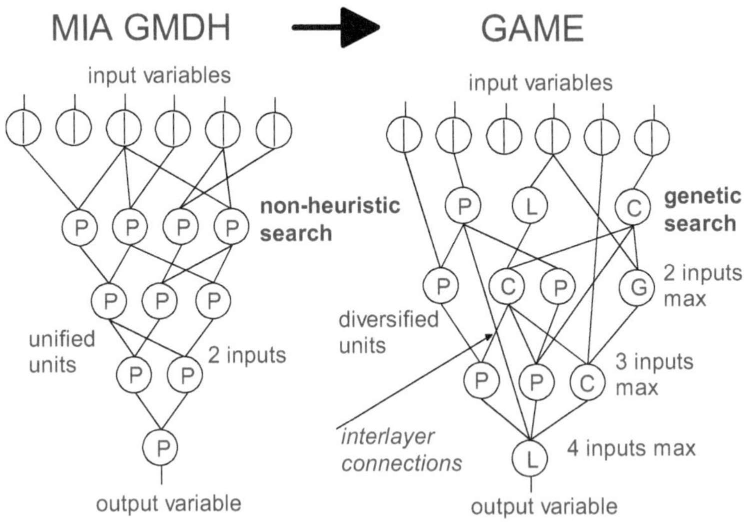

Fig. 8 Comparison: original MIA GMDH network and the GAME network

The Fig. 8 illustrates the difference between models produced by the two algorithms.

The GAME model (see Fig. 8, right) has more degrees of freedom (neurons with more inputs, interlayer connections, transfer functions etc.) than MIA GMDH models. To search the huge state space of model's possible topologies, the GAME algorithm incorporates the niching genetic algorithm in each layer.

Improvements to the MIA GMDH are discussed below in more detailed form.

- *Heterogeneous neurons* - several types of neurons compete to survive in GAME models.
- *Optimization of neurons* - Efficient gradient based training algorithm developed for hybrid networks.
- *Heterogeneous learning methods* - Several optimization methods compete to build the most successful neurons.
- *Structural innovations* - Growth from a minimal form, interlayer connections etc.
- *Regularization* - Regularization criteria are employed to reduce the complexity of transfer functions.
- *Genetic algorithm* - A heuristic construction of GAME models. Inputs of neurons are evolved.
- *Niching methods* - Diversity promoted to maintain less fit but more useful neurons.

- *Evolving neurons (active neurons)* - Neurons such as the CombiNeuron evolve their transfer functions.
- *Ensemble of models generated* - Ensemble improves accuracy; the credibility of models can be estimated.

2.2.1 Heterogeneous Neurons

In MIA GMDH models, all neurons have the same polynomial transfer function. The Polynomial Neural Networks (PNN) [36] models supports multiple types of polynomials used within a single model.

Our previous research showed that employing heterogeneous neurons within a model gives better results when using neurons of a single type only [27]. Hybrid models are often more accurate than homogeneous ones, even if the homogeneous model has a suitable transfer function appropriate for modelled system.

In GAME models, neurons within a single model can have several types of transfer functions (Hybrid Inductive Model). Transfer functions can be linear, polynomial, logistic, exponential, Gaussian, rational, perceptron network etc. (see Table 1 and Fig. 9).

The motivation, for implementing so many different neurons was as follows. Each problem or data set is unique. Our previous experiments showed [27] that for simple problems, models with simple neurons were superior, whereas for complex

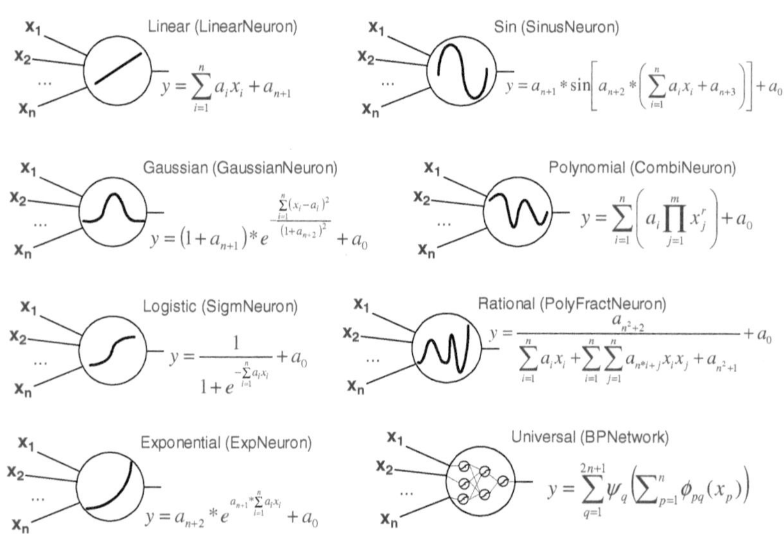

Fig. 9 Neurons are building blocks of GAME models. Transfer functions of neurons can be combined in a single model (then we call it a hybrid model with heterogeneous neurons). The list of neurons includes some neurons implemented in the FAKE GAME environment.

Table 1 Summary of neuron types appearing in GAME networks

Name of neuron	Transfer function.	Learning method
LinearNeuron	Linear	- any method -
LinearGJNeuron	Linear	Gauss-Jordan method
CombiNeuron	Polynomial	- any method -
PolySimpleNeuron	Polynomial	- any method -
PolySimpleGJNeuron	Polynomial	Gauss-Jordan method
PolyHornerNeuron	Polynomial	- any method -
PolySimpleNRNeuron	Polynomial	- any method + GL5 -
SigmNeuron	Sigmoid	- any method -
ExpNeuron	Exponential	- any method -
PolyFractNeuron	Rational	- any method -
SinusNeuron	Sinus	- any method -
GaussNeuron	Gaussian	- any method -
MultiGaussNeuron	Gaussian	- any method -
GaussianNeuron	Gaussian	- any method -
BPNetwork	Universal	BackPropagation algorithm
NRBPNetwork	Universal	BP alg. + GL5 stop.crit.

problems, the winning models were those with neurons having a complex transfer function. The best performance on all tested problems was achieved by models where the neurons were mixed.

2.2.2 Experiments with Heterogeneous Neurons

To prove our assumptions and to support our preliminary results [27], we designed and conducted the following experiments. We used several real world data sets of various complexity and noise levels. For each data set, we built simple ensembles [13] of 10 models. Each ensemble had a different configuration. In ensembles of homogeneous models, there was just a single type of neuron allowed (e.g. **Exp** stands for an ensemble of 10 models consisting of ExpNeuron neurons only). Ensembles of heterogeneous inductive models, where all types of neurons are allowed to participate in the evolution, are labelled **all**, **all-simple** and **all-fast** respectively. In the all-simple configuration, Linear, Sigmoid and Exponential functions were enabled, in the all-fast configurations, Linear, Sigmoid, Polynomial, Exponential, Sine, Rational and Gaussian transfer functions were employed.

For all experiments in this section, we used only one optimization method (Quasi-Newton) to avoid biased results.

The first experiment was performed on the Building data set. This data set has three output variables. One of these variables is considerably noisy (Energy consumption) and the other two output variables have low noise levels. The results are consistent with this observation. The **Combi** and the **Polynomial** ensembles perform very well on variables with low noise levels, but for the third, “noisy” variable, they both overfitted the training data (having a huge error on the testing data set).

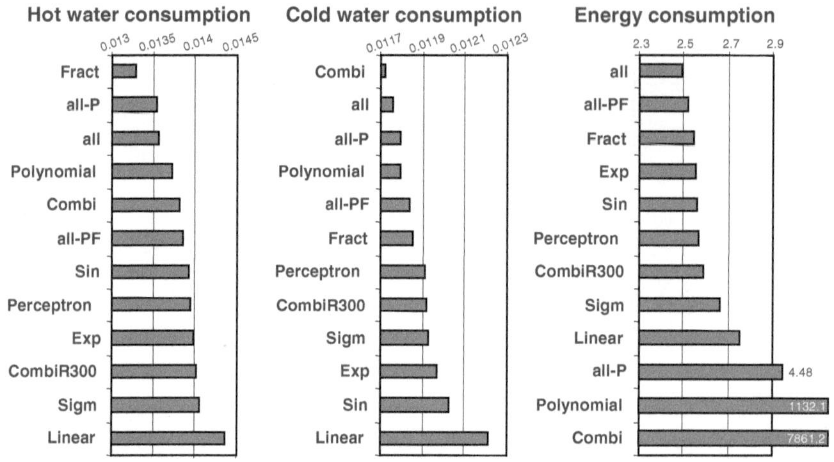

Fig. 10 Performance comparison of GAME neurons on the Building data set. In the all-PF configuration all neurons except the Perceptron and Fract neurons were enabled; similarly in all-P only the Perceptron neuron was excluded.

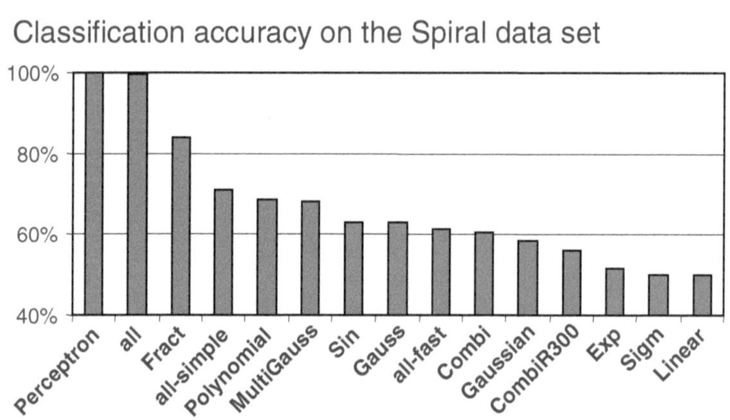

Fig. 11 Performance comparison of GAME neurons on the Spiral data set.

Notice that the configuration **all** has an excellent performance for all three variables, no matter what the level of noise (Fig. 10).

In the Fig. 11, we present the results of the experiment on the Spiral data set [20]. As you can see, the **Perceptron** ensemble learned to tell two spirals apart without any mistake. The second best performing configuration was **all** with almost one hundred percent accuracy[1]. The worst performing ensembles were **Linear** and **Sigm** (neurons with linear and logistic transfer functions). Their 50% classification

[1] We have to mention that building the ensemble of **all** models took just a fraction of time needed to build the ensemble of **Perceptron** models (consisting of BPnetwork neurons).

accuracy signifies that these neurons absolutely failed to learn the Spiral data set. The failure of neurons with logistic transfer function signifies that the GAME algorithm is not as efficient on this problem as the Cascade Correlation algorithm using 16 sigmoid neurons on average to solve this problem.

We performed a number of similar experiments with other real world data sets. We can conclude that the **all** ensemble performed extremely well for almost all data sets under investigation.

The conclusion of our experiments is that for best results we recommend enabling all neurons which have been so far implemented in the GAME engine. The more types of transfer function we have, the more diverse relationships we can model. The type of selected neurons depends only on the nature of the data modelled. The main advantage of using neurons of various types in a single model is that models are adapted to the character of the modelled system. Only neurons with appropriate transfer function survive. Hybrid models also better approximate relationships that can be expressed by the superposition of different functions (e.g. polynomial * sigmoid * linear).

These results are a significant step towards automated data mining, where exhaustive experiments with optimal configuration of data mining methods are no longer necessary.

In the same sense, we also use several types of optimization methods.

2.3 *Optimization of GAME Neurons*

The process of learning aims to minimize the error of each neuron (distance of output from target variable for every training instance).

$$E = \sum_{j=0}^{m} (y_j - d_j)^2, \tag{1}$$

where $y_j$is the output of the model for the jth training vector and $d_j$is the corresponding target output value.

The optimal values of parameters are values minimizing the difference in behavior between a real system and its model. This difference is typically measured by a root mean squared error.

The aim of the learning process is to find values of transfer function coefficients $a_1, a_2, \cdots, a_n$ in order to minimize the error of the neuron.

Most of the coefficients are continuous without constraints. When the transfer function of the neuron is differentiable, we can derive the gradient of the error. The analytic gradient helps the optimization method to adjust coefficients efficiently, providing faster convergence.

Optimal values of coefficients cannot be determined in one step. After random initialization of their values, the error of the neuron is computed (e.g. 1) and the optimization method proposes new values of coefficients, after that, error is computed again (see Fig. 12a). This single optimization step is called iteration. If the

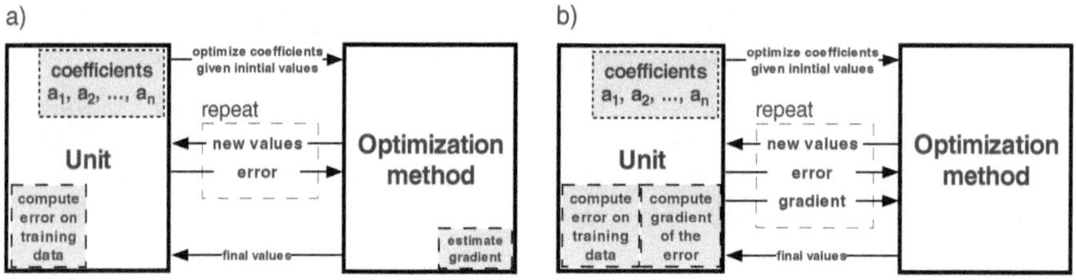

Fig. 12 Optimization of the coefficients can be performed without the analytic gradient a) or with the gradient supplied b). Utilization of the analytic gradient significantly reduces the number of iterations needed for the optimization of coefficients.

analytical gradient of the error can be computed, the number of iterations would be significantly reduced, because we know in which direction coefficients should be adjusted (see Fig. 12b). The gradient of the error $\nabla \mathbf{E}$ in the error surface of a GAME neuron can be written as

$$\nabla \mathbf{E} = \left(\frac{\partial \mathbf{E}}{\partial a_1}, \frac{\partial \mathbf{E}}{\partial a_2}, \cdots, \frac{\partial \mathbf{E}}{\partial a_n} \right), \tag{2}$$

where $\frac{\partial \mathbf{E}}{\partial a_i}$ is a partial derivative of the error in the direction of the coefficient a_i. It tell us how to adjust the coefficient to get a smaller error E on the training data. This partial derivative can be computed as

$$\frac{\partial \mathbf{E}}{\partial a_i} = \sum_{j=0}^{m} \frac{\partial \mathbf{E}}{\partial y_j} * \frac{\partial y_j}{\partial a_i}, \tag{3}$$

where m is the number of training vectors. The first part of the summand can be is easily derived from the Equation 1 as

$$\frac{\partial \mathbf{E}}{\partial y_j} = 2 \sum_{j=0}^{m} (y_j - d_j). \tag{4}$$

The second part of the summand from the Equation 3 is unique for each neuron, because it depends on its transfer function. We demonstrate the computation of the analytic gradient for the Gaussian neuron. For other neurons the gradient is computed in a similar manner.

2.3.1 The Analytic Gradient of the Gaussian Neuron

Gaussian functions are very important and can be found almost everywhere. The most common distribution in nature follows the Gaussian probability density function $f(x) = \frac{1}{2\pi\sigma} * e^{-\frac{(x-\mu)^2}{2\sigma^2}}$. Neurons with Gaussian transfer function are typically used in Radial Basis Function Networks. We have modified the function for our purposes. We added coefficients to be able to scale and shift the function. The first version of the transfer function as implemented in *GaussianNeuron* is the following:

$$y_j = (1 + a_{n+1}) * e^{-\frac{\sum_{i=1}^{n}\left(x_{ij}-a_i\right)^2}{\left(1+a_{n+2}\right)^2}} + a_0 \tag{5}$$

The second version (*GaussNeuron*) proved to perform better on several low dimensional real world data sets:

$$y_j = (1 + a_{n+1}) * e^{-\frac{\sum_{i=1}^{n}\left(a_i * x_{ij}-a_{n+3}\right)^2}{\left(1+a_{n+2}\right)^2}} + a_0 \tag{6}$$

Finally, the third version (*MultiGaussNeuron*), as the combination of the transfer functions above showed the best performance, but sometimes exhibited almost fractal behavior.

$$y_j = (1 + a_{2n+1}) * e^{\underbrace{-\frac{\sum_{i=1}^{n}\left(a_i * x_{ij}-a_{n+i}\right)^2}{\left(1+a_{2n+2}\right)^2}}_{\rho_j}} + a_0 \tag{7}$$

We computed gradients for all these transfer functions. Below, we derive the gradient of the error (see Equation 2) for the third version of the Gaussian transfer function (Equation 7). We need to derive partial derivatives of the error function according to Equation 3. The easiest partial derivative to compute is the one in the direction of the a_0 coefficient. The second term $\frac{\partial y_j}{\partial \rho_j}$ is equal to 1. Therefore we can write $\frac{\partial \mathbf{E}}{\partial a_0} = 2\sum_{j=0}^{m}(y_j - d_j)$. In the case of the coefficient a_{2n+1}, the equation becomes more complicated

$$\frac{\partial \mathbf{E}}{\partial a_{2n+1}} = 2\sum_{j=0}^{m}\left[(y_j - d_j) * e^{-\frac{\sum_{i=1}^{n}\left(a_i * x_{ij}-a_{n+i}\right)^2}{\left(1+a_{2n+2}\right)^2}}\right]. \tag{8}$$

Remaining coefficients are in the exponential part of the transfer function. Therefore the second summand in the Equation 3 cannot be formulated directly. We have to rewrite the Equation 3 as

$$\frac{\partial \mathbf{E}}{\partial a_i} = \sum_{j=0}^{m}\left[\frac{\partial \mathbf{E}}{\partial y_j} * \frac{\partial y_j}{\partial \rho_j} * \frac{\partial \rho_j}{\partial a_i}\right], \tag{9}$$

where ρ_j is the exponent of the transfer function 7. Now we can formulate partial derivatives of remaining coefficients as

$$\frac{\partial \mathbf{E}}{\partial a_{2n+2}} = 2\sum_{j=0}^{m}\left[(y_j - d_j) * (1 + a_{2n+1})e^{\rho_j} * 2\frac{\sum_{i=1}^{n}\left(a_i * x_{ij}-a_{n+i}\right)^2}{\left(1+a_{2n+2}\right)^3}\right] \tag{10}$$

$$\frac{\partial \mathbf{E}}{\partial a_i} = 2\sum_{j=0}^{m}\left[(y_j - d_j) * (1 + a_{2n+1})e^{\rho_j} * -2\frac{a_i^2 * x_{ij}^2 - a_{n+i} * x_{ij}}{(1 + a_{2n+2})^2}\right] \quad (11)$$

$$\frac{\partial \mathbf{E}}{\partial a_{n+i}} = 2\sum_{j=0}^{m}\left[(y_j - d_j) * (1 + a_{2n+1})e^{\rho_j} * -2\frac{a_{n+i} - a_i * x_{ij}}{(1 + a_{2n+2})^2}\right]. \quad (12)$$

We derived the gradient of error on the training data for the Gaussian transfer function neuron. An optimization method often requires these partial derivatives' every iteration to adjust parameters in the proper direction. This mechanism (as described in Fig. 12b) can significantly save the number of error evaluations needed (see Fig. 13).

2.3.2 The Experiment: Analytic Gradient Saves Error Function Evaluations

We performed an experiment to evaluate the effect of analytic gradient computation.

The Quasi-Newton optimization method was used to optimize the SigmNeuron neuron (a logistic transfer function). In the first run the analytic gradient was provided and in the second run, the gradient was not provided so the QN method was forced estimate the gradient itself. We measured the number of function evaluation calls and for the first run we recorded also the number of gradient computation requests.

The results are displayed in the Fig. 13 and in the Table 2. In the second run, without the analytic gradient provided, the number of error function evaluation calls

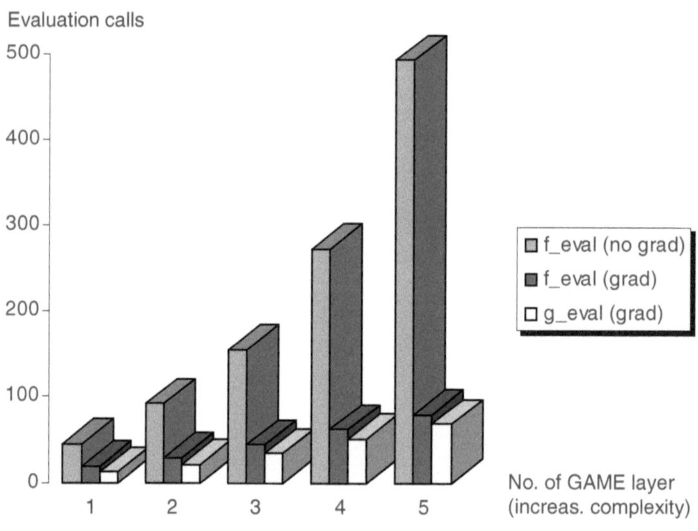

Fig. 13 When the gradient have to be estimated by the optimization method, number of function evaluation calls grows exponentially with an increasing complexity of the problem. When the analytic gradient is computed, the growth is almost linear.

Table 2 Number of evaluations saved by supplying gradient depending on the complexity of the energy function.

Complexity energy fnc.	Avg. evaluations without grad.	Avg. evals. with grad.	Avg. gradient calls	Evaluations saved	Computation time saved
1	45.825	20.075	13.15	56.19%	13.15%
2	92.4	29.55	21.5	68.02%	33.12%
3	155.225	44.85	34.875	71.11%	37.41%
4	273.225	62.75	51.525	77.03%	48.75%
5	493.15	79.775	68.9	83.82%	62.87%

increased exponentially with rising complexity of the error function. For the first run, when the analytic gradient is provided, number of error function evaluation calls increases just linearly and the number of gradient computations grows also linearly. The computation of gradient is almost equally time-consuming as the error function evaluation. When we sum up these two numbers for the first run, we still get growth increasing linearly with the number of layer (increasing complexity of the error surface). This is perfect result, because some models of complex problems can have 20 layers, the computational time saved by providing the analytic gradient is huge. Unfortunately some optimization methods such as genetic algorithms and swarm methods are not designed to use the analytic gradient of the error surface. On the other hand, for some data sets, the usage of analytic gradient can worsen convergence characteristic of optimization methods (getting stuck in local minima).

The training algorithm described in this Section enables possibility of efficient training of hybrid neural networks. The only problem that remains is to select appropriate optimization method.

2.4 *Optimization Methods (Setting Up Coefficients)*

The question "Which optimization method is the best for our problem?" has not a simple answer. There is no method superior to others for all possible optimization problems. However there are popular methods performing well on whole range of problems.

Among these popular methods, we can include gradient methods - the Quasi Newton method, the Conjugate Gradient method and the Levenberg-Marquardt method. They use an analytical gradient (or its estimation) of the problem error surface. The gradient brings them faster convergence, but in cases when the error surface is jaggy, they are likely to get stuck in local optima.

Other popular optimization methods are genetic algorithms. They search the error surface by jumping on it with several individuals. Such search is usually slower, but more prone to get stuck in local minima. The Differential Evolution (DE) perform genetic search with an improved crossover scheme.

The search performed by swarm methods can be imagined as a swarm of birds flying over the error surface, looking for food in deep valleys. You can also imagine that for certain types of terrain, they might miss the deepest valley. Typical examples of swarm methods are Particle Swarm Optimization (PSO) and Ant Colony Optimization (ACO) that mimics the behavior of real ants and their communication using pheromone.

Optimization methods with different behavior are often combined in one algorithm such as Hybrid of the Genetic Algorithm and the Particle Swarm Optimization (HGAPSO).

In our case, we use optimization methods to adjust coefficients of neurons – building blocks of inductive model. The inductive model is created from particular data set. The character of the data set influences which transfer functions will be used and also the complexity of error surface. The surface of a model's RMS error depends heavily on the data set, transfer functions of optimized neuron and also on preceding neurons in the model. The problem is to decide which optimization method should be used to minimize the error. Each data set has different complexity. Therefore we might expect there is no universal optimization method performing optimally on all data sets. We decided to implement several different methods and test their performance on data sets of various complexities.

Below, you can find short description of optimization methods used in GAME algorithm.

2.4.1 Optimization Methods Used in GAME

Optimization methods we have so far implemented to the GAME engine are summarized in the Table3.

There are three different classes of optimization methods named after type of search, they utilize – gradient, genetic and swarm. We will shortly describe particular algorithms we experiment with.

Table 3 Optimization methods summary

Abbrevation	Search	Optimization method
QN	Gradient	Quasi-Newton method
CG	Gradient	Conjugate Gradient method
PCG	Gradient	Powell CG method
PalDE	Genetic	Differtial Evolution ver. 1
DE	Genetic	Differtial Evolution ver. 2
SADE	Genetic	SADE genetic method
PSO	Swarm	Particle Swarm Optimization
CACO	Swarm	Cont. Ant Colony Opt.
ACO*	Swarm	Ext. Ant Colony Opt.
DACO	Swarm	Direct ACO
AACA	Swarm	Adaptive Ant Colony Opt.
API	Swarm	ACO with API heur.
HGAPSO	Hybrid	Hybrid of GA and PSO
SOS	Other	Stoch. Orthogonal Search
OS	Other	Orthogonal Search

Gradient based methods

The most popular optimization method of nonlinear programming is the Quasi-Newton method (QN) [39]. It computes search directions using gradients of an energy surface. To reduce their computational complexity, second derivatives (Hessian matrix) are not computed directly, but estimated iteratively using so called updates [38].

The Conjugate gradient method (CG) [51], a non-linear iterative method, is based on the idea that the convergence can be improved by considering also all previous search directions, not only the actual one. Several variants of the direction update are available (Fletcher-Reeves, Polak-Ribiere, Beale-Sorenson, Hestenes-Stiefel) and bounds are respected. Restarting (previous search direction are forgotten) often improves properties of CG method [42].

Genetic search

Genetic Algorithms (GA) [15] are inspired by Darwin's theory of evolution. Population of individuals are evolved according simple rules of evolution. Each individual has a *fitness* that is computed from its genetic information. Individuals are crossed and mutated by genetic operators and the most fit individuals are selected to survive. After several generations the mean fitness of individuals is maximized.

Niching methods [31] extend genetic algorithms to domains that require location of multiple solutions. They promote the formation and maintenance of stable subpopulations in genetic algorithms (GAs). The GAME engine uses the *Deterministic Crowding* (DC) [30] niching method to evolve structure of models. There exist several other niching strategies such as fitness sharing, islands, restrictive competition, semantic niching, etc.

The Differential Evolution (DE) [47] is a genetic algorithm with special crossover scheme. It adds the weighted difference between two individuals to a third individual. For each individual in the population, an offspring is created using the weighted difference of parent solutions. The offspring replaces the parent in case it is fitter. Otherwise, the parent survives and is copied to the next generation. The pseudocode, how offsprings are created, can be found e.g. in [50].

The Simplified Atavistic Differential Evolution (SADE) algorithm [16] is a genetic algorithm improved by one crossover operator taken from differential evolution. It also prevents premature convergence by using so called radiation fields. These fields have increased probability of mutation and they are placed to local minima of the energy function. When individuals reach a radiation field, they are very likely to be strongly mutated. At the same time, the diameter of the radiation field is decreased. The global minimum of the energy is found when the diameter of some radiation field descend to zero.

Swarm methods

The Particle Swarm Optimization method (PSO) use a swarm of particles to locate the optimum. According to [19] particles "communicate" information they find about each other by updating their velocities in terms of local and global bests; when a new best is found, the particles will change their positions accordingly so that the new information is "broadcast" to the swarm. The particles are always drawn back both to their own personal best positions and also to the best position of the entire swarm. They also have stochastic exploration capability via the use of the random constants.

The Ant colony optimization (ACO) algorithm is primary used for discrete problems (e.g. Traveling Salesman Problem, packet routing). However many modifications of the original algorithm for continuous problems have been introduced recently [48]. These algorithms mimic the behavior of real ants and their communication using pheromone. We have so far implemented the following ACO based algorithms:

The Continuous Ant colony optimization (CACO) was proposed in [8] and it works as follows. There is an ant nest in a center of a search space. Ants exits the nest in a direction given by quantity of pheromone. When an ant reaches the position of the best ant in the direction, it moves randomly (the step is limited by decreasing diameter of search. If the ant find better solution, it increases the quantity of pheromone in the direction of search [28].

The Ant Colony Optimization for Continuous Spaces (ACO*) [7] was designed for the training of feed forward neural networks. Each ant represents a point in the search space. The position of new ants is computed from the distribution of existing ants in the state space.

Direct Ant Colony Optimization (DACO) [23] uses two types of pheromone - one for mean values and one for standard deviation. These values are used by ants to create new solutions and are updated in the ACO way.

The Adaptive Ant Colony Algorithm (AACA) [29] encodes solutions into binary strings. Ants travel from least significant bit to the most significant bit and back. After finishing the trip, the binary string is converted to the solution candidate. The probability of change decreases with significance of bit position by boosting pheromone deposits.

The API algorithm [33] is named after Pachycondyla apicalis and it simulates the foraging behavior of these ants. Ants move from nest to its neighborhood and randomly explore the terrain close to their hunting sites. If an improvement occurs, next search leads to the same hunting site. If the hunt is unsuccessful for more than p times for one hunting site, the hunting site is forgotten and ant randomly generates a new one.

Hybrid search

The Hybrid of the GA and the PSO (HGAPSO) algorithm was proposed in [19]. PSO works based on social adaptation of knowledge, and all individuals are considered to be of the same generation. On the contrary, GA works based on evolution from generation to generation, so the changes of individuals in a single generation are not considered. In nature, individuals will grow up and become more suitable to the environment before producing offspring. To incorporate this phenomenon into GA, PSO is adopted to enhance the top-ranking individuals on each generation.

Other methods

The Orthogonal Search (OS) optimizes multivariate problem by selecting one dimension at a time, minimizing the error at each step. The OS can be used [6] to train single layered neural networks.

We use minimization of a real-valued function of several variables without using gradient, optimizing variables one by one. The Stochastic Orthogonal Search (SOS) differs from OS just by random selection of variables.

2.4.2 Benchmark of Optimization Methods

Following experiments are to demonstrate performance of individual methods applied to optimize models of several real world data sets.

For each data set, we generated models, where neurons with simple transfer functions (Linear, Sigm, Combi, Polynomial, Exp) were enabled. Coefficients of these neurons were optimized by a single method from the Table 3. The configuration **all**, will be explained later. Because these experiments were computationally expensive (optimization methods not utilizing the analytic gradient need many more iterations to converge), we built the ensemble of 5 models for each configuration and data set.

Results on Boston data set from UCI repository are summarized in the Fig.14. For all optimization methods the difference between their error on training and testing data set was almost the same. It signifies that this data set is not very noisy so the overfitting did not occur. The best performance showed the Conjugate Gradient method, but all methods except the worst performing one (Orthogonal Search) achieved similar results. Both training and testing errors of models optimized by OS were significantly higher.

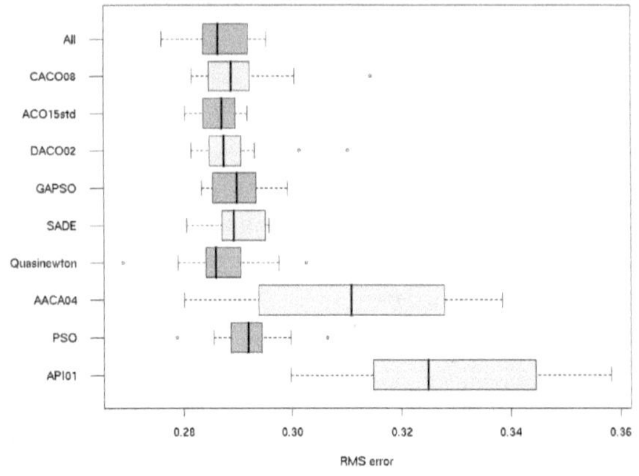

Fig. 14 The RMS error of models on the Boston data set. Neurons of models were optimized by individual methods.

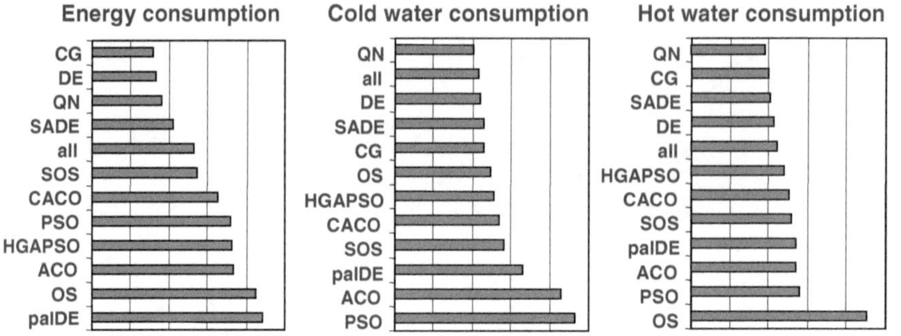

Fig. 15 The performance comparison of optimization methods on the Building data set. The size of bars for individual methods is proportional to the average testing RMS error of models generated using these methods on the Building data set. Models were generated individually for each output variable.

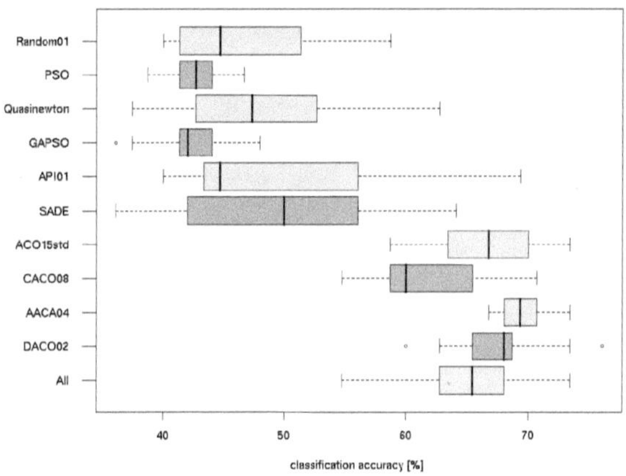

Fig. 16 The classification accuracy of models optimized by individual methods on two intertwined spirals problem.

The results on the Building data set for it's three output variables are shown in the Fig.15. There is no significant difference between results for the noisy variable (Energy consumption) and the other two. We can divide optimization methods into the good and bad performing classes. Good performers are Conjugate Gradient, Quasi Newton, SADE genetic algorithm, Differential Evolution, and the all configuration standing for all methods participation in models evolution. On the other hand badly performing optimization methods for the Building data set are Particle Swarm Optimization, PAL- Differential Evolution[2] and the Ant Colony Optimization.

[2] The palDE is the second version of the Differential Evolution algorithm implemented in the GAME engine. The result when the first version of DE performed well and the second version badly is pellicular. It signifies that the implementation and the proper configuration of a method is of crucial importance.

In accordance with results published in [50], our version of differential evolution outperformed swarm optimization methods. On the other hands, experiment with the Spiral data (telling apart two intertwined spirals) showed different results. Fig. 16 shows that Ant Colony based methods trained better models than methods based on Differential Evolution or gradient based methods. Spiral data set is very hard classification problem and it is difficult to solve it. Error surface has plenty local optima and ant colony methods were able to locate diverse solutions of high quality. Combining them provided increased accuracy.

Conclusion of our experiment with several different data sets is in accordance with our expectations. There is no universal optimization method applicable to arbitrary problem (data set). This statement is also one of the consequences of so called "No Free Lunch Theorem" [14].

2.5 Combining Optimization Methods

We assumed that for each data set, some optimization methods are more efficient than others. If we select appropriate method to optimize coefficients of each neuron within single GAME network, the accuracy will increase. The problem is to find out which method is appropriate (and most effective).

In the "**all**" configuration, we used simple strategy. When new neuron was initialised, random method was assigned to optimize the coefficients of neurons. In case the optimization method was inappropriate, coefficients were not set optimally and neuron did not survived in the genetic algorithm evolving neurons in the layer of the GAME model. Only appropriate optimization methods were able to generate fittest neurons.

2.5.1 Evolution of Optimization Methods

The question is if it is better to assign optimization method randomly or inherit it from parent neurons.

The type of optimization method can be easily inherited from parents, because neurons are evolved by means of niching genetic algorithm. This genetic algorithm can also assign appropriate optimization methods to neurons being evolved. We added the type of the optimization into the chromosome (see Fig. 18). When new neurons are generated by crossover to the next generation, they also inherit type of optimization from their parent neurons. The result should be that methods, training successful neurons, ale selected more often than methods, training poor performers on a particular data set.

Again, an experiment was designed to prove this assumption. We prepared configurations of the GAME engine with several different inheritance settings. In the configuration *p0%* new neurons inherit their optimization method from their paren neurons. In the configuration *p50%* offsprings have 50% chance to get random method assigned. In the configuration *p100%* nothing is inherited, all optimization methods are set randomly.

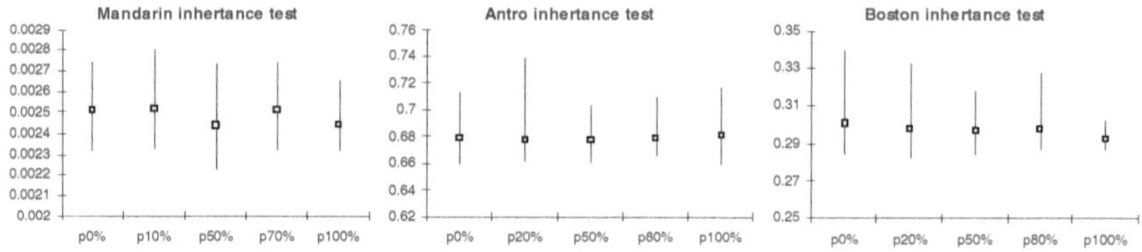

Fig. 17 The experiments with the inheritance of transfer function and learning method. For all three data sets, the fifty percent inheritance level is a reasonable choice.

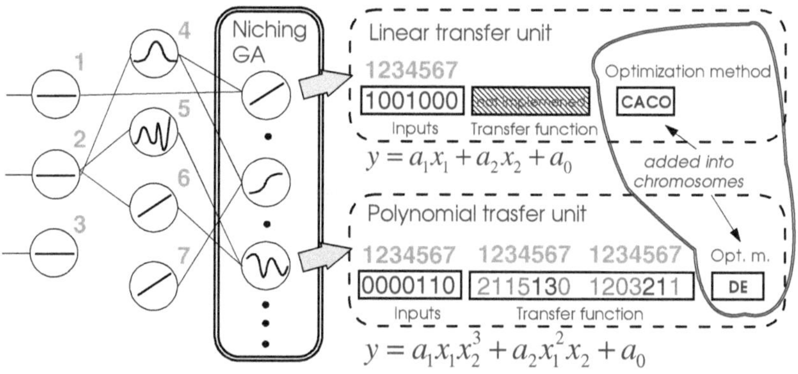

Fig. 18 The example of chromosomes for GAME neurons with linear and polynomial transfer function. Chromosomes contain encoded input connections and for some neurons, the structure of the transfer function is also encoded to be able to evolve it. The type of the optimization method was appended to the chromosome.

We have been experimenting with the Mandarin, Antro and Boston data sets. The description of these data sets can be found in [25].

For each configuration 30 models were evolved. The maximum, minimum and mean of their RMS errors for each configuration are displayed in the Fig. 17. Results are very similar for all configurations and data sets. There is no configuration significantly better than others. For all data sets we can observe that the $p50\%$ and the $p100\%$ configuration have slightly better mean error values and lower dispersion of errors. We chose the $p50\%$ configuration to be default in the GAME engine. It means offspring neurons have 50% chance to get random optimization method assigned otherwise their methods are inherited from parent neurons.

The result of this approach is that methods, training successful neurons, ale used more often than methods, training poor performers on a particular data set. Also, advantages of several different optimization approaches are combined within single model making possible to obtain diverse solutions of high quality.

Again, we demonstrate the advantage of our approach by experimental results.

2.5.2 Combined Optimization Outperforms Individual Methods

We measured the performance of individual methods and the combined optimization (all) on several data sets.

We used the same methodology as for previous experiments on Building data set, Boston and Spiral data sets. Along with these data sets, we used diverse real world data sets described in [25].

Optimization methods are ranked according to the accuracy of their models on several data sets. Fig. 19 displays the results.

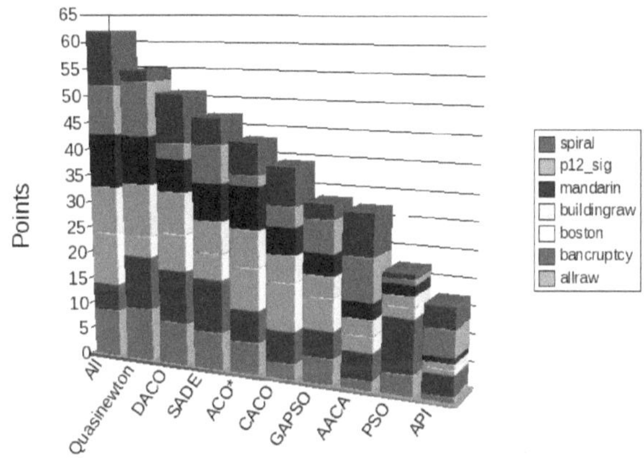

Fig. 19 Final comparison of all tested methods. Points are derived from the ranking for each data test - better position means more points.

Final ranking shows, that the Quasi-Newton optimization method was the most successful from individual methods. It was also the fastest. Combined optimization (**All**) clearly outperformed all individual methods, but it was much slower than Quasi-Newton method. The reason is that computing time was wasted by inefficient methods that do not use analytic gradient of the error surface (such as PSO). Possible solution is to exclude the least efficient methods (accuracy will decrease just marginally), or to enhance these methods by hybridizing them with gradient based methods.

The experiments in this section showed that gradient methods like Quasi Newton and Conjugate Gradients performed very well for all data sets we have been experimenting with. When **All** methods are used, superb performance is guarantied, but the computation is significantly slower (some methods need many iterations to converge). At this stage of the research and implementation, we recommend using the Quasi Newton (QN) optimization method only, because it is the fastest and very reliable. If the computing time is not important for you, the evolution of optimization methods is the best choice.

Another modification of the MIA GMDH algorithm is the topology of models produced.

2.6 Structural Innovations

As stated in the introductional part of this chapter, the topology of the original MIA GMDH was adapted to computational capabilities of early seventies. Experiments that can be nowadays run on every personal computer were intractable even on the most advanced supercomputer.

To make the computation of an inductive model possible, several restrictions on the structure of the model had to be imposed. Because of growing computational power and the development of heuristic methods capable of approximative solutions np-hard problems, we can leave out some of these restrictions.

The most restrictive rule of the original MIA GMDH is the fixed number of neurons' inputs (two) and a polynomial transfer function that is constant (except coefficients) for all neurons in a model.

The next restriction is the absence of layer breakthrough connections. In the original version inputs to a neuron can be from previous layer only.

2.6.1 Growth from a Minimal Form

The GAME models grow from a minimal form. There is a strong parallel with state of the art neural networks as the NEAT [44]. In the default configuration of the GAME engine, neurons are restricted to have at least one input and the maximum number of inputs must not exceed the number of the hidden layer, the neuron belongs to.

The number of inputs to the neuron increases together with the depth of the neuron in the model. Transfer functions of neurons reflect growing number of inputs. We showed [27] that increasing number of neuron's inputs and allowing interlayer connections plays significant role in improving accuracy of inductive models. The growing limit of inputs a neuron is allowed to have is crucial for inductive networks. It helps to overcome the curse of dimensionality. According to the induction principle it is easier to decompose problem to one-dimensional interpolation problems and then combine solutions in two and more dimensions, than to start with multi-dimensional problems (for full connected networks - dimensionality is proportional to the number of input features).

To improve the modeling accuracy of neural networks, artificial input features are often added to the training data set. These features are synthesized from original features by math operations and can possibly reveal more about the modelled output. This is exactly what is GAME doing automatically (neurons of the first hidden layer serve as additional synthesized features for the neurons deeper in the network).

Our additional experiments showed that the restriction on the maximum number of inputs to neurons has moderately negative effect on the accuracy of models. However when the restriction is enabled[3], the process of model generation is much faster. The accuracy of produced models is also more stable than without the restriction. Without the restriction we would need many more epochs of the genetic

[3] No restriction on the maximal number of inputs does not mean a fully connected network!

algorithm evolving neurons in a layer (models accuracy would be stable and the feature ranking algorithm deriving significance from proportional numbers of neurons connected to a particular feature would work properly (feature ranking algorithm will be described later in this chapter).

2.6.2 Interlayer Connections

Neurons have no longer inputs just from previous layer. Inputs can be connected to the output of any neuron from previous layers as well as to any input feature. This modification greatly increases the state space of possible model topologies, but the improvement in accuracy of models is rather high [27].

The GMDH algorithm implemented in the KnowledgeMiner software [34] can also generate models with layer breakthrough connections.

Expanded search space of possible model topologies requires methods of efficient heuristic search.

2.7 Genetic Algorithm

The genetic algorithm is frequently used to optimize a topology of neural networks [32, 41, 45]. Also in GMDH related research, recent papers [36, 35] report improving the accuracy of models by employing genetic search to identify their optimal structure.

In the GAME engine, we also use genetic search to optimize the topology of models and also the configuration and shapes of transfer functions within their neurons. The model is constructed layer by layer as in the MIA GMDH. In each layer, genetic is executed and after several generations, most fit and diverse neurons are selected to form the layer. After that, the construction process continues with next layers. Neurons from these layers can be connected to input variables and also to neurons from preceding layers.

If the optimal structure of model is to be identified, we need to find optimal interconnection of neurons, types and structure of their transfer functions and size of the model (number of neurons in each layer and number of layers). Connections of neurons and structure of their transfer functions can be optimized by genetic algorithm. The example of encoding into chromosomes are depicted in the Fig. 20.

The individual in the genetic algorithm represents one particular neuron of the GAME network. Inputs of a neuron are encoded into a binary string chromosome. The structure of transfer function can be also added into the chromosome. The chromosome can also include type of the optimization method and configuration options such as stopping criteria, strategies utilized during optimization of parameters, etc.

The length of the "Inputs" part of the chromosome equals to the number of input variables plus number of neurons from previous layers, the neuron can be connected to. The existing connection is represented by "1" in the corresponding gene. The number of ones is restricted to maximal number of neuron's inputs. The example how the transfer function can be encoded is in the Fig. 20.

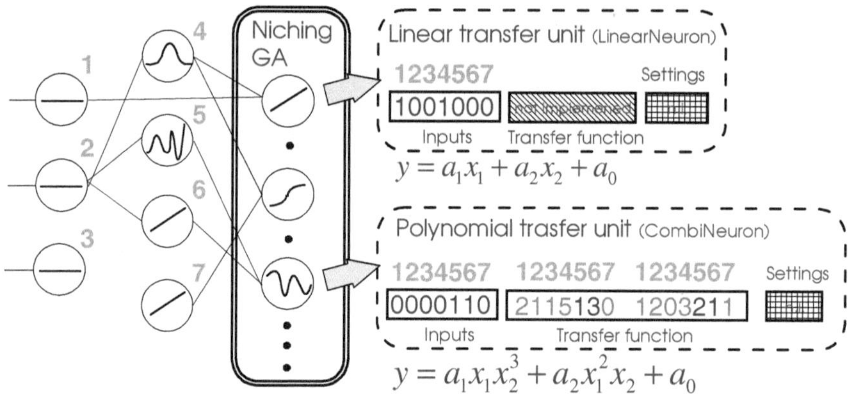

Fig. 20 Neurons are encoded into chromosomes and evolution identifies optimal inputs of neurons and structure of their transfer functions.

If two neurons of different types are crossed, just the "Inputs" part of the chromosome comes into play. If two Polynomial neurons cross over, also the second part encoding transfer function is involved.

Note that coefficients of the transfer functions (a_0,a_1,. . . ,a_n) are not encoded in the chromosome Fig. 20. These coefficients are adjusted separately by optimization methods as described in previous section. This is crucial difference from the Topology and Weight Evolving Artificial Neural Network (TWEANN) approach [45].

The fitness of the individual is inversely proportional to the error of the individual computed on the validation data set.

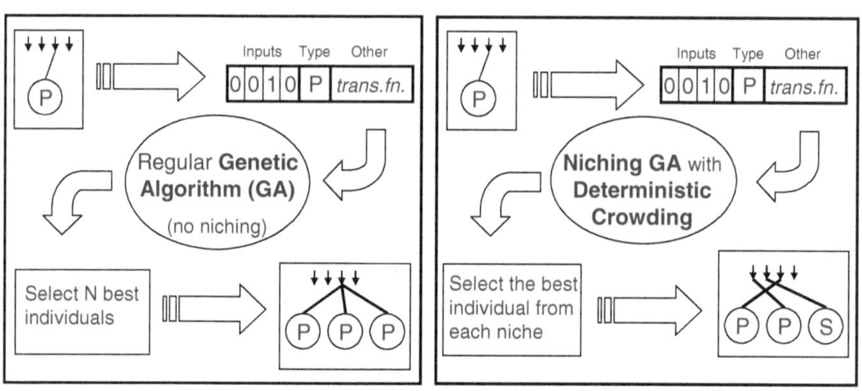

Fig. 21 GAME neurons in the first layer are encoded into chromosomes, then GA is applied to evolve the best performing neurons. After few epochs all neurons will be connected to the most significant input and therefore correlated. When the Niching GA is used instead of the basic variant of GA, neurons connected to different inputs survive.

The application of the genetic algorithm in the GAME engine is depicted in the Fig. 21. The left schema describes the process of single GAME layer evolution when the standard genetic algorithm [15] is applied.

Neurons randomly initialized and encoded into chromosomes. Then the genetic algorithm is executed. After several epochs of the evolution, individuals with the highest fitness (neurons connected to the most significant input) dominate the population. The best solution represented by the best individual is found.

Whole population have very similar (or the same) chromosomes as the winning individual has. This is also the reason why all neurons surviving in the population (after several epochs of evolution by the regular genetic algorithm) are highly correlated.

The regular genetic algorithm found one best solution. We want to find also multiple suboptimal solutions (e.g. neurons connected to the second and the third most important input). By using less significant features we get more additional information than by using several best individuals connected to the most significant feature, which are in fact highly correlated (as shown on Fig. 22.). Therefore we employ a niching method described below. It maintains diversity in the population and therefore neurons connected to less significant inputs are allowed to survive, too (see Fig. 21 right).

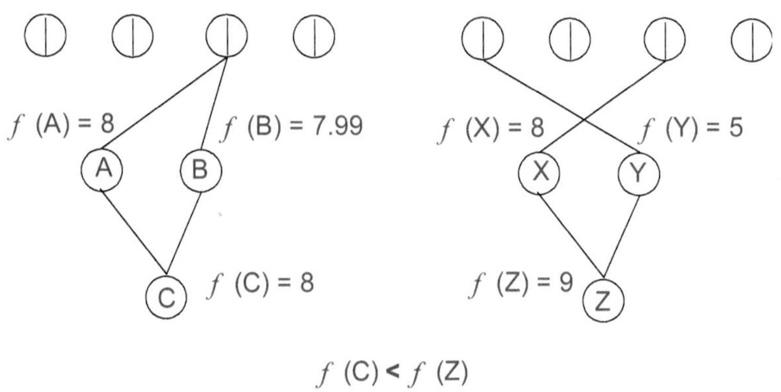

Fig. 22 Fitness of neuron Z is higher than that of neuron C, although Z has less fit inputs.

2.7.1 Experiments with Deterministic Crowding

The major difference between the regular genetic algorithm and a niching genetic algorithm is that in the niching GA the distance among individuals is defined.

The distance of two individuals from the pseudocode of Deterministic Crowding [30] can be based on the phenotypic or genotypic difference of neurons. In the GAME engine, the distance of neurons is computed from both differences. Fig. 23 shows that the distance of neurons is partly computed from the correlation of their errors and partly from their genotypic difference. The genotypic difference consists the obligatory part "difference in inputs", then some neurons add "difference in transfer functions" and also "difference in configurations" can be defined.

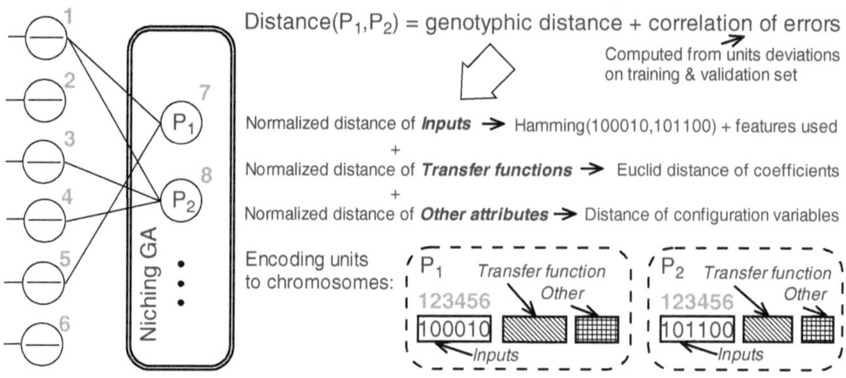

Fig. 23 The distance of two neurons in the GAME network.

Neurons that survive in layers of GAME networks are chosen according to the following algorithm. After the niching genetic algorithm finished the evolution of neurons, a multi-objective algorithm sorts neurons according to their RMS error, genotypic distance and the correlation of errors. Surviving neurons have low RMS errors, high mutual distances and low correlations of errors.

Niches in GAME are formed by neurons with similar inputs, similar transfer functions, similar configurations and high correlation of errors.

The next idea is that neurons should inherit their type and the optimization method used to estimate their coefficients. This improvement allows reducing time wasted with optimizing neurons with an improper transfer function by optimization methods not suitable to processed data.

Evaluation of the distance computation

The GAME engine enables the visual inspection of complex processes that are normally impossible to control. One of these processes is displayed in the Fig. 2.7.1. From left we can see the matrix of genotypic distances computed from chromosomes of individual neurons during the evolution of the GAME layer. Note that this distance is computed as a sum of three components: distance of inputs, transfer functions and configuration variables, where last two components are optional.

The darker color of background signifies the higher distance of corresponding individuals and vice versa. The next matrix visualizes distances of neurons based on the correlation of their errors. Darker background signifies less correlated errors. The next graph shows deviations of neurons output from the target value of individual training vectors. From these distances the correlation is computed. The most right graph of the Fig. 2.7.1 shows a normalized RMS error of neurons on the training data.

All these graphs are updated as the evolution proceeds from epoch to epoch. When the niching genetic algorithm finishes, you can observe how neurons are sorted (multi objective sorting algorithm based on the Bubble sort algorithm) and

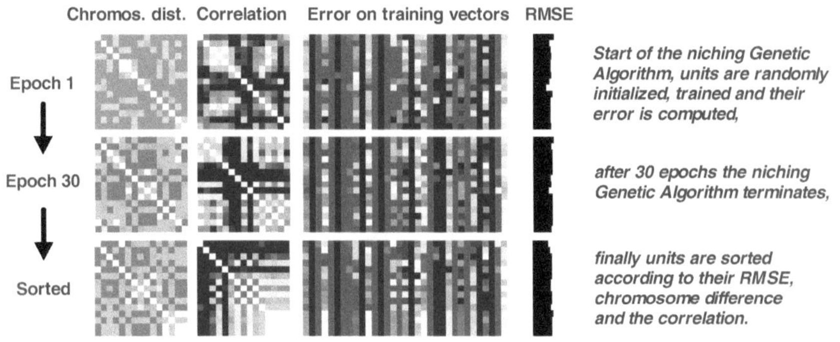

Fig. 24 During the GAME layer evolution, distances of neurons can be visually inspected. The first graph shows their distance based on the genotypic difference. The second graph derives distance from their correlation. Third graph shows deviations of neurons on individual training vectors and the most right graph displays their RMS error on the training data.

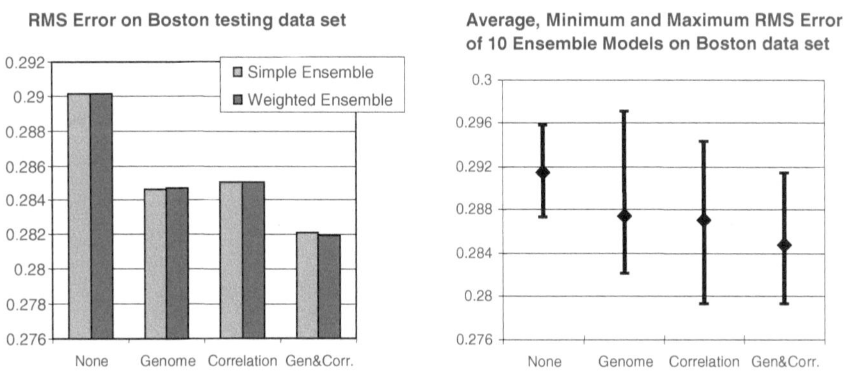

Fig. 25 The best results were obtained when the distance of neurons is computed as a combination of their genotypic distance and the correlation of their errors on training vectors.

which neurons are finally selected to survive in the layer. Using this visual inspection tool, we have evaluated and tuned the distance computation in the niching genetic algorithm.

The next goal was to evaluate if the distance computation is well defined. The results in the Fig. 2.7.1 show that the best performing models can be evolved with the proposed combination of genotypic difference and correlation as the distance measure. The worst results are achieved when the distance is set to zero for all neurons. Medium accuracy models are generated by either the genotypic difference based distance or the correlation of errors based distance.

In the Fig. 26, there is a comparison of the regular genetic algorithm and the niching GA with the Deterministic Crowding scheme. The data set used to model the output variable (Mandarin tree water consumption) has eleven input features.

Neurons in the first hidden layer of the GAME network have a single input, so they are connected to a single feature.

The population of 200 neurons in the first layer was initialized randomly (genes are uniformly distributed - approx. the same number of neurons connected to each feature). After 250 epochs of the regular genetic algorithm the fittest individuals (neurons connected to the most significant feature) dominated the population. On the other hand the niching GA with DC maintained diversity in the population. Individuals of three niches survived. As Fig. 26 shows, the functionality of niching genetic algorithm in the GAME engine is evident.

When you look at the Fig. 26 you can also observe that the number of individuals (neurons) in each niche is proportional to the significance of the feature, neurons are connected to. From each niche the fittest individual is selected and the construction goes on with the next layer. The fittest individuals in next layers of the GAME network are these connected to features which brings the maximum of additional information. Individuals connected to features that are significant, but highly correlated with features already used, will not survive. By monitoring which individuals endured in the population we can estimate the significance of each feature for the output variable modelling. This information can be subsequently used for the feature ranking.

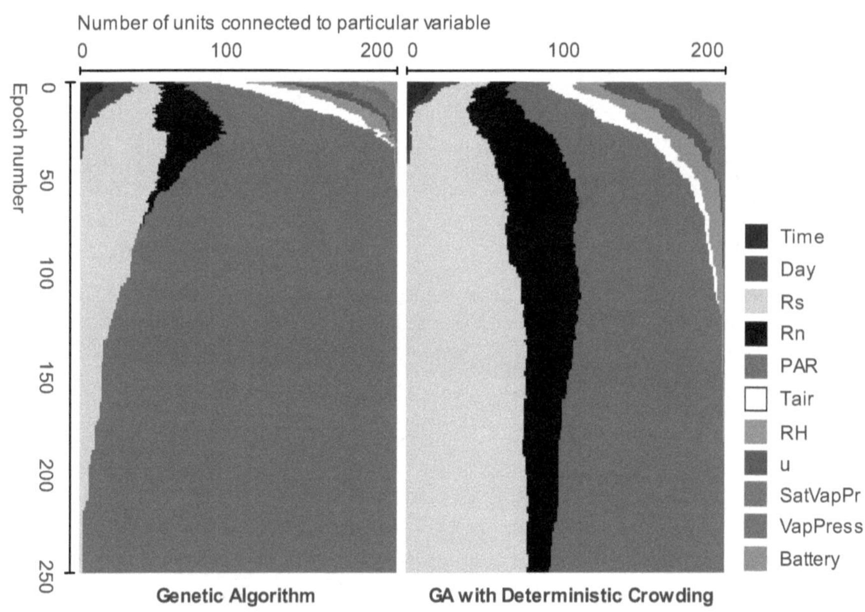

Fig. 26 The experiment demonstrated that the regular Genetic Algorithm approaches an optimum relatively quickly. Niching preserves different neurons for many more iterations so we can chose the best neuron from each niche at the end. Niching also increases a probability of the global minimum not being missed.

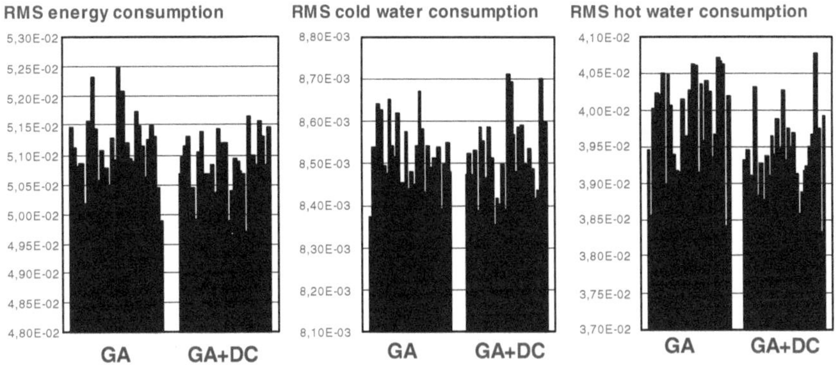

Fig. 27 RMS error of GAME models evolved by means of the regular GA and the GA with the Deterministic Crowding respectively (on the complex data). For the the hot water and the energy consumption, the GA with DC is significantly better than the regular GA

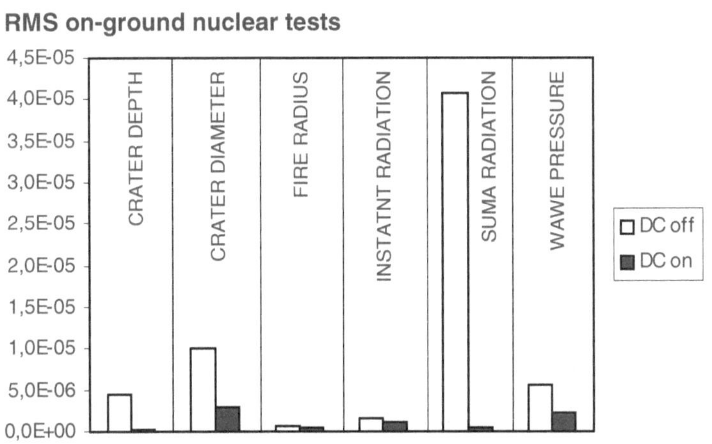

Fig. 28 Average RMS error of GAME models evolved by means of simple GA (DC off) and GA with Deterministic Crowding (DC on) respectively (on the simple data). Here for all variables, the Deterministic Crowding attained the superior performance.

We also compared the performance (the inverse of RMS error on a testing data) of GAME models evolved by means of the regular GA and the niching GA with Deterministic Crowding respectively. Extensive experiments were executed on the complex data (Building dataset) and on the small simple data (On-ground nuclear tests dataset).

The statistical test proved that on the level of significance 95%, the GA with DC performs better than simple GA for the energy and hot water consumption. The Fig. 27 shows RMS errors of several models evolved by means of the regular GA and the GA with Deterministic Crowding respectively.

The results are more significant for the On-ground nuclear dataset. The Fig. 28 shows the average RMS error of 20 models evolved for each output attribute. Leaving out models of the fire radius attribute, the performance of all other models is significantly better with Deterministic Crowding enabled.

We can conclude, than niching strategies significantly improved the evolution of GAME models. Generated models are more accurate than models evolved by the regular GA as showed our experiments with real world data.

2.8 Ensemble Techniques in GAME

The GAME method generates on the training data set models of similar accuracy. They are built and validated on random subsets of the training set (this technique is known as bagging [17]). Models have also similar types of neurons and similar complexity. It is difficult to choose the best model - several models have the same (or very similar) performance on the testing data set. We do not choose one best model, but several optimal models - ensemble models [9].

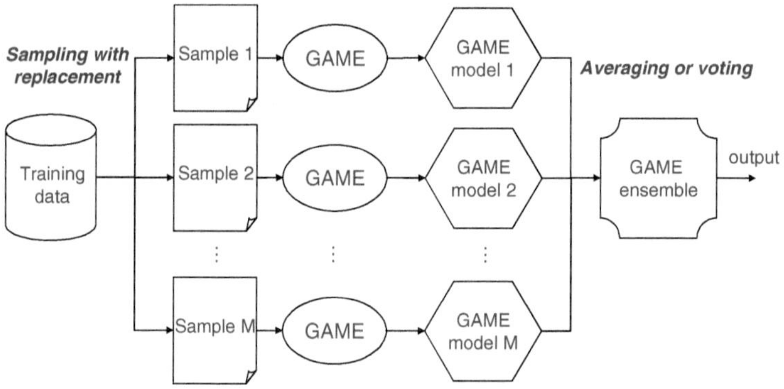

Fig. 29 The Bagging approach is used to build an ensemble of GAME models, models are then combined by the Simple or Weighted averaging.

The Fig. 29 illustrates the principle how GAME ensemble models are generated using bootstrap samples of training data and later combined into a simple ensemble or a weighted ensemble. This technique is called Bagging and it helps that member models demonstrate diverse errors on a testing data.

Other techniques that promote diversity in the ensemble of models play significant role in increasing the accuracy of the ensemble output. The diversity in the ensemble of GAME models is supported by following techniques:

- Input data varies (Bagging)
- Input features vary (using subset of features)
- Initial parameters vary (random initialization of weights)
- Model architecture varies (heterogeneous neurons used)

- Training algorithm varies (several training methods used)
- Stochastic method used (niching GA used to evolve models)

We assumed that the ensemble of GAME models will be more accurate than any of individual models. This assumption appeared to be true just for GAME models whose construction was stopped before they reached the optimal complexity (Fig. 30 left).

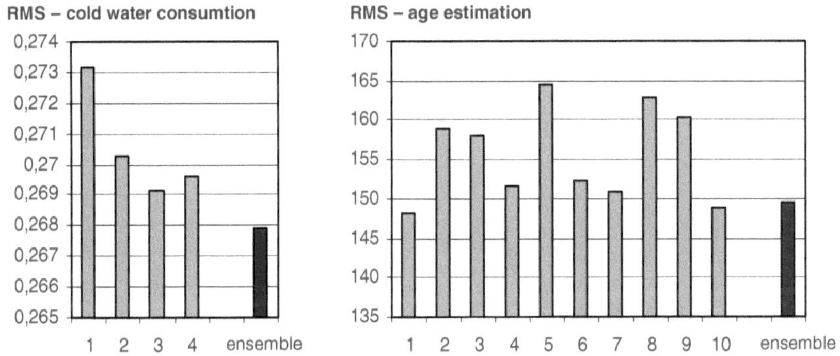

Fig. 30 The Root Mean Square error of the simple ensemble is significantly lower than RMS of individual suboptimal models on testing data (left graph). For optimal GAME models it is not the case (right).

We performed several experiments on both synthesized and real world data sets. These experiments demonstrated that ensemble of optimal GAME models is seldom significantly better then single the best performing model from the ensemble Fig. 30 right).

The problem is, we cannot say in advance which single model will perform the best on testing data. The best performing model on training data can be the worst performing one on testing data and vice versa.

Usually, models badly performing on training data perform badly also on testing data. Such models can impair the accuracy of ensemble model. To limit the influence of bad models on the output of ensemble, models can be weighted according to their performance on training data set. Such ensemble is called the weighted ensemble and we discuss its performance below.

Contrary to the approach introduced in [12], we do not use the whole data set to determine performances (Root Mean Square Errors) of individual models in the weighted ensemble.

In Fig. 31 you can see that weighted ensemble has tendency to overfit the data - stronger than simple ensemble. While its performance is superior on the training and validation data, on the testing data there are several individual models performing better.

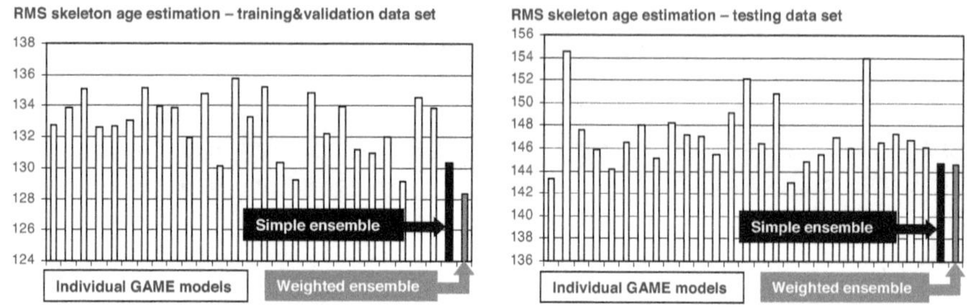

Fig. 31 Performance of the simple ensemble and weighted ensemble on very noisy data set (Skeleton age estimation based on senescence indicators).

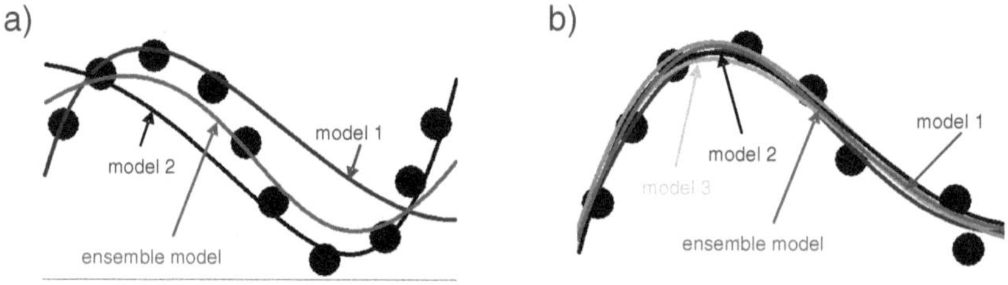

Fig. 32 Ensemble of two models exhibiting diverse errors can provide significantly better result.

The theoretical explanation for such behavior might be the following. Fig. 32 a shows ensemble of two models that are not complex enough to reflect the variance of data (weak learner). The error of the ensemble is lower than that of individual models, similarly like in the first experiment mentioned above.

In Fig. 32 b, there is an ensemble of three models having the optimal complexity. It is apparent that the accuracy of the ensemble cannot be significantly better, than those of individual models. The negative result of second experiment is therefore caused by the fact, that the bias of optimal models cannot be further reduced.

We can conclude that by using the simple ensemble, instead of single GAME model, we can in some cases improve the accuracy of modeling. The accuracy improvement is not only advantage of using ensembles. There is highly interesting information encoded in the ensemble behavior. It is the information about the credibility of member models.

These models approximate data similarly and their behavior differ outside of areas where system can be successfully modelled (insufficient data vectors present, etc.). In well defined areas all models have compromise response. We use this fact for models' quality evaluation purposes and for estimation of modeling plausibility in particular areas of the input space.

3 Benchmarking the GAME Method

In this section we benchmark the regression and classification performance of the GAME method against the performance of methods implemented in the Weka machine learning environment.

We performed experiments on the A-EGM data set, described in [26]. At first, we studied the regression performance of GAME models produced by different configurations of the GAME algorithm. The target variable was the average A-EGM signal ranking by three experts (the A-EGM-regression data set). We found out, and it is also apparent in the boxplot charts, that comparison of the 10-fold cross validation error is not stable enough to decide, which configuration is better. Therefore we repeated the 10-fold cross validation ten times, each time with different fold splitting. For each box plot it was necessary to generate and validate one hundred models.

For all experiments we used three default configurations of the GAME algorithm available in FAKE GAME environment [5]. The *std* configuration uses just subset of neurons (those with implemented analytic gradient for faster optimization). It evolves 15 neurons for 30 epochs in each layer. The *quick* configuration is the same as *std* except that it do not use the niching genetic algorithm (just 15 neurons in the initial population). The *linear* restricts type of neurons that can be used to linear transfer function neurons. The *all* configuration is the same as *std*, in addition it uses all neurons available in the FAKE GAME environment. This configuration is more computationally expensive, because it also optimizes complex neurons such as BPNetwork containing standard MLP neural network with the back-propagation of error [32].

The GAME algorithm also allows to generate ensemble of models [13, 9]. Ensemble configurations contain digit (number of models) in their name.

Fig. 33 shows that the regression of the AER output is not too difficult task. All basic GAME configurations performed similarly (left chart) and ensembling of

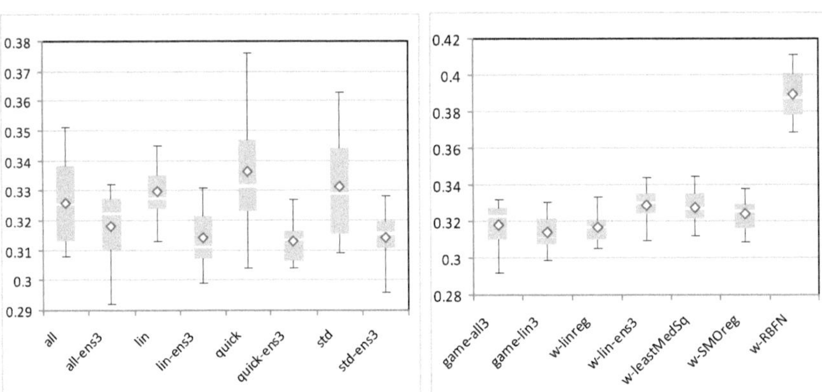

Fig. 33 The comparison of RMS cross validation errors for several configuration configuration of the GAME engine(left). Selected GAME models compared with models generated in Weka environment(right).

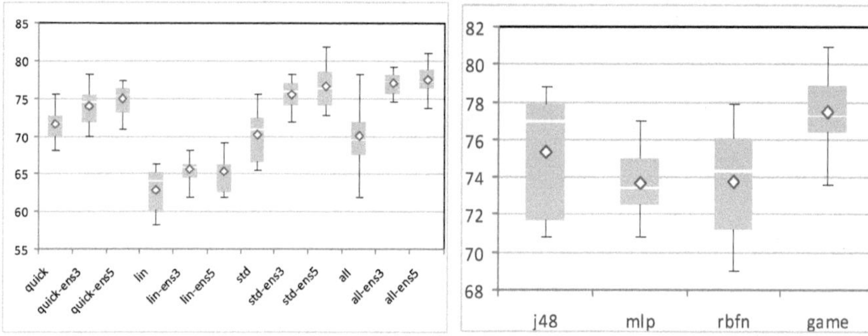

Fig. 34 Classification accuracy in percent for several GAME configurations (left) and comparison with Weka classifiers (right).

three models further improved their accuracy. The ensemble of three linear models performed best in average, but the difference from $all - ens3$ configuration is not significant.

In Weka data mining environment, *LinearRegression* with embedded feature selection algorithm was the best performing algorithm. Ensembling (bagging) did not improved results of generated model, quite the contrary. The Radial Basis Function Network (RBFN) failed to deliver satisfactory results in spite of experiments with its optimal setting (number of clusters).

Secondly, our experiments were performed on the A-EGM-classification data set. The methodology remained the same as for regression data. Additionally we tested classification performance of 5 models ensembles. Fig. 34 left shows that the classes are not linearly separable - *linear* configuration generates poor classifiers and ensembling does not help. Combining models in case of all other configurations improve the accuracy. For *all* configuration the dispersion of cross validation errors is quite high. The problem is in the configuration of the genetic algorithm - with 15 individuals in the population some "potentially useful" types of neurons do not have chance to be instantiated. Ensembling models generated by this configuration improves their accuracy significantly.

Comparison with Weka classifiers (Fig. 34 right) shows that GAME ensemble significantly outperforms Decision Trees (j48), MultiLayered Perceptron (mlp) and Radial Basis Function network (rbfn) implemented in Weka data mining environment.

The last experiment (Fig. 35) showed that the best split of the training and validation data set is 40%/60% (training data are used by optimization method to adjust parameters of GAME neurons transfer functions, whereas from validation part, the fitness of neurons is computed).

Implicitly, and in all previous experiments, training and validation data set was divided 70%/30% in the GAME algorithm. Changing the implicit setting to 40%/60% however involves additional experiments on different data sets.

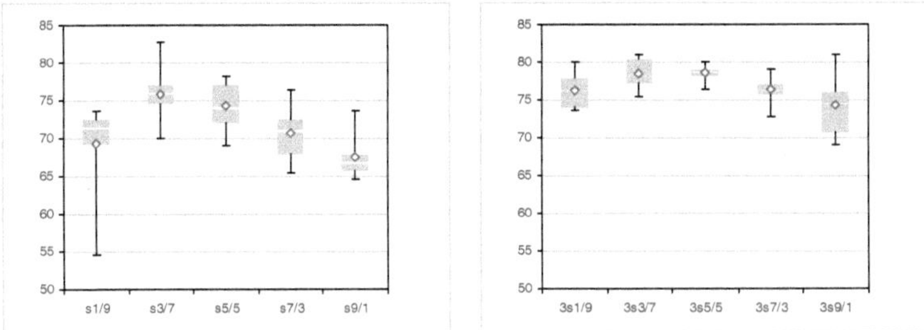

Fig. 35 Classification performance for different ratios of training/validation data split. Left - results for single game models generated by *std* configuration. Right - results for GAME ensemble ($std - ens3$).

3.1 Summary of Results

For this data set, the GAME algorithm outperforms well established methods in both classification and regression accuracy. What is even more important, both winning configurations were identical $all - ens$. Natural selection evolved optimal models for very different tasks - that is in accordance with our previous experiments and with our aim to develop automated data mining engine.

4 Case Studies – Data Mining Using GAME

4.1 Fetal Weight Prediction Formulae Extracted from GAME

An accurate model of ultrasound estimation of fetal weight (EFW) can help in decision if the cesarean childbirth is necessary. We collected models from various sources and compared their accuracy. These models were mostly obtained by standard techniques such as linear and nonlinear regression. The best performing model, from 14 we have been experimenting with, was the equation published by Hadlock et al:

$$\begin{aligned}\log_{10} EFW &= 1.326 - 0.00326 \times AC \times FL + 0.0107 \times HC \\ &\quad + 0.0438 \times AC + 0.158 \times FL \end{aligned} \tag{13}$$

Alternatively, we generated several linear and non-linear models by using the GAME algorithm. GAME models can be serialized into simple equations that are understandable by domain experts.

We generated several models (see Eq. 14, 15, 16) by the GAME algorithm and compare them with well known EFW models, which has been found by linear and nonlinear regression methods by various authors in the past.

We loaded the data into the FAKE GAME open source application [5] and generated models by using standard configuration (if not indicated differently) of the GAME engine.

All generated models are simple and we also checked regression graphs of each model in GAME toolkit and see that every model has smooth progression (see Fig. 36) and approximate the output data set by hyperplane. Because the error is measured on testing data and the regression hyperplane is smooth, we can see that models are not overtrained and have good generalization ability.

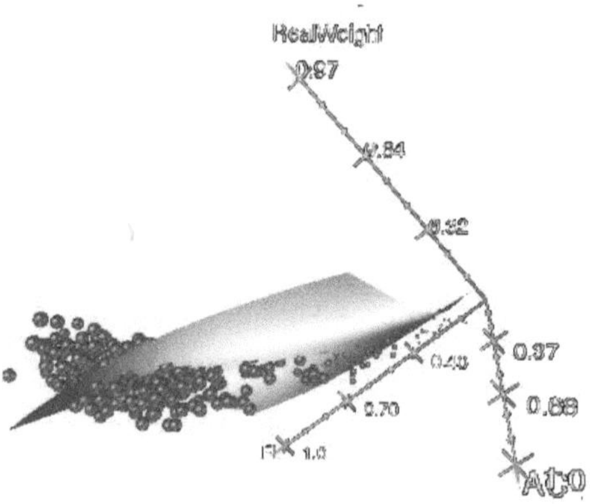

Fig. 36 An example of GAME model evolved on FL data. Regression hyperplane is smooth as expected.

The first model was serialized into polynomial formula (just polynomial neurons were enabled and the penalization for complexity of the transfer function was applied to get simple formulae). The error of the model is therefore higher (Tab. 5) than that of models with more complex formulas obtained with the standard configuration of the GAME engine:

$$EFW = 0.0504 \times AC^2 - 16.427 \times AC + 38.867 \times FL + 284.074 \tag{14}$$

$$\begin{aligned} EFW &= -7637.17 + 7870.09 \times e^{3.728\times 10^{-6}\times (AC-163)^2 + 0.0002\times HC} \\ &\quad \times e^{\frac{1}{10.676+40011\times e^{-0.096\times BPD}} + \frac{1}{7.557+6113.68\times e^{-0.102\times FL}}} \end{aligned} \tag{15}$$

Note that exponential and sigmoid neurons are very successful on this data set. Observed relationship of variables (Fig. 36) is apparently nonlinear. To simplify generated equations, we transformed the output into logarithmic scale for the last model. Model produced by GAME does not contain exponential terms any more, but neurons with sine transfer function were more successful than polynomial neurons:

$$\begin{aligned} \log_{10} EFW &= 2.18 + 0.0302 \times BPD + 0.0293 \times FL \\ &\quad -0.603 \sin(0.524 - 0.0526 \times AC) \\ &\quad -0.344 \sin(-0.029 \times AC - 0.117 \times FL + 0.946) \end{aligned} \tag{16}$$

In case that experts prefer polynomial equation, Sine neurons can be easily disabled in the configuration of the GAME engine.

Table 4 Basic statistic characteristics of models, Major Percentiles [g]

Method	5%	10%	50%	90%	95%
Hadlock (13)	894	1401	3168	3678	3836
GAME (14)	950	1462	3149	3623	3779
GAME (15)	937	1424	3145	3633	3741
GAME (16)	886	1394	3173	3625	3720

Table 5 Model Correlation with Actual Birth Weight R^2, Mean absolute Error ± Standard deviation, RMS Error

Method	R^2	Mean Abs. Error [g] ± SD	RMS Error [g]
Hadlock (13)	0.91	199 ± 171	261
GAME (15)	0.91	199 ± 168	261
GAME (16)	0.91	203 ± 173	266
GAME (14)	0.91	209 ± 174	272

4.1.1 Statistical Evaluation of Models

All Fake Game models are at least good as best models found by statistical approach. We succeeded to find models with the same R^2, lower mean absolute error, lower RMS error and lower standard deviation than models found by traditional techniques. We also decreased mean absolute error, standard deviation and RMS error by using ensemble of three models which increases accuracy of estimation of fetal weight (see [43]).

5 The FAKE GAME Project

Knowledge discovery and data mining are popular research topics in recent times. It is mainly due to the fact that the amount of collected data significantly increases. Manual analysis of all data is no longer possible. This is where the data mining and the knowledge discovery (or extraction) can help.

The process of knowledge discovery [11] is defined as the non-trivial process of finding valid, potentially useful, and ultimately understandable patterns. The problem is that this process still needs a lot of human involvement in all its phases in order to extract some useful knowledge. Our research focuses on methods aimed at significant reduction of expert decisions needed during the process of knowledge extraction. Within the FAKE GAME environment we develop methods for automatic data preprocessing, adaptive data mining and for the knowledge extraction (see Fig. 37). The data preprocessing is very important and time consuming phase of the knowledge extraction process. According to [37] it accounts for almost 60% of total time of the process. The data preprocessing involves dealing with non-numeric variables (alpha values coding), missing values replacement (imputing), outlier detection, noise reduction, variables redistribution, etc. The data preprocessing phase

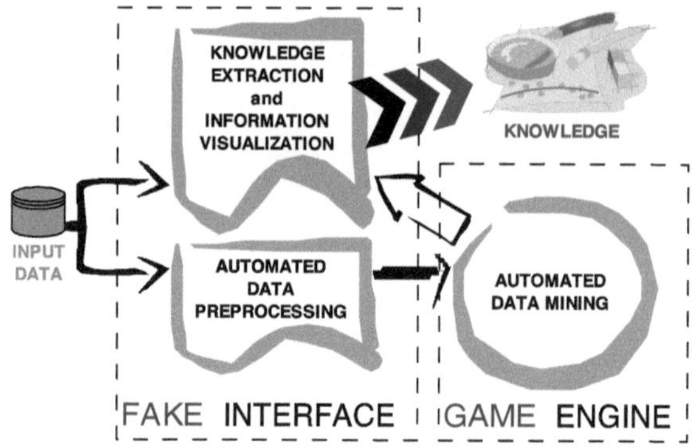

Fig. 37 FAKE GAME environment for the automated knowledge extraction.

cannot be fully automated for every possible data set. Each data have unique character and each data mining method requires different preprocessing.

Existing data mining software packages support just very simple methods of data preprocessing [3]. There are new data mining environments [4, 1] trying to focus more on data preprocessing, but their methods are still very limited and give no hint which preprocessing would be the best for your data. It is mainly due to the fact that the theory of data preprocessing is not very developed. Although some preprocessing methods seem to be simple, to decide which method would be the most appropriate for some data might be very complicated. Within the FAKE interface we develop more sophisticated methods for data preprocessing and we study which methods are most appropriate for particular data. The final goal is to automate the data preprocessing phase as much as possible.

In the knowledge extraction process, the data preprocessing phase is followed by the phase of data mining. In the data mining phase, it is necessary to choose appropriate data mining method for your data and problem. The data mining method usually generates a predictive, regressive model or a classifier on your data. Each method is suitable for different task and different data. To select the best method for the task and the data, the user has to experiment with several methods, adjust parameters of these methods and often also estimate suitable topology (e.g. number of neurons in a neural network). This process is very time consuming and presumes strong expert knowledge of data mining methods by the user.

In the new version of one commercial data mining software [46], an evolutionary algorithm is used to select the best data mining method with optimal parameters for actual data set and a problem specified. This is really significant step towards the automation of the data mining phase. We propose a different approach. The ensemble of predictive, regressive models or classifiers is generated automatically using the GAME engine. Models adapt to the character of a data set so that they have an optimal topology. We develop methods eliminating the need of parameters adjustment so that the GAME engine performs independently and optimally on bigger range of different data.

The results of data mining methods can be more or less easily transformed into the knowledge, finalizing the knowledge extraction process. Results of methods such as simple decision tree are easy to interpret. Unfortunately majority of data mining methods (neural networks, etc.) are almost black boxes - the knowledge is hidden inside the model and it is difficult to extract it.

Almost all data mining tools bound the knowledge extraction from complex data mining methods to statistical analysis of their performance. More knowledge can be extracted using the techniques of information visualization. Recently, some papers [49] on this topic had been published. We propose techniques based on methods such as scatterplot matrix, regression plots, multivariate data projection, etc. to extract additional useful knowledge from the ensemble of GAME models. We also develop evolutionary search methods to deal with the state space dimensionality and to find interesting projections automatically.

5.1 The Goal of the FAKE GAME Environment

The ultimate goal of our research is to automate the process of knowledge extraction from data. It is clear that some parts of the process still need the involvement of expert user. We build the FAKE GAME environment to limit the user involvement during the process of knowledge extraction. To automate the knowledge extraction process, we research in the following areas: data preprocessing, data mining, knowledge extraction and information visualization (see Fig. 38).

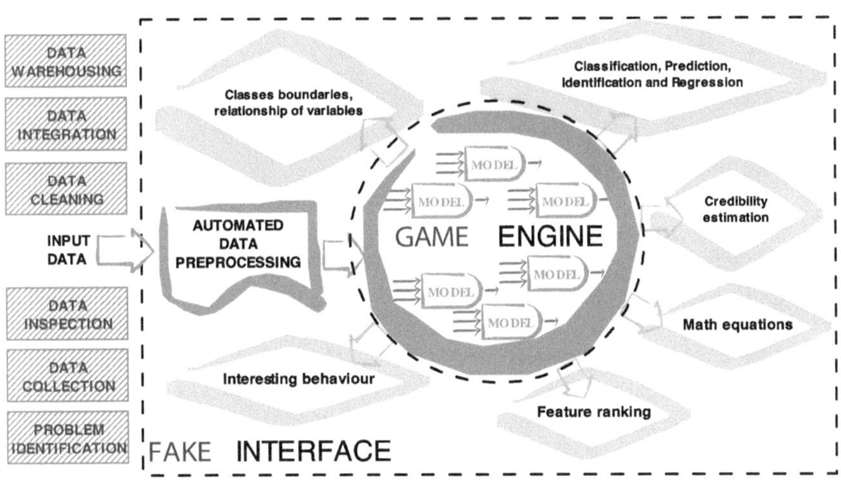

Fig. 38 Fully Automated Knowledge Extraction (FAKE) using Group of Adaptive Models Evolution (GAME)

5.1.1 Research of Methods in the Area of Data Preprocessing

In order to automate the data preprocessing phase, we develop more sophisticated methods for data preprocessing.We focus on data imputing (missing values replacement), that is in existing data mining environments [4, 1] realized by zero or mean value replacement although more sophisticated methods already exist [37]. We also developed a method for automate nonlinear redistribution of variables. FAKE GAME is not focusing on data warehousing, because this process is very difficult to automate in general. It is very dependent on particular conditions (structure of databases, information system, etc.) We assume that source data are already collected cleansed and integrated (Fig. 38).

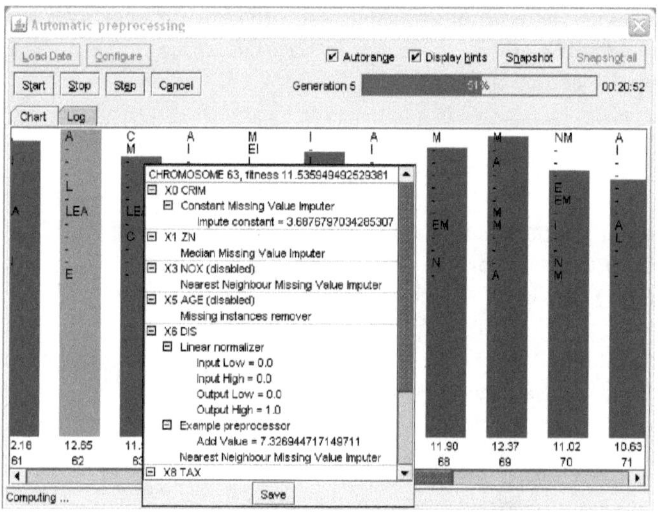

Fig. 39 Automated preprocessing module implemented in the FAKE GAME.

5.1.2 Automated Data Mining

To automate the data mining phase, we develop an engine that is able to adapt itself to the character of data. This is necessary to eliminate the need of parameter tuning. The GAME engine autonomously generates the ensemble of predictive, regressive models or classifiers. Models adapt to the character of data set so that they have optimal topology. Unfortunately, the class of problems where the GAME engine performs optimally is still limited. To make the engine more versatile, we need to add more types of building blocks, more learning algorithms, improve the regularization criteria, etc.

5.1.3 Knowledge Extraction and Information Visualization

To extract the knowledge from complex data mining models is very difficult task. Visualization techniques are promising way how to achieve it. Recently, some papers [49] on this topic had been published. In our case, we need to extract information from an ensemble of GAME inductive models. To do that we enriched methods such as scatterplot matrix, regression plots by the information about the behavior of models. For data with many features (input variables) we have to deal with curse of dimensionality. The state space is so big, that it is very difficult to find some interesting behavior (relationship of system variables) manually. For this purpose, we developed evolutionary search methods to find interesting projections automatically.

Along with the basic research, we implement proposed methods in Java programming language and integrate it into the FAKE GAME environment [5] so we can directly test the performance of proposed methods, adjust their parameters, etc.

Based on the research and experiments performed within this dissertation, we are developing the open source software FAKE GAME. This software should be able to automatically preprocess various data, to generate regressive, predictive models and classifiers (by means of GAME engine), to automatically identify interesting relationships in data (even in high-dimensional ones) and to present discovered knowledge in a comprehensible form. The software should fill gaps which are not covered by existing open source data mining environments [3, 4]. You can download the application to experiment with your data or join our community at Sourceforge [5].

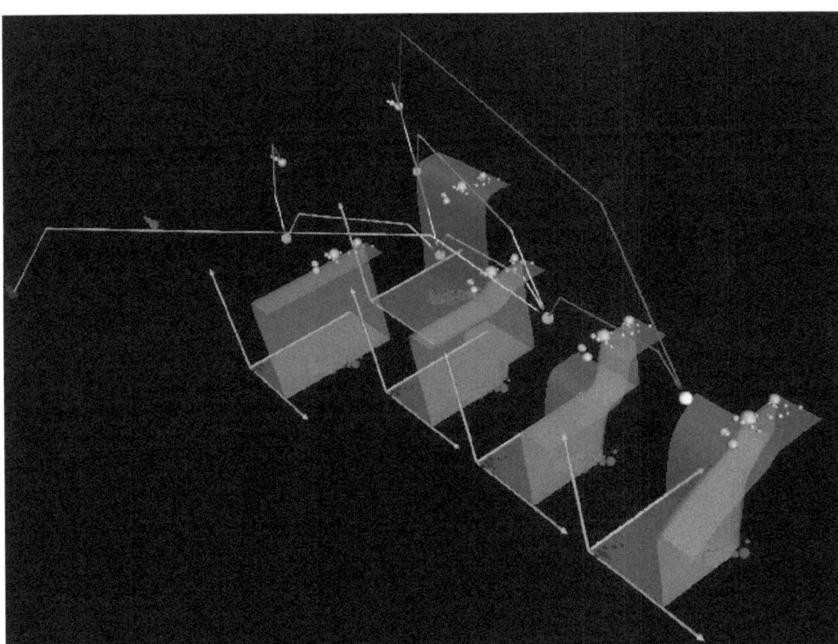

Fig. 40 3D inspection of GAME model topology and behavior (Iris Versicolor class).

Acknowledgement

I would like to thank to my collaborators Miroslav Cepek, Jan Drchal, Ales Pilny, Oleg Kovarik, Jan Koutnik, Tomas Siegl, members of the Computational Intelligence Research Group and all students participating in the FAKE GAME project. Thanks to head and former head of our research group Miroslav Skrbek and Miroslav Snorek.

This research is partially supported by the grant Automated Knowledge Extraction (KJB201210701) of the Grant Agency of the Academy of Science of the Czech Republic and the research program "Transdisciplinary Research in the Area of Biomedical Engineering II" (*MSM*6840770012) sponsored by the Ministry of Education, Youth and Sports of the Czech Republic.

References

1. The sumatra tt data preprocessing tool (September 2006), `http://krizik.felk.cvut.cz/sumatra/`
2. Uci machine learning repository (September 2006), `http://www.ics.uci.edu/~mlearn/MLSummary.html`
3. Weka open source data mining software (September 2006), `http://www.cs.waikato.ac.nz/ml/weka/`
4. The yale open source learning environment (September 2006), `http://www-ai.cs.uni-dortmund.de/SOFTWARE/YALE/intro.html`
5. The fake game environment for the automatic knowledge extraction (November 2008), `http://www.sourceforge.net/projects/fakegame`
6. Adeney, K., Korenberg, M.: An easily calculated bound on condition for orthogonal algorithms. In: IEEE-INNS-ENNS International Joint Conference on Neural Networks (IJCNN 2000), vol. 3, p. 3620 (2000)
7. Bilchev, G., Parmee, I.C.: The ant colony metaphor for searching continuous design spaces. In: Selected Papers from AISB Workshop on Evolutionary Computing, pp. 25–39. Springer, London (1995)
8. Blum, C., Socha, K.: Training feed-forward neural networks with ant colony optimization: An application to pattern classification. In: Proceedings of Hybrid Intelligent Systems Conference, HIS 2005, pp. 233–238. IEEE Computer Society, Los Alamitos (2005)
9. Brown, G.: Diversity in Neural Network Ensembles. PhD thesis, The University of Birmingham, School of Computer Science, Birmingham B15 2TT, United Kingdom (January 2004)
10. Fahlman, S.E., Lebiere, C.: The cascade-correlation learning architecture. Technical Report CMU-CS-90-100, Carnegie Mellon University Pittsburgh, USA (1991)
11. Fayyad, U., Shapiro, G., Smyth, P.: From data mining to knowledge discovery in databases. AI Magazine 17(3), 37–54 (1996)
12. Granitto, P., Verdes, P., Ceccatto, H.: Neural network ensembles: evaluation of aggregation algorithms. Artificial Intelligence 163, 139–162 (2005)
13. Hansen, L., Salamon, P.: Neural network ensembles. IEEE Trans. Pattern Anal. Machine Intelligence 12(10), 993–1001 (1990)
14. Ho, Y.-C., Pepyne, D.: Simple explanation of the no free lunch theorem of optimization. In: Proceedings of the 40th IEEE Conference on Decision and Control, December 4-7, vol. 5, pp. 4409–4414 (2001)

15. Holland, J.: Adaptation in Neural and Artificial Systems. University of Michigan Press (1975)
16. Hrstka, O., Kučerová, A.: Improvements of real coded genetic algorithms based on differential operators preventing premature convergence. Advances in Engineering Software 35(3-4), 237–246 (2004)
17. Islam, M., Yao, X., Murase, K.: A constructive algorithm for training cooperative neural network ensembles. IEEE Transitions on Neural Networks 14(4) (July 2003)
18. Ivakhnenko, A.G.: Polynomial theory of complex systems. IEEE Transactions on Systems, Man, and Cybernetics SMC-1(1), 364–378 (1971)
19. Juang, C.-F., Liou, Y.-C.: On the hybrid of genetic algorithm and particle swarm optimization for evolving recurrent neural network. In: Proceedings of the IEEE International Joint Conference on Neural Networks, Dept. of Electr. Eng., Nat. Chung-Hsing Univ., Taichung, Taiwan, July 25-29, vol. 3, pp. 2285–2289 (2004)
20. Juille, H., Pollack, J.B.: Co-evolving intertwined spirals. In: Lawrence, P.J.A., Fogel, J., Baeck, T. (eds.) Proceedings of the Fifth Annual Conference on Evolutionary Programming. Evolutionary Programming V, pp. 461–467. MIT Press, Cambridge (1996)
21. Kohavi, R.: A study of cross-validation and bootstrap for accuracy estimation and model selection. In: Proceedings of International Joint Conference on Artificial Intelligence (1995)
22. Kohonen, T.: Self-Organizing Maps. Springer, Heidelberg (2001)
23. Kong, M., Tian, P.: A direct application of ant colony optimization to function optimization problem in continuous domain. In: Dorigo, M., Gambardella, L.M., Birattari, M., Martinoli, A., Poli, R., Stützle, T. (eds.) ANTS 2006. LNCS, vol. 4150, pp. 324–331. Springer, Heidelberg (2006)
24. Kordík, P.: Game - group of adaptive models evolution. Technical Report DCSE-DTP-2005-07, Czech Technical University in Prague, FEE, CTU Prague, Czech Republic (2005)
25. Kordík, P.: Fully Automated Knowledge Extraction using Group of Adaptive Models Evolution. PhD thesis, Czech Technical University in Prague, FEE, Dep. of Comp. Sci. and Computers, FEE, CTU Prague, Czech Republic (September 2006)
26. Kordík, P., Křemen, V., Lhotská, L.: The game algorithm applied to complex fractionated atrial electrograms data set. In: Koutník, J., Kůrková, V., Neruda, R. (eds.) ICANN 2008, Part II. LNCS, vol. 5164, pp. 859–868. Springer, Heidelberg (2008)
27. Kordík, P., Náplava, P., Šnorek, M., Genyk-Berezovskyj, M.: The Modified GMDH Method Applied to Model Complex Systems. In: International Conference on Inductive Modeling - ICIM 2002, Lviv, pp. 150–155. State Scientific and Research Institute of Information Infrastructure (2002)
28. Kuhn, L.: Ant Colony Optimization for Continuous Spaces. PhD thesis, The Department of Information Technology and Electrical Engineering The University of Queensland (October 2002)
29. Li, Y.-J., Wu, T.-J.: An adaptive ant colony system algorithm for continuous-space optimization problems. J. Zhejiang Univ. Sci. 4(1), 40–46 (2003)
30. Mahfoud, S.W.: A comparison of parallel and sequential niching methods. In: Sixth International Conference on Genetic Algorithms, pp. 136–143 (1995)
31. Mahfoud, S.W.: Niching methods for genetic algorithms. Technical Report 95001, Illinois Genetic Algorithms Laboratory (IlliGaL), University of Ilinios at Urbana-Champaign (May 1995)
32. Mandischer, M.: A comparison of evolution strategies and backpropagation for neural network training. Neurocomputing (42), 87–117 (2002)

33. Monmarché, N., Venturini, G., Slimane, M.: On how pachycondyla apicalis ants suggest a new search algorithm. Future Gener. Comput. Syst. 16(9), 937–946 (2000)
34. Muller, J.A., Lemke, F.: Self-Organising Data Mining, Berlin (2000) ISBN 3-89811-861-4
35. Nariman-Zadeh, N., Darvizeh, A., Jamali, A., Moeini, A.: Evolutionary design of generalized polynomial neural networks for modelling and prediction of explosive forming process. Journal of Materials Processing Technology (165), 1561–1571 (2005)
36. Oh, S.-K., Pedrycz, W., Park, B.-J.: Polynomial neural networks architecture: analysis and design. Computers and Electrical Engineering 29(29), 703–725 (2003)
37. Pyle, D.: Data Preparation for Data Mining. Morgan Kaufman, Fondi di Ricerca Salvatore Ruggieri - Numero 421 d'inventario (1999)
38. Salane, Tewarson: A unified derivation of symmetric quasi-newton update formulas. Applied Math. 25, 29–36 (1980)
39. Schnabel, R., Koontz, J., Weiss, B.: A modular system of algorithms for unconstrained minimization. Technical Report CU-CS-240-82, Comp. Sci. Dept., University of Colorado at Boulder (1982)
40. Seiffert, U., Michaelis, B.: Adaptive three-dimensional self-organizing map with a two-dimensional input layer. In: Australian and New Zealand Conference on Intelligent Information Systems, November 18-20, pp. 258–263 (1996)
41. Sexton, R.S., Gupta, J.: Comparative evaluation of genetic algorithm and backpropagation for training neural networks. Information Sciences (129), 45–59 (2000)
42. Shewchuk, J.R.: An introduction to the conjugate gradient method without the agonizing pain. Technical report, School of Computer Science Carnegie Mellon University, Pittsburgh, PA 15213 (August. 1994)
43. Siegl, T., Kordík, P., Šnorek, M., Calda, P.: Fetal weight prediction models: Standard techniques or computational intelligence methods? In: Koutník, J., Kůrková, V., Neruda, R. (eds.) ICANN 2008, Part I. LNCS, vol. 5163, pp. 462–471. Springer, Heidelberg (2008)
44. Stanley, K., Bryant, B., Miikkulainen, R.: Real-time neuroevolution in the nero video game. IEEE Transactions on Evolutionary Computation 9(6), 653–668 (2005)
45. Stanley, K.O., Miikkulainen, R.: Evolving neural networks through augmenting topologies. Evolutionary Computation 10(2), 99–127 (2002)
46. Statsoft. Statistica neural networks software (September 2006), `http://www.statsoft.com/products/stat_nn.html`
47. Storn, R., Price, K.: Differential evolution - a simple and efficient heuristic for global optimization over continuous spaces. Journal of Global Optimization 11, 341–359 (1997)
48. Tsutsui, S., Pelikan, M., Ghosh, A.: Performance of aggregation pheromone system on unimodal and multimodal problems. In: The IEEE Congress on Evolutionary Computation, 2005 (CEC 2005), September 2-5, vol. 1, pp. 880–887. IEEE, Los Alamitos (2005)
49. Tzeng, F.-Y., Ma, K.-L.: Opening the black box - data driven visualization of neural networks. In: Proceedings of IEEE Visualization 2005 Conference, Minneapolis, USA, pp. 23–28 (October 2005)
50. Vesterstrom, J., Thomsen, R.: A comparative study of differential evolution, particle swarm optimization, and evolutionary algorithms on numerical benchmark problems. In: Proceedings of the 2004 Congress on Evolutionary Computation, vol. 2, pp. 1980–1987 (2004)
51. Wade, J.G.: Convergence properties of the conjugate gradient method (September 2006), `www-math.bgsu.edu/~gwade/tex_examples/example2.txt`
52. Wickera, D., Rizkib, M.M., Tamburinoa, L.A.: E-net: evolutionary neural network synthesis. Neurocomputing 42, 171–196 (2002)

Author Index

Ali, Jamali 99

Hitoshi, Iba 27

Kordík, Pavel 233

Nariman-zadeh, Nader 99

Onwubolu, Godfrey 1, 139, 193

Sharma, Anurag 193

FSC
www.fsc.org
MIX
Papier | Fördert gute Waldnutzung
FSC® C083411

Kolibri 360 GmbH
Ferdinand-Jühlke-Straße 7
99095 Erfurt, Deutschland
produktsicherheit@kolibri360.de